BIOCHEMISTRY RESOURCE BOOK

$$\frac{[H^+][COO^-]}{[COOH]}$$

BIOCHEMISTRY RESOURCE BOOK

Laurence A. Moran

University of Toronto

K. Gray Scrimgeour

University of Toronto

Contributing Authors:

R. Roy Baker	*University of Toronto*
H. Robert Horton	*North Carolina State University*
Willy Kalt	*Agriculture Canada*
Robert N. Lindquist	*San Francisco State University*
Robert K. Murray	*University of Toronto*
Frances J. Sharom	*University of Guelph*
Malcolm Watford	*The State University of New Jersey, Rutgers*

NEIL PATTERSON PUBLISHERS
PRENTICE HALL
Englewood Cliffs NJ 07632

Biochemistry Resource Book

© 1994 by Neil Patterson Publishers/Prentice-Hall, Inc.
ISBN 0–13–816679–X

Printed in the United States of America: February, 1994

Publisher: Neil Patterson
Editorial Director: Sherri Foster
Principal Editor: Charlotte Pratt
Contributing Editors: John Challice, Terri O'Quin, Morgan Ryan
Production Manager: Donna F. Young
Artist: George Sauer

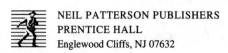 NEIL PATTERSON PUBLISHERS
PRENTICE HALL
Englewood Cliffs, NJ 07632

Prentice-Hall International (UK) Limited, *London*
Prentice-Hall of Australia Pty, Limited, *Sydney*
Prentice-Hall Canada Inc., *Toronto*
Prentice-Hall Hispanoamerica, S.A., *Mexico*
Prentice-Hall of India Private Limited, *New Delhi*
Prentice-Hall of Japan, Inc., *Tokyo*
Simon & Schuster Asia Pte. Ltd., *Singapore*
Editora Prentice-Hall do Brasil, Ltda., *Rio de Janeiro*
Prentice-Hall, Inc., *Englewood Cliffs, New Jersey*

Preface

This *Biochemistry Resource Book* is designed to serve as a study guide to accompany Moran·Scrimgeour *Biochemistry* and as a source of useful data that students can use long after completing the course. The book consists of several parts. The study guide portion is arranged by chapters that correspond to chapters in Moran·Scrimgeour *Biochemistry*. Next is an appendix of useful biochemical information for general reference. Following that is a list of common abbreviations. Finally, there is a dictionary of biochemical terms, including complete definitions for over 800 terms. A stereo viewer comes with each copy of the *Resource Book* and can be used for viewing the stereo images that appear in Moran·Scrimgeour *Biochemistry*. An alphabetical listing of the contents of all figures and tables in the *Resource Book* (beginning on page xi) allows easy access to information.

Each chapter in the *Resource Book* includes a summary of the corresponding chapter in Moran·Scrimgeour *Biochemistry* and a selection of figures and tables that warrant careful study. The figures include important molecular structures, major metabolic pathways, and summary figures of various kinds. Summary tables and tables of important data are included for easy reference. The problems for each chapter are thought-provoking, a test not of memorization skills but of the student's ability to apply principles. For each problem, a complete, step-by-step solution is provided, including illustrations when appropriate.

The appendix contains 12 tables of scientific data, with emphasis on material most useful for students of biochemistry. Following the appendix is a list of common biochemical abbreviations.

The dictionary of biochemistry contains over 930 cross-referenced entries and over 800 precise definitions. This material can be used to test mastery of particular terms or can be consulted whenever an unfamiliar term is encountered.

We intend this *Resource Book* to be valuable not only as a supplement to Moran·Scrimgeour *Biochemistry* but also as a reference work and review manual that will become a permanent part of the student's library.

Charlotte W. Pratt
Principal Editor
February 1994

Contents

Tables and Figures Listed by Topic

2

Cells

Summary

The unit of life is the cell. Cells are surrounded by a plasma membrane consisting of lipids and protein molecules. Chemical traffic between the cell interior and exterior is strictly controlled by selective transport across the plasma membrane. Cells are of two types, eukaryotic or prokaryotic. Eukaryotic cells contain internal membranous structures termed organelles that divide the cell interior into compartments, and they possess a cytoskeleton consisting of fibers formed from protein subunits. Prokaryotic cells are generally much smaller than eukaryotic cells, and they do not possess internal compartments.

Bacteria are prokaryotes. Most bacteria possess a cell wall surrounding the plasma membrane. If the cell wall is surrounded by an outer membrane, the bacteria are Gram-negative; prokaryotic cells that do not possess an outer membrane surrounding their cell wall are Gram-positive. Much of the biochemical diversity of nature is found in the prokaryotic realm, reflecting adaptation to an enormous range of different environments.

The organelles of eukaryotic cells have specialized functions. The nucleus, surrounded by a double membrane, contains the genetic material. The endoplasmic reticulum is an extensive membrane system continuous with the outer membrane of the nucleus. The rough endoplasmic reticulum is studded with membrane-bound ribosomes. Synthesis of proteins to be exported or proteins that will remain embedded in membranes is one of the principal metabolic activities of the rough endoplasmic reticulum. Ribosomes are absent in regions called smooth endoplasmic reticulum, where much of the lipid synthesis of the cell occurs.

Material newly synthesized in the endoplasmic reticulum is packaged in vesicles that merge with the Golgi apparatus, where modification and sorting of materials occurs. Vesicles then bud off on the far side of the Golgi and carry sorted material to specific cellular destinations.

Lysosomes and peroxisomes contain enzymes that catalyze potentially destructive reactions. Sequestration of these enzymes in lysosomes and peroxisomes protects the cell.

Mitochondria and chloroplasts are organelles involved in energy metabolism. Mitochondria are the centers of oxidative respiration and are the main sites of ATP formation. Chloroplasts are large organelles specialized for the conversion of light energy into chemical energy by the process of photosynthesis. Both organelles almost certainly arose endosymbiotically.

Eukaryotic cells contain elaborate networks of fibrous proteins collectively termed the cytoskeleton. The principal components of the cytoskeleton are actin filaments, microtubules, and intermediate filaments. Actin filaments and microtubules contribute to cell structure and are capable of directed motion. The role of intermediate filaments appears to be primarily structural.

Viruses are genetic parasites. They do not carry out independent metabolism, but depend on the metabolic capacities of host cells to reproduce themselves.

Study Information

Table 2·1 Comparison of prokaryotes and eukaryotes

	Prokaryotes	Eukaryotes
Organisms	Eubacteria, archaebacteria	Animals, plants, fungi, protists
Organization	Unicellular, some colonial	Unicellular, multicellular
Cell size (diameter)	~1–10 μm	~10–100 μm
Membranous organelles	No	Yes
Cytoskeleton	No	Yes
Peptidoglycan cell walls	Yes	No
Endo- and exocytosis	No	Yes
Chromosomes		
Number	1	>1
Location	Nucleoid	Nucleus
Topology	Circular	Linear
Chromosome segregation	Mechanism uncertain	Mitotic spindle

[Adapted from Neidhardt, F. C., Ingraham, J. L., and Schaecter, M. (1990). *Physiology of the Bacterial Cell: A Molecular Approach* (Sunderland, Massachusetts: Sinauer Associates), p. 4, and Stanier, R. Y., Ingraham, J. L., Wheelis, M. L., and Painter, P. R. (1986). *The Microbial World*, 5th ed. (Englewood Cliffs, New Jersey: Prentice Hall), p. 74.]

Problems

1. List three major criteria that distinguish bacterial cells from animal cells.

2. Match the organelles with their principal function.

	Organelle:	**Function:**
d	Nucleus	(a) Photosynthesis
e	Mitochondrion	(b) Synthesis of phospholipids; synthesis of membrane and secretory proteins
b	Endoplasmic reticulum	(c) Intracellular digestion
f	Golgi apparatus	(d) Contains genetic material
c	Lysosome	(e) Aerobic energy metabolism
a	Chloroplast	(f) Modification and sorting of protein products

Solutions

1. (a) Bacterial cells are prokaryotic; animal cells are eukaryotic.
    (b) Bacterial cells are generally much smaller than animal cells.
    (c) Bacterial cells reproduce at a much faster rate than animal cells.
    (d) Bacterial cells have cells walls; animal cells do not.

2. | | |
    |---|---|
    | Nucleus | (d) Contains genetic material |
    | Mitochondrion | (e) Aerobic energy metabolism |
    | Endoplasmic reticulum | (b) Synthesis of phospholipids; synthesis of membrane and secretory proteins |
    | Golgi apparatus | (f) Modification and sorting of protein products |
    | Lysosome | (c) Intracellular digestion |
    | Chloroplast | (a) Photosynthesis |

3

Water

Summary

Water is a nonlinear molecule whose H—O—H bond angle is 104.5°. Because oxygen is more electronegative than hydrogen, an uneven distribution of charge occurs within each O—H bond of the water molecule (that is, each O—H bond is polar). The polarity of the covalent bonds of the water molecule and the geometry of the molecule are such that the molecule has a positive end and a negative end and thus has a permanent dipole.

A water molecule forms four hydrogen bonds in ice and up to four hydrogen bonds in liquid water. In biological systems, intermolecular hydrogen bonds form between water molecules and many types of biomolecules. Some biomolecules also form intramolecular hydrogen bonds.

Ionic substances, such as sodium chloride, and highly polar nonionic compounds, such as short-chain alcohols and glucose, readily dissolve in water and are said to be hydrophilic. Ionic and polar molecules are surrounded by water molecules that form a solvation sphere. Nonpolar substances, such as hydrocarbons, are essentially insoluble in water and are called hydrophobic. The phenomenon of exclusion of nonpolar substances by water is called the hydrophobic effect. Detergents are amphipathic, meaning they have both hydrophobic and hydrophilic groups. These compounds may form monolayers on the surface of aqueous solutions and micelles when dispersed in aqueous media. Chaotropes enhance the solubility of nonpolar compounds in water.

The major noncovalent interactions in cells are electrostatic attractions, hydrogen bonds, van der Waals forces, and hydrophobic interactions. These weak forces stabilize the structures of proteins, nucleic acids, and membranes.

Although water is nucleophilic and present in large amounts, cells use several strategies to prevent degradative hydrolytic reactions and to allow condensation reactions. Unwanted hydrolysis of biopolymers is prevented by storing some hydrolases in inactive forms or sequestering them in organelles. Cells use the chemical potential energy of ATP to overcome the unfavorable equilibria of biosynthetic reactions. Furthermore, water is often excluded from the active sites of enzymes, where biosynthetic reactions are catalyzed.

Pure water, which ionizes slightly, contains 10^{-7} M protons and 10^{-7} M hydroxide ions. The concentration of H$^\oplus$ in aqueous solutions is measured on the logarithmic pH scale. Neutral solutions have a pH value of 7.0, acidic solutions have pH values less than 7.0, and basic (or alkaline) solutions have pH values greater than 7.0. Weak acids only partially dissociate when dissolved in water. The pK_a values of weak acids are determined by titration. These values are the pH values at the midpoints of titrations. The Henderson-Hasselbalch equation quantitatively describes the relationship between the pH of a solution, the pK_a of a weak acid, and the ratio of concentrations of the weak acid and its conjugate base.

A solution that resists changes in pH when small amounts of acid or base are added to it is said to be buffered. Maximum buffering is afforded when a weak acid and its conjugate base are present in equal concentrations (in other words, when the pH is equal to the pK_a). In humans, the pH of blood, 7.4 (sometimes called physiological pH), is maintained by the carbon dioxide–carbonic acid–bicarbonate buffer system, which depends upon both the concentration of plasma bicarbonate and the partial pressure of carbon dioxide in the lungs. Proteins and inorganic phosphate contribute to intracellular buffering. Natural and synthetic buffer compounds are used to maintain a constant pH during biochemical experiments.

Problems

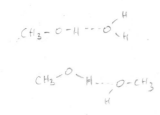

1. Draw the hydrogen bond(s) that are likely to form between the following pairs of molecules:

   (a) Methanol and water

   (b) Two molecules of ethanol

   (c) Acetone and

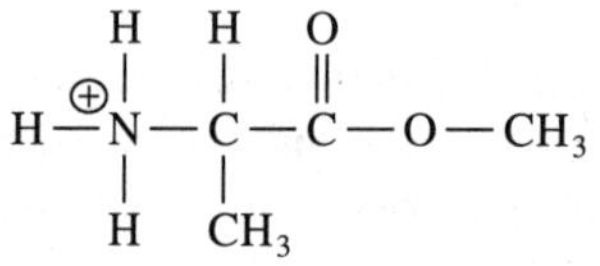

   (d) Two molecules of

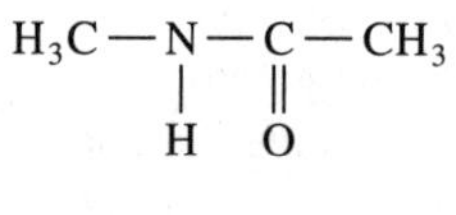

   (e) 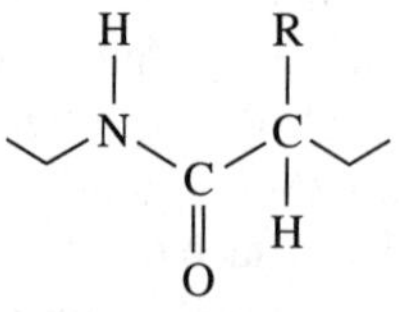 ___ and water

   Peptide chain

2. (a) Sketch the titration curve for carbonic acid (H_2CO_3). See Table 3·4 for pK_a values.
   (b) Referring to Figure 3·18, draw the structures of the major species of the carbon dioxide–carbonic acid–bicarbonate buffer that are present at pH 2.0, pH 7.0, and pH 11.0.
   (c) Referring to the titration curve in part (a) and to Figure 3·18, determine whether the carbon dioxide–carbonic acid–bicarbonate system of the blood is a more effective buffer against acid or against base between pH 7.0 and pH 8.0.

3. The pK_a of ascorbic acid (vitamin C) is 4.2 at 24°C. At what pH is the ratio of the unprotonated to protonated forms of ascorbic acid (a) 1:1, (b) 1:10, (c) 10:1, (d) 1:3? Which form predominates at physiological pH?

$[HA] : [A^-] = 1:1$

4. Many phosphorylated sugars (phosphate esters of sugars) are metabolic intermediates in the cellular degradation of carbohydrates. The two ionizable —OH groups of the phosphate group of the monophosphate ester of ribose (ribose 5-phosphate) have pK_a values of 1.2 and 6.6. The fully protonated form of α-D-ribose 5-phosphate has the structure shown on the right.
   (a) Draw, in order, the ionic species formed upon titration of this phosphorylated sugar from pH 0.0 to pH 10.0.
   (b) Sketch the titration curve for ribose 5-phosphate.
   (c) What is the predominant ionic form of ribose 5-phosphate at physiological pH?
   (d) Would ribose 5-phosphate be a good physiological buffer for cells?

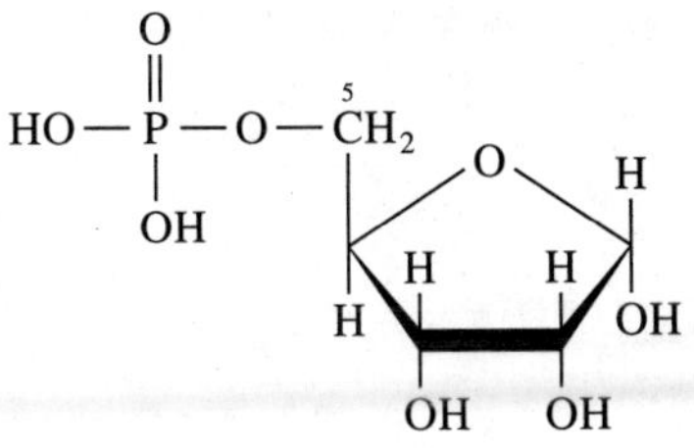

α-D-Ribose 5-phosphate

5. What is the pH of a solution of (a) 0.1 M HCl, (b) 0.01 M NaOH?

6. What is the approximate pH of a 1 M solution of a weak acid (HCOOH) that has a pK_a of about 4.0?

Solutions

1.

(a)

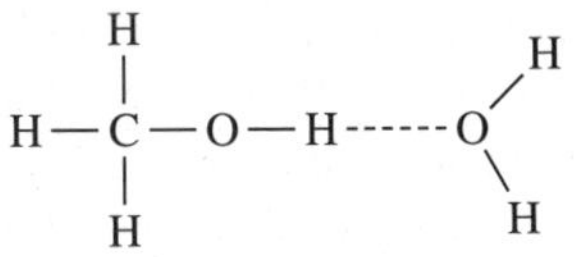

and

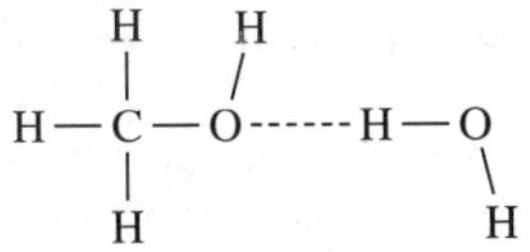

(b)

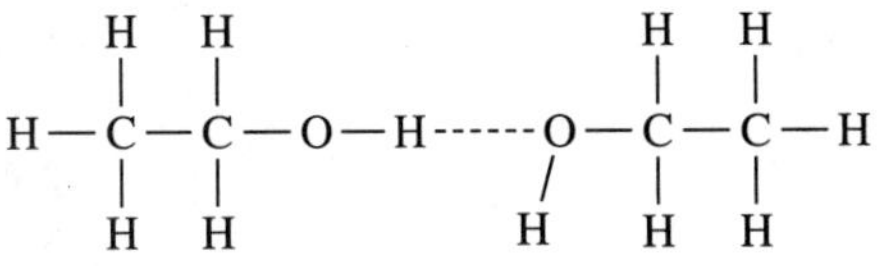

(c)

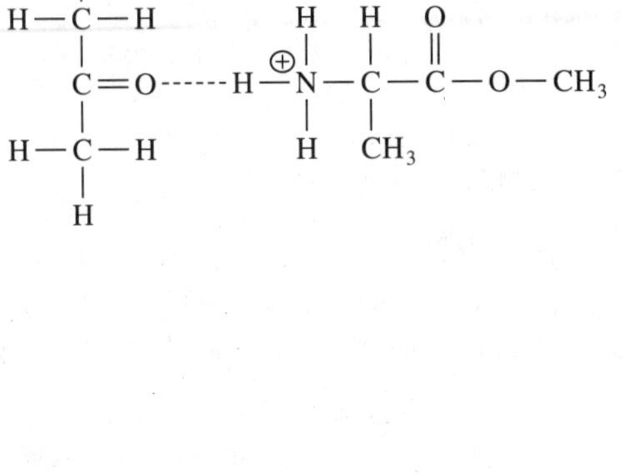

(d)

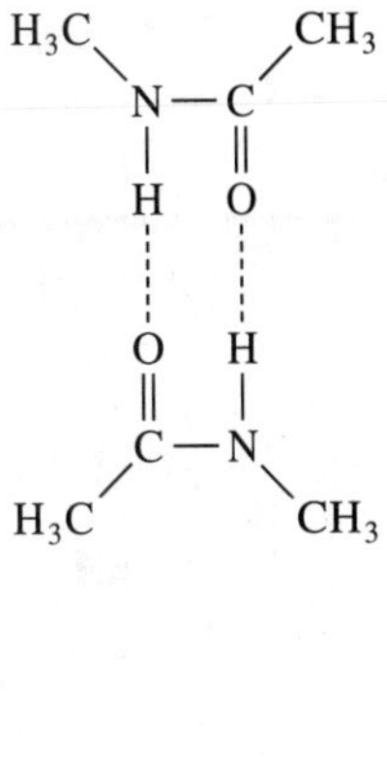

(e)

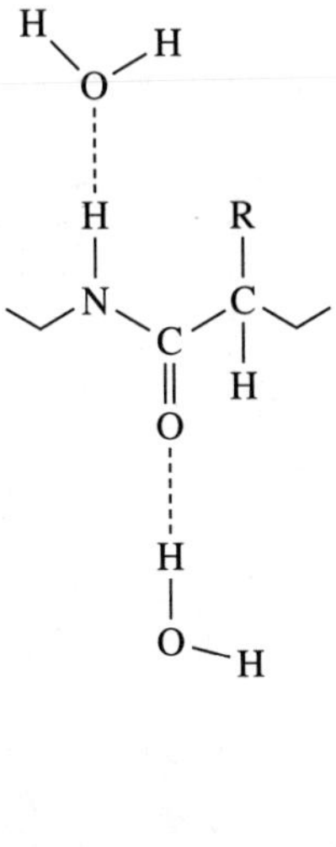

2. (a)

Second midpoint
$[HCO_3^{\ominus}] = [CO_3^{2\ominus}]$

Second endpoint
$CO_3^{2\ominus}$

First midpoint
$[H_2CO_3] = [HCO_3^{\ominus}]$

First endpoint
$HCO_3^{\ominus}$

pH

Equivalents of $OH^{\ominus}$

(b)

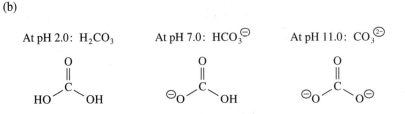

At pH 2.0: H_2CO_3 At pH 7.0: $HCO_3^{\ominus}$ At pH 11.0: $CO_3^{2\ominus}$

(c) Whether the blood is more effectively buffered against acid or more effectively buffered against base depends on which species of the buffer
system, a conjugate acid or a conjugate base, is more abundant. At a pH
equal to the pK_a (6.4), equal concentrations of H_2CO_3 and $HCO_3^{\ominus}$ are
present. Between pH 7.0 and 8.0, which is above the pK_a of 6.4, the conjugate base ($HCO_3^{\ominus}$) predominates. Since $HCO_3^{\ominus}$ buffers blood against
acid ($HCO_3^{\ominus} + H^{\oplus} \longrightarrow H_2CO_3$), the blood is more effectively
buffered against acid than against base in this pH range.

However, as explained in the text, CO_2(aqueous) dissolved in the
blood and CO_2(gaseous) in the lungs form a large reservoir of CO_2 that is
in equilibrium with H_2CO_3; thus, any H_2CO_3 that is used to buffer
against a base is rapidly replenished. Consequently, in vivo, the carbon
dioxide–carbonic acid–bicarbonate system is effective as a buffer against
both acids and bases.

3. (a) The Henderson-Hasselbalch equation (Equation 3·17) defines the relationship between pH, pK_a, and the ratio of concentrations of unprotonated ($A^{\ominus}$) and protonated (HA) species present in solution. When the
ratio of ascorbate anion to ascorbic acid is 1:1, $[A^{\ominus}]/[HA] = 1.0$. If
$[A^{\ominus}]/[HA] = 1.0$, then $\log(1.0) = 0$, and pH $= pK_a$. Thus, pH $= 4.2$.

(b) By the same principle, when the ratio of the unprotonated to protonated
forms is 1:10, $\log[A^{\ominus}]/[HA] = \log(0.10) = -1.0$; thus, pH $= 4.2 - 1.0 = 3.2$.

(c) pH $= 4.2 + \log(10) = 4.2 + 1.0 = 5.2$

(d) pH $= 4.2 + \log\dfrac{1}{3} = 4.2 - 0.5 = 3.7$

Since physiological pH (7.4) is greater than the pK_a of ascorbic acid
(4.2), the conjugate base (ascorbate) predominates at physiological pH.

4. (a)

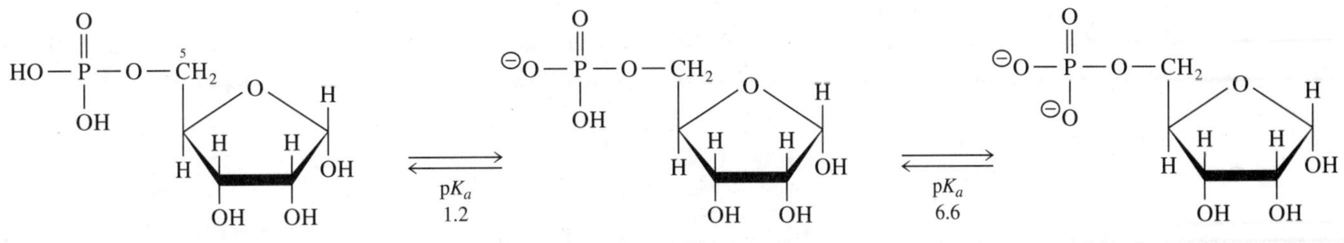

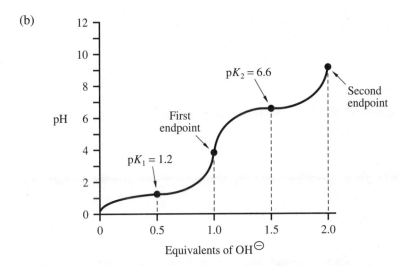

Fully protonated   Partially ionized   Fully ionized
                         (monoanion)   (dianion)

(b)

(c) Since physiological pH (7.4) is greater than the pK_a of the monoanionic compound (6.6), the fully ionized (dianion) form predominates.

(d) Ribose 5-phosphate would be a good physiological buffer. The buffering capacity of an ionizable molecule is greatest in a solution in which the pH is equal to the pK_a, but considerable buffering capacity exists at pH values within one unit of the pK_a. Thus, ribose 5-phosphate would be a good buffer from about pH 5.6 to pH 7.6.

5. (a) Since $[H^{\oplus}] = 1 \times 10^{-1}$ M, the pH is 1.0.
   (b) Since $[OH^{\ominus}] = 1 \times 10^{-2}$ M, $[H^{\oplus}] = 10^{-12}$ M. Therefore the pH is 12.0.

6. The dissociation of the acid can be written as

$$HCOOH \; \underset{\longleftarrow}{\overset{\longrightarrow}{}} \; H^{\oplus} + HCOO^{\ominus}$$

If the pK_a for this ionization is 4, then

$$K_a = 10^{-4} \text{ M} = \frac{[H^{\oplus}][HCOO^{\ominus}]}{[HCOOH]}$$

Since $[H^{\oplus}]$ is equal to $[HCOO^{\ominus}]$,

$$\frac{[H^{\oplus}]^2}{[HCOOH]} = 10^{-4} \text{ M}$$

Since HCOOH is a weak acid, it is only slightly ionized and the concentration of $HCOO^{\ominus}$ is small compared to the concentration of HCOOH. Thus, the initial concentration of HCOOH (1 M) is not significantly diminished.

$$\frac{[H^{\oplus}]^2}{1 \text{ M}} = 10^{-4}$$

Thus, the approximate $[H^{\oplus}]$ is 10^{-2} M and the pH is around 2.0.

4

Amino Acids and the Primary Structures of Proteins

Summary

Proteins are made from 20 amino acids, some of which may be modified after synthesis of the protein. Differences in the properties of amino acids reflect differences among their side chains, or R groups. Except for glycine, which has no chiral carbon, all amino acids in proteins are of the L configuration. The side chains of amino acids can be classified according to their chemical structures: aliphatic, aromatic, sulfur-containing, alcohols, bases, acids, and amides. Some amino acids are further classified as highly hydrophobic or highly polar. The properties of the side chains of amino acids are important factors in stabilizing the conformations and determining the functions of proteins.

The ionic state of the acidic and basic groups of amino acids and polypeptides depends on the pH. At pH 7, the α-carboxyl group is anionic ($-COO^{\ominus}$) and the α-amino group is cationic ($-NH_3^{\oplus}$). The charges of ionizable side chains depend on both the pH and their pK_a values. Differences in charges can be used to separate amino acids and proteins.

Amino acid residues in proteins are linked by peptide bonds. The sequence of residues is called the primary structure. Peptides and small proteins can be chemically synthesized. Proteins from biological sources are purified by methods that take advantage of the differences in solubility, net charge, size, and binding properties of individual proteins. The amino acid composition of a protein can be determined quantitatively by hydrolyzing the peptide bonds and analyzing the hydrolysate chromatographically.

The sequence of a polypeptide chain can be determined by the Edman degradation procedure. In each cycle of this procedure, the N-terminal residue of the protein reacts with phenylisothiocyanate to form a phenylthiocarbamoyl-peptide. The modified N-terminal residue is then cleaved and treated with aqueous acid to form a phenylthiohydantoin derivative, which can be identified chromatographically. The polypeptide chain, now one amino acid shorter, is again treated with phenylisothiocyanate. Often, sequences of 30 or more residues can be determined by using the Edman degradation procedure.

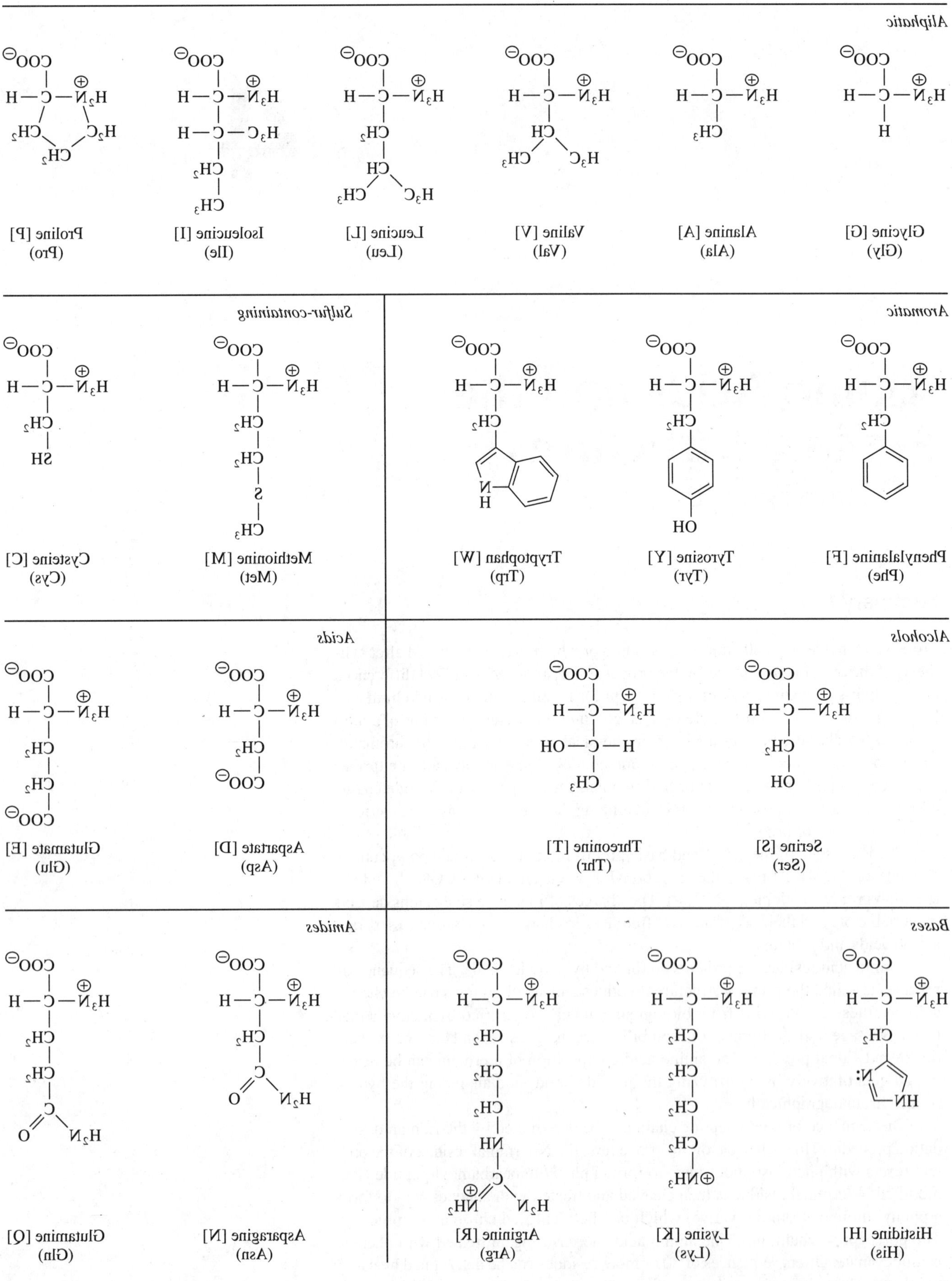

Aliphatic

Glycine [G] (Gly)   Alanine [A] (Ala)   Valine [V] (Val)   Leucine [L] (Leu)   Isoleucine [I] (Ile)   Proline [P] (Pro)

Aromatic   Sulfur-containing

Phenylalanine [F] (Phe)   Tyrosine [Y] (Tyr)   Tryptophan [W] (Trp)   Methionine [M] (Met)   Cysteine [C] (Cys)

Alcohols   Acids

Serine [S] (Ser)   Threonine [T] (Thr)   Aspartate [D] (Asp)   Glutamate [E] (Glu)

Bases   Amides

Histidine [H] (His)   Lysine [K] (Lys)   Arginine [R] (Arg)   Asparagine [N] (Asn)   Glutamine [Q] (Gln)

Sequences of larger polypeptides can be determined by selective cleavage of peptide bonds using proteases or chemical reagents, followed by Edman degradation of the resulting fragments. The amino acid sequence of the entire polypeptide is then deduced by lining up matching sequences of overlapping peptide fragments.

Comparisons of the primary structures of proteins reveal evolutionary relationships. As species diverge, the primary structures of their common proteins also diverge. By comparing differences in protein structures, we may reach a clearer understanding of evolutionary history.

Study Information

Table 4·2 pK_a values of acidic and basic constituents of free amino acids at 25°C*

Amino acid	pK_a values		
	α-Carboxyl group	α-Amino group	Side chain
Glycine	2.4	9.8	
Alanine	2.4	9.9	
Valine	2.3	9.7	
Leucine	2.3	9.7	
Isoleucine	2.3	9.8	
Methionine	2.1	9.3	
Proline	2.0	10.6	
Phenylalanine	2.2	9.3	
Tryptophan	2.5	9.4	
Serine	2.2	9.2	
Threonine	2.1	9.1	
Cysteine	1.9	10.7	8.4
Tyrosine	2.2	9.2	10.5
Asparagine	2.1	8.7	
Glutamine	2.2	9.1	
Aspartic acid	2.0	9.9	3.9
Glutamic acid	2.1	9.5	4.1
Lysine	2.2	9.1	10.5
Arginine	1.8	9.0	12.5
Histidine	1.8	9.3	6.0

*Values have been rounded.
[Values from Dawson, R. M. C., Elliott, D. C., Elliott, W. H., and Jones, K. M. (1986). *Data for Biochemical Research*, 3rd ed. (Oxford: Clarendon Press).]

Figure 4·5 (Page 12)
Fischer projections of the 20 common amino acids at pH 7, their names, and one- and three-letter abbreviations. Amino acids are classified by their hydrophobic or polar characteristics and by the functional groups of their side chains.

Problems

1. Amino acid side chains are important in the structure and biological function of a peptide or protein. Using both three- and one-letter abbreviations, list amino acids whose side chains 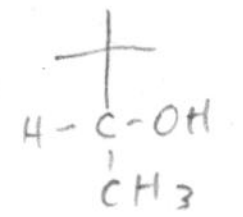
   (a) contain a hydroxyl group
   (b) contain an amide group
   (c) contain a group with aromatic characteristics
   (d) contain branched aliphatic hydrocarbons
   (e) contain sulfur
   (f) contain a group or atom than can act as a nucleophile in the pH range of 7 to 10. Identify the nucleophilic group or atom.

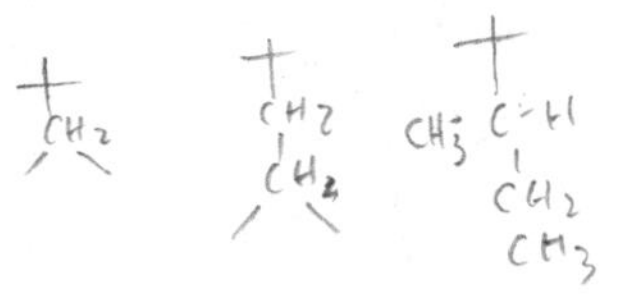

13

2. Based on the pK_a values given in Table 4·2, draw the predominant ionic structure of each of the following amino acids at pH 3.0, at pH 7.0, and at pH 13.0.
   (a) Glycine
   (b) Glutamate
   (c) Asparagine
   (d) Arginine
   (e) Histidine
   (f) Proline

3. Calculate the isoelectric point (pI) of each amino acid in Problem 2.

4. A solution contains a mixture of three tripeptides, Tyr–Arg–Ser, Glu–Met–Phe, and Asp–Pro–Lys. The α-COOH groups have a pK_a of 3.8; the α-NH$_3^{\oplus}$ groups have a pK_a of 8.5. At which pH value (2.0, 6.0, or 13.0) would electrophoresis provide the best resolution of the three tripeptides in the mixture?

5. Draw the structures of the PTH–amino acids that would be generated during the first three cycles of Edman degradation of the pentapeptide Phe–Pro–Arg–Ser–Met.

6. A tridecapeptide (an oligopeptide with 13 amino acid residues) has the following amino acid composition: Ala, Arg, Asp$_2$, Glu$_2$, Gly$_3$, Leu, Val$_3$. After partial acid hydrolysis of the tridecapeptide, the following peptides were isolated, and their sequences were determined by Edman degradation. Deduce the sequence of the original tridecapeptide.
   (a) Asp–Glu–Val–Gly–Gly–Glu–Ala
   (b) Val–Asp–Val–Asp–Glu
   (c) Val–Asp–Val
   (d) Glu–Ala–Leu–Gly–Arg
   (e) Val–Gly–Gly–Glu–Ala–Leu–Gly–Arg
   (f) Leu–Gly–Arg

7. Referring to Figure 4·29, answer the following questions.
   (a) From the peptide sequence alone, how can you tell that peptides 4 and 6 originate from the N-terminus of cytochrome b_5?
   (b) How can you tell that peptides 7 and 9 represent the C-terminus?
   (c) Which peaks would be visible if the peptides were detected by absorbance at 280 nm?

Solutions

1. (a) Ser (S), Thr (T), Tyr (Y)
   (b) Asn (N), Gln (Q)
   (c) Phe (F), Trp (W), Tyr (Y)
   (d) Ile (I), Leu (L), Val (V)
   (e) Cys (C), Met (M)
   (f) Side-chain groups or atoms that can act as nucleophiles are —OH in Ser (S), Thr (T), and Tyr (Y); sulfur in Cys (C) and Met (M); —COO$^{\ominus}$ in Asp (D) and Glu (E); and nitrogen in His (H) and Lys (K).

2. As the Henderson-Hasselbalch equation (Equation 4·3) shows, when the pH equals the pK_a of an ionizable group, one-half of the group is dissociated, and equal concentrations of the basic and acidic forms exist. At a pH *below* the pK_a, a majority of the species are in the protonated form; at a pH *above* the pK_a, a majority of the species are in the unprotonated (conjugate base) form.

Amino acid	pK_a value	Predominant form at pH 3.0	Predominant form at pH 7.0	Predominant form at pH 13.0
(a) Glycine	2.4, 9.8	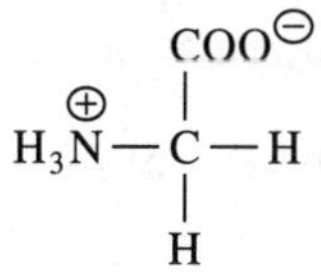 $COO^{\ominus}$ / $H_3\overset{\oplus}{N}-C-H$ / H	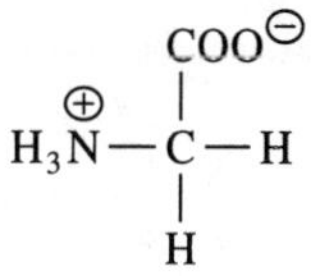 $COO^{\ominus}$ / $H_3\overset{\oplus}{N}-C-H$ / H	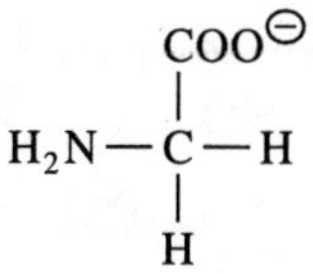 $COO^{\ominus}$ / H_2N-C-H / H
(b) Glutamate	2.1, 4.1, 9.5	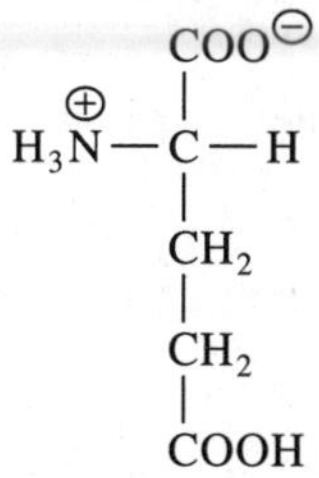 $COO^{\ominus}$ / $H_3\overset{\oplus}{N}-C-H$ / CH_2 / CH_2 / $COOH$	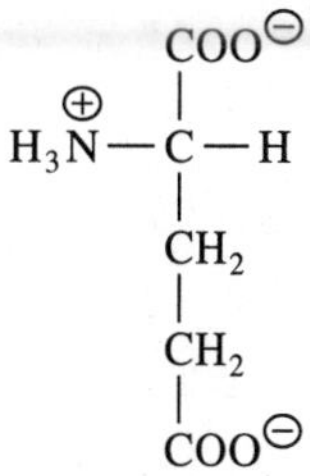 $COO^{\ominus}$ / $H_3\overset{\oplus}{N}-C-H$ / CH_2 / CH_2 / $COO^{\ominus}$	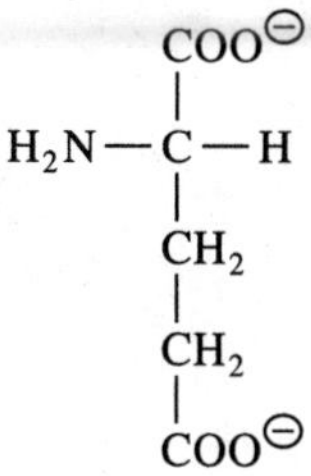 $COO^{\ominus}$ / H_2N-C-H / CH_2 / CH_2 / $COO^{\ominus}$
(c) Asparagine	2.1, 8.7	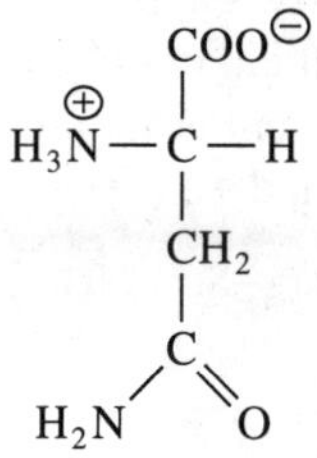 $COO^{\ominus}$ / $H_3\overset{\oplus}{N}-C-H$ / CH_2 / $C(=O)$ / H_2N	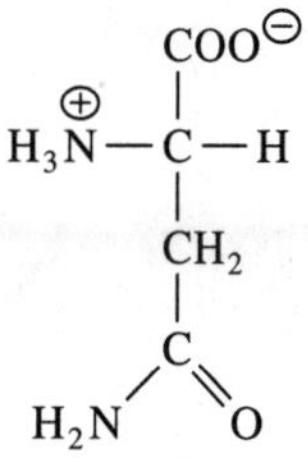 $COO^{\ominus}$ / $H_3\overset{\oplus}{N}-C-H$ / CH_2 / $C(=O)$ / H_2N	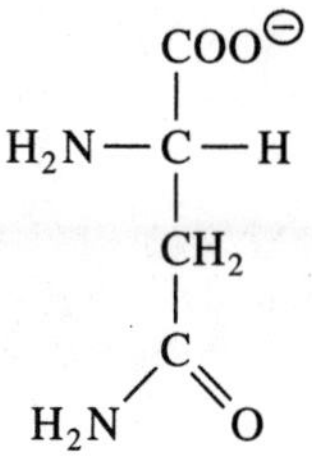 $COO^{\ominus}$ / H_2N-C-H / CH_2 / $C(=O)$ / H_2N
(d) Arginine	1.8, 9.0, 12.5	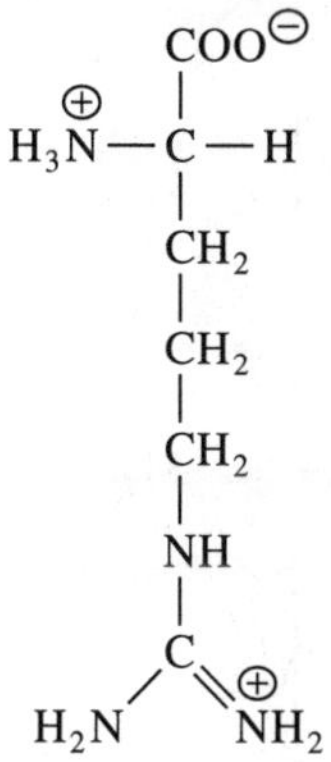 $COO^{\ominus}$ / $H_3\overset{\oplus}{N}-C-H$ / CH_2 / CH_2 / CH_2 / NH / C / $H_2N \quad \overset{\oplus}{NH_2}$	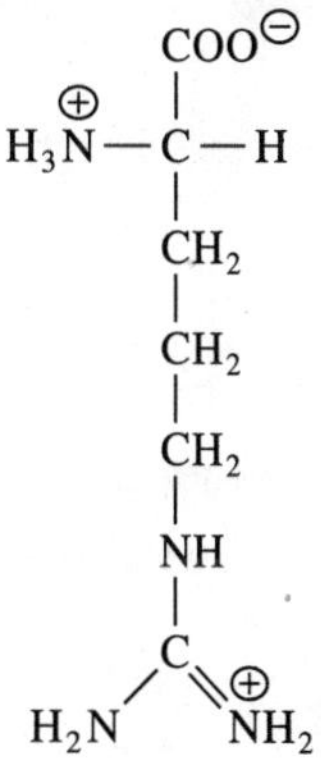 $COO^{\ominus}$ / $H_3\overset{\oplus}{N}-C-H$ / CH_2 / CH_2 / CH_2 / NH / C / $H_2N \quad \overset{\oplus}{NH_2}$	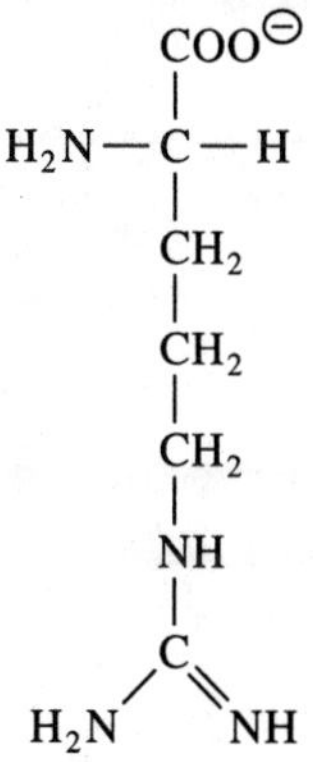 $COO^{\ominus}$ / H_2N-C-H / CH_2 / CH_2 / CH_2 / NH / C / $H_2N \quad NH$
(e) Histidine	1.8, 6.0, 9.3	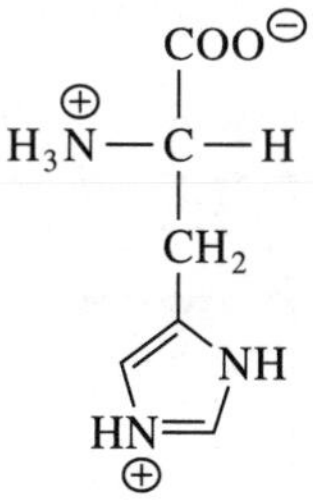 $COO^{\ominus}$ / $H_3\overset{\oplus}{N}-C-H$ / CH_2 / imidazole ($HN\overset{\oplus}{=}$, NH)	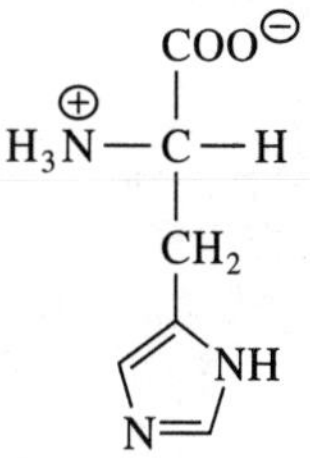 $COO^{\ominus}$ / $H_3\overset{\oplus}{N}-C-H$ / CH_2 / imidazole (N, NH)	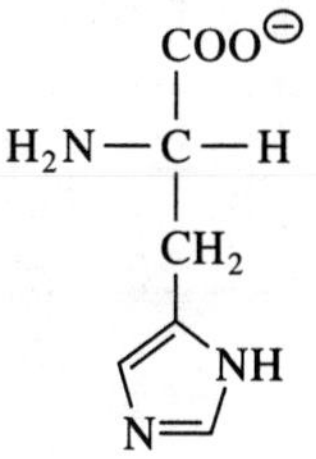 $COO^{\ominus}$ / H_2N-C-H / CH_2 / imidazole (N, NH)
(f) Proline	2.0, 10.6	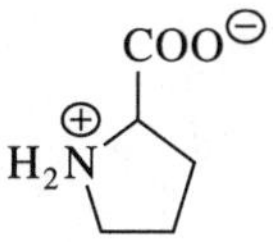 $COO^{\ominus}$ / $H_2\overset{\oplus}{N}$ pyrrolidine ring	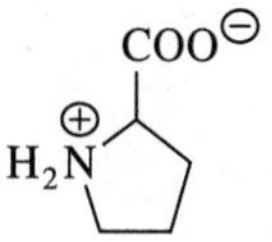 $COO^{\ominus}$ / $H_2\overset{\oplus}{N}$ pyrrolidine ring	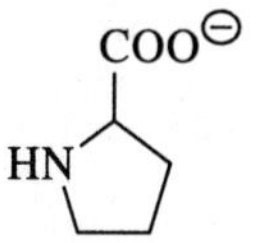 $COO^{\ominus}$ / HN pyrrolidine ring

15

3. The isoelectric point lies midway between the two pK_a values that indicate the protonation and deprotonation of the isoionic (electrically neutral) form.

(a) Glycine: $\quad pI = \dfrac{2.4 + 9.8}{2} = 6.1$

(b) Glutamate: $pI = \dfrac{2.1 + 4.1}{2} = 3.1$

(c) Asparagine: $pI = \dfrac{2.1 + 8.7}{2} = 5.4$

(d) Arginine: $\quad pI = \dfrac{9.0 + 12.5}{2} = 10.8$

(e) Histidine: $\quad pI = \dfrac{6.0 + 9.3}{2} = 7.6$

(f) Proline: $\quad pI = \dfrac{2.0 + 10.6}{2} = 6.3$

4. Electrophoresis at pH 6.0 would provide better resolution of the three peptides than at either pH 2.0 or pH 13.0 because at pH 6.0 each species would have a different net charge (+1, −1, and 0). At pH 2.0, the net charges would be +2, +1, and +2 ; at pH 13.0, they would be −2, −2, and −2.

5. Phe−Pro−Arg−Ser−Met

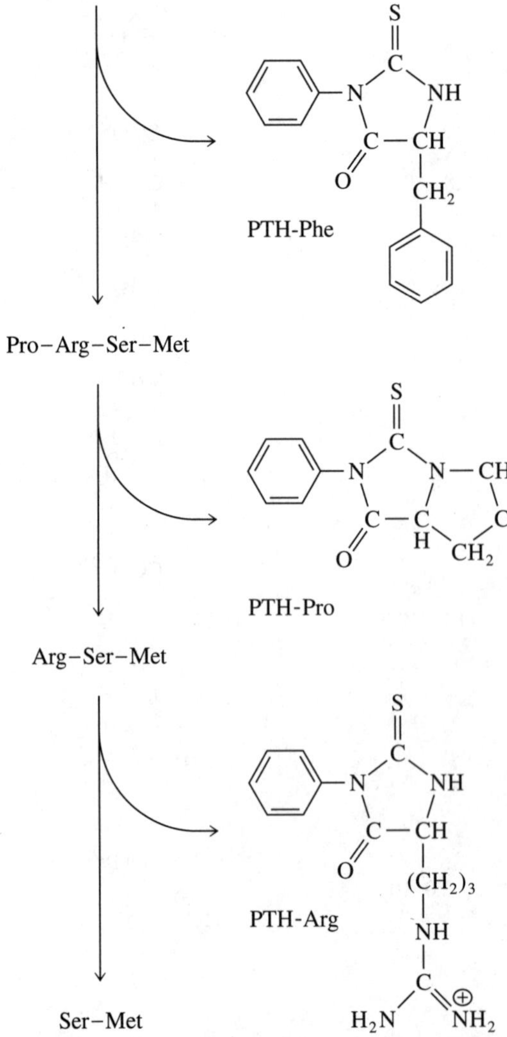

PTH-Phe

Pro−Arg−Ser−Met

PTH-Pro

Arg−Ser−Met

PTH-Arg

Ser−Met

6. The peptide sequence can be determined by aligning the overlapping identical sequences of the peptide fragments.

Peptide
fragment

(a)	Asp – Glu – Val – Gly – Gly – Glu – Ala
(b)	Val – Asp – Val – Asp – Glu
(c)	Val – Asp – Val
(d)	Glu – Ala – Leu – Gly – Arg
(e)	Val – Gly – Gly – Glu – Ala – Leu – Gly – Arg
(f)	Leu – Gly – Arg

Entire peptide: Val – Asp – Val – Asp – Glu – Val – Gly – Gly – Glu – Ala – Leu – Gly – Arg

7. (a) Since only the N-terminal residue of proteins can normally be acetylated, the acetylated peptides 4 and 6 must originate from the N-terminus of the protein.

(b) The C-terminal residue of both peptides 7 and 9 is aspartate. Because the protein fragments were produced by trypsin digestion, all other peptides have either lysine or arginine as C-terminal residues.

(c) Since only tryptophan and tyrosine absorb appreciable light at 280 nm, only peptides 7, 9, 10, 12, 13, and 14 would be easily detected.

Proteins: Three-Dimensional Structure and Function

Summary

Proteins can be classified as fibrous or globular. Fibrous proteins are generally water insoluble, physically tough, and built of repetitive structures. They usually have static functions. Globular proteins, most of which are soluble in aqueous solutions, are compact, roughly spherical macromolecules whose polypeptide chains are tightly folded. They act dynamically by transiently binding specific molecules.

Adjacent amino acid residues are joined by peptide bonds. Peptide bonds are polar and planar and have some double-bond character. Because of steric restraints, peptide groups are usually in the *trans* conformation. Rotation around the N—C_α and C_α—C bonds gives polypeptide chains conformational flexibility.

Proteins may have up to four levels of structure: primary (sequence of amino acid residues), secondary (local conformation, stabilized by hydrogen bonds), tertiary (the compacted globular structure of an entire polypeptide chain), and quaternary (assembly of two or more polypeptide chains into a multisubunit protein). The highly complex three-dimensional structures of proteins, many of which have been determined by X-ray crystallography, must be preserved to maintain biological activity.

The right-handed α helix is a common secondary structure found in some fibrous and globular proteins. It contains 3.6 amino acid residues per turn and has a pitch of 0.54 nm. The other major type of secondary structure is the β sheet, either parallel or antiparallel, in which the polypeptide chain is extended.

A different helical structure is found in collagen, a fibrous protein of connective tissue. A collagen molecule consists of three left-handed polypeptide helices intertwined to form a right-handed supercoil. Interchain hydrogen bonding and covalent cross-linking through modification of proline and lysine residues stabilizes the protein.

The elastic protein elastin can rapidly extend and relax. Its strong cross-linked structural network is required for the proper functions of organs such as heart and lungs.

Folding of a globular protein into its biologically active state is a sequential, cooperative process involving the hydrophobic effect and, to a lesser extent, hydrogen bonding, van der Waals interactions, and ion pairing. In cells, enzymes and chaperones assist folding. The native conformation of a protein can be disrupted by the addition of denaturing agents.

The compact, folded structures of globular proteins allow them to selectively bind other molecules. For example, the globular structures of the heme-containing proteins myoglobin and hemoglobin allow these proteins to bind and release oxygen in a manner that facilitates delivery of oxygen to respiring tissues.

A monomeric protein, myoglobin contains a single polypeptide chain of 153 residues that is folded into a compact globular structure composed of eight α helices. Its heme prosthetic group, which binds oxygen, is shielded from water in a cleft, or hydrophobic cage, formed by the protein.

Most globular proteins have considerable stretches of residues in nonrepeating conformations. These regions include turns and loops needed to connect α helices and β strands. The secondary structural elements are often connected into recognizable combinations called supersecondary structures, or motifs. Larger globular units called domains, or lobes, are usually associated with a particular function.

Hemoglobin consists of four chains (two α and two β chains in adult hemoglobin), each similar to the globin chain of myoglobin. The deoxy (T) and oxy (R) conformations of hemoglobin differ in their affinity for oxygen. Due to structural interactions associated with its tertiary and quaternary structure, hemoglobin displays positive cooperativity in the binding of oxygen and is subject to allosteric regulation.

Slight differences in the primary structures of hemoglobin molecules can result in significant functional differences: substitution of a serine residue in fetal hemoglobin (Hb F) for a histidine residue found in normal adult hemoglobin (Hb A) increases the affinity of Hb F for oxygen, thereby increasing the efficiency of oxygen delivery from maternal blood to the fetus. A mutation leading to replacement of a glutamate residue at position 6 in the β chains of normal adult hemoglobin with a valine residue results in Hb S, the hemoglobin responsible for sickle-cell anemia.

Antibodies are multidomain proteins that bind foreign substances. The variable domains at the ends of the heavy and light chains of the antibody interact with the antigen. Immunoassays take advantage of the binding specificity of antibodies.

Study Information

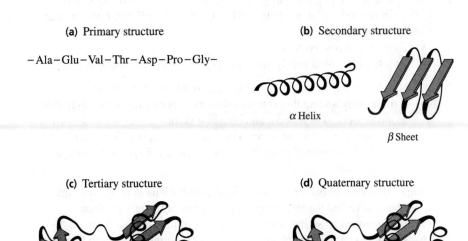

(a) Primary structure

−Ala−Glu−Val−Thr−Asp−Pro−Gly−

(b) Secondary structure

α Helix

β Sheet

(c) Tertiary structure

(d) Quaternary structure

Figure 5·2
Levels of protein structure. **(a)** The linear sequence of amino acid residues defines the primary structure. **(b)** Secondary structure consists of regions of regularly repeating conformations of the peptide chain, such as α helices and β sheets. **(c)** Tertiary structure refers to the folding of a polypeptide chain into its compact, globular shape. Two domains are shown. **(d)** Quaternary structure refers to the arrangement of two or more separate polypeptide chains into a multi-subunit molecule.

Table 5·3 Examples of hydrogen bonds in proteins

Type of hydrogen bond		Typical distance between donor and acceptor atom (nm)
Hydroxyl-hydroxyl	—O—H------O— H	0.28
Hydroxyl-carbonyl	—O—H------O=C	0.28
Amide-carbonyl	N—H------O=C	0.29
Amide-hydroxyl	N—H------O— H	0.30
Amide-imidazole nitrogen	N—H------N NH	0.31

Problems

1. The synthetic polypeptide polyglutamate, or $(Glu)_n$, forms regions of α helices in aqueous solutions below pH 3.0, but above pH 5.0, it assumes an extended conformation.
   (a) Explain this observation.
   (b) At what pH would polylysine, or $(Lys)_n$, be likely to form α helices?

2. An inhibitor of prolyl hydroxylase has been isolated from a plant.
   (a) Explain why adding this compound to the diet of a rat results in fragile blood vessels, skin lesions, and bleeding gums.
   (b) Symptoms similar to those described in part (a) appear in humans who suffer from vitamin C (ascorbic acid) deficiency. Explain the similarity of symptoms.

3. Studies of naturally occurring amino acid substitutions in proteins (mutations) have revealed the following. Explain each of the observations.
   (a) Replacement of serine is least likely to eliminate protein function.
   (b) Replacement of tryptophan is most likely to eliminate protein function.
   (c) Changes such as Lys $\longrightarrow$ Arg and Ile $\longrightarrow$ Leu often have little effect on protein function.
   (d) Glycine residues are often conserved at certain positions within protein chains despite changes at other positions.

4. Spectroscopic investigations of the molecule below indicate that, in aqueous solution, it remains a monomer up to the limits of its solubility. In methylene chloride (CH_2Cl_2), however, it forms dimers even at low concentration.

   (a) Explain these observations.
   (b) What do these findings imply about interactions in protein folding?

5. How does the reaction of carbon dioxide with water help explain the Bohr effect?
   (a) Include the equation for the formation of bicarbonate ion from CO_2 and water, and explain the effects of $H^\oplus$ and CO_2 on hemoglobin oxygenation.
   (b) Explain the physiological basis for the intravenous administration of bicarbonate to shock victims.

6. A number of mutant hemoglobins in addition to Hb S have single amino acid–residue replacements in either the α or β chain. In the mutant hemoglobin known as Hb Providence, an asparagine residue in the β chain replaces Lys-81, which projects into the central cavity of the tetrameric protein.
   (a) Predict the effect of this Lys $\longrightarrow$ Asn mutation on the affinity of Hb Providence for 2,3-*bis*phospho-D-glycerate.
   (b) What effect would this mutation have on the oxygenation of Hb Providence?

7. Hemoglobin is a tetramer of two α and two β subunits. Given that the dissociation of hemoglobin ($\alpha_2\beta_2 \rightleftharpoons 2\,\alpha\beta$) does not yield α_2 or β_2 fragments, show schematically how the four subunits might be symmetrically arranged.

8. If lysine labelled with radioactive ^{14}C is added to the diet of an animal, will ^{14}C appear in the animal's collagen? If hydroxylysine labelled with ^{14}C is introduced, will collagen containing ^{14}C be synthesized?

9. One technique for studying subunit composition (quaternary structure) and determining the molecular weight of proteins is the cross-linking of subunits with dimethylsuberimidate (DMSI). DMSI possesses two imido ester groups that react with amines. Suggest how DMSI might interact with proteins and what information could be obtained from treating a protein with DMSI.

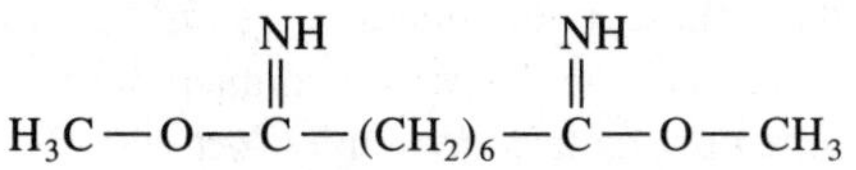

$$H_3C-O-\underset{\substack{\|\\ C}}{\overset{\substack{NH\\ \|}}{}}-(CH_2)_6-\underset{\substack{\|\\ C}}{\overset{\substack{NH\\ \|}}{}}-O-CH_3$$

Dimethylsuberimidate
(DMSI)

Solutions

1. (a) α Helices containing amino acid residues with ionizable side chains are sensitive to changes in pH because the pH of the solution determines whether the side chains are charged. Polymers of a single amino acid form helices only when the side chains are uncharged. Identical charges on side chains of adjacent residues create electrostatic repulsion that precludes packing of a polypeptide chain into an α-helical conformation. The side chain of glutamate has a pK_a of 4.1 (Table 4·2). When the pH is sufficiently below 4.1 (about 3), nearly all of the side chains of poly-aspartate are uncharged, and the polypeptide chain can form α helices. At pH 5 or above, however, nearly all of the side chains are anionic; electrostatic repulsion between neighboring charges thus prevents formation of the helices and causes the homopolymer to assume an extended conformation.

   (b) The pK_a of the side chain of lysine is 10.5 (Table 4·2). When the pH is sufficiently above 10.5, most of the side chains of polylysine are uncharged, and the polypeptide is likely to form an α helix. At lower pH values, however, the polycationic molecule is likely to assume an extended conformation.

2. (a) Prolyl hydroxylase catalyzes the conversion of proline residues to hydroxyproline residues, which stabilize the collagen triple helix and collagen fibers by participating in interchain hydrogen bonding. Since collagen fibers are structural components of blood vessels, inhibition of proline hydroxylation leads to greater fragility of blood vessels, bleeding, and skin lesions.

   (b) Ascorbic acid is necessary for the enzymatic activities of both prolyl hydroxylase and lysyl hydroxylase (Figure 5·28). Unlike most mammals, humans and other primates require a dietary source of ascorbic acid (vitamin C) because they lack the ability to synthesize it from carbohydrates such as glucose. Therefore, a deficiency of vitamin C leads to less hydroxylation of proline and lysine residues, resulting in the synthesis of defective collagen fibers and the same physiological effects as described in (a). These effects of vitamin C deficiency constitute the disease scurvy.

3. (a) A radical change in size, charge, or polarity of an important amino acid residue is likely to lead to a loss of protein function, whereas a small change in size or polarity may have little effect on function. Serine has a medium-sized side chain that, although polar, is not charged. It is possible for either a polar, uncharged residue (such as Asn, Gln, or Thr) or a

polar, charged residue (such as Arg, Asp, Glu, or Lys) to replace serine without sterically disrupting the protein conformation or greatly altering hydrogen bonding with other molecules.

(b) Tryptophan is an aromatic residue with a large, primarily hydrophobic side chain. Tryptophan is therefore usually found in the nonpolar interior of globular proteins. Although phenylalanine side chains are also hydrophobic and aromatic, phenylalanine may not be a suitable substitute for tryptophan because van der Waals contacts with the benzene ring of phenylalanine may be less favorable than with the indole ring system of tryptophan, resulting in subtle but significant alteration of protein structure and function.

(c) Lysine and arginine both possess cationic side chains (Lys $pK_a = 10.5$; Arg $pK_a = 12.5$) of similar lengths, and isoleucine and leucine are similar in size and are both hydrophobic. In either case, a substitution of one residue for the other results in only small steric changes and causes little alteration in the overall conformation of a protein.

(d) Glycine has the smallest side chain of any amino acid (a hydrogen atom) and the greatest conformational flexibility. Glycine can therefore fit into regions of protein structure that accommodate no other residue. For example, glycine is frequently found in the most common types of turns, type I and type II turns, and is the only residue that can occupy the third position of a type II turn. Substitution of any other amino acid for glycine in such instances necessitates significant structural changes that could have profound effects on protein function. Glycine residues that are not subject to such conformational constraints are more likely to be substituted by other residues.

4. (a) In aqueous solution, this thioamide remains a monomer stabilized by hydrogen bonding to water (Structure 1). In hydrophobic methylene chloride, however, the polar thioamide group is not stable as a solvated species; the more stable form is a dimer in which intermolecular hydrogen bonds provide stable interactions between the polar groups (Structure 2).

(b) When extrapolated to interactions in proteins, these findings imply that specific patterns of hydrogen bonding between peptide groups, such as those in an α helix or a β sheet, are more stable within the hydrophobic interior of a globular protein than they are in an aqueous environment, where peptide groups must compete with water molecules for hydrogen bonding.

5. The reaction of carbon dioxide with water explains why there is a concomitant lowering of pH when the concentration of CO_2 increases. Carbon dioxide produced by rapidly metabolizing tissue reacts with water to produce bicarbonate ions and $H^{\oplus}$.

(a) $$CO_2 + H_2O \rightleftharpoons H_2CO_3 \rightleftharpoons HCO_3^{\ominus} + H^{\oplus}$$

The $H^{\oplus}$ generated in this reaction decreases the pH of the blood and thus stabilizes the deoxy form (T conformation) of hemoglobin. The net effect is an increase in the P_{50}, that is, a lower affinity of hemoglobin for oxygen, so that more oxygen is released to the tissue (Figure 5·63). Carbon dioxide also lowers the affinity of hemoglobin for oxygen by forming carbamate adducts with the N-termini of the four chains (Figure 5·64). These adducts contribute to the stability of the deoxy (T) conformation, thereby further increasing the P_{50} and promoting the release of oxygen to the tissue.

(b) Shock victims suffer a critical deficit of oxygen supply to their tissues. Bicarbonate administered intravenously provides a source of carbon dioxide to the tissues. By lowering the affinity of hemoglobin for oxygen, carbon dioxide facilitates release of oxygen from oxyhemoglobin to the tissues.

6. (a) The substitution of asparagine, which has a neutral, polar side chain, for lysine, which has a positively charged side chain, would eliminate the ionic bonding that normally occurs between this lysine and the anionic 2,3BPG molecule in the central cavity of hemoglobin. This lack of ionic bonding would decrease the affinity of Hb Providence for 2,3BPG relative to that of normal hemoglobin (Hb A). (A similar phenomenon is seen in fetal hemoglobin, Hb F, as discussed in Section 5·18.)

(b) Binding of 2,3BPG stabilizes the deoxy (T) conformation of hemoglobin. Since Hb Providence does not bind 2,3BPG as tightly as Hb A does, the P_{50} of Hb Providence would be lower than that of Hb A in the presence of 2,3BPG. The result would be a less efficient oxygen-transport system due to the diminished ability of Hb Providence to unload oxygen to the tissues.

7. The dissociation data suggest that the interaction between α and β subunits is greater than the interaction between two α subunits or between two β subunits. Several arrangements for the quaternary structure of hemoglobin are possible. In fact, the four subunits of hemoglobin are packed in a tetrahedral array.

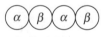
Linear

(1)

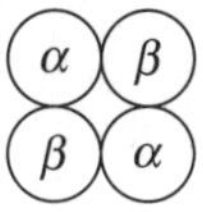

(2)
Square planar

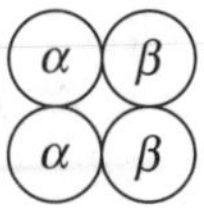

Tetrahedral

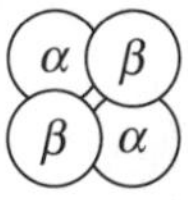

8. Because ingested lysine is incorporated into proteins, including collagen, the ^{14}C label will appear in the animal's collagen. However, labelled hydroxylysine will not be incorporated into collagen, because hydroxylysine is produced by hydroxylation of lysine residues in collagen polypeptides (Figure 5·28).

9. DMSI reacts primarily with lysine residues, most of which are on the protein surface because they are polar. Since each molecule of DMSI has two reactive imido ester groups, one molecule of DMSI can react with lysine residues in separate subunits (provided they are in close proximity), cross-linking two polypeptides. The untreated (a) and DMSI-treated (b) proteins can be denatured and analyzed by SDS-polyacrylamide gel electrophoresis (Section 4·7). If the two subunits are identical, the cross-linked molecule (b) will be twice as large as an unreacted subunit (a). DMSI cross-linking is useful for analyzing the quaternary structures of proteins containing many subunits of different molecular weights.

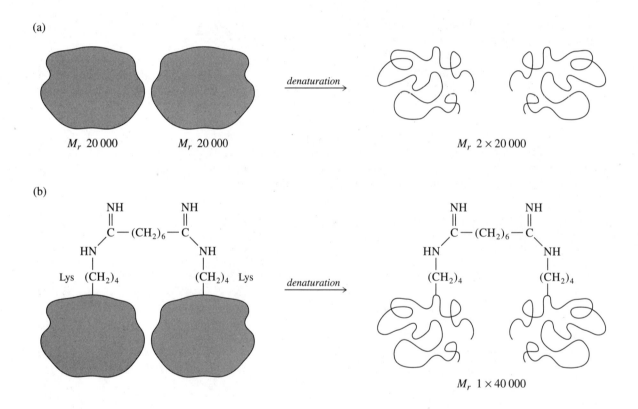

6

Properties of Enzymes

Summary

Enzymes, the catalysts of living organisms, are remarkable for their catalytic efficiency and their substrate and reaction specificity. With the exception of certain catalytic RNA molecules, enzymes are proteins, or proteins plus auxiliary compounds or ions called cofactors. Enzymes can be grouped into six major classes according to the nature of the reactions they catalyze: oxidoreductases (dehydrogenases), transferases, hydrolases, lyases, isomerases, and ligases (synthetases).

Chemical-kinetic experiments involve the systematic variation of reaction conditions, especially the concentration of substrate, and measurement of the alteration in the rate of formation of the product. Enzyme-kinetic measurements are usually made by measuring initial velocities, which are obtained from progress curves that show the amount of product formed over time. The first step in an enzyme-catalyzed reaction is the formation of a noncovalent enzyme-substrate complex. As a result, enzymatic reactions are characteristically first order with respect to enzyme concentration and typically show hyperbolic dependence on substrate concentration. Maximum velocity (V_{max}) is reached when the substrate concentrations are saturating. The Michaelis-Menten equation describes such kinetic behavior. The Michaelis constant (K_m) is equal to the concentration of substrate that gives half-maximum reaction velocity—that is, half saturation of E with S. Values for the kinetic constants K_m and V_{max} can be determined by computer analysis of kinetic data, or they can be estimated graphically. The double-reciprocal, or Lineweaver-Burk, plot of $1/v_0$ versus $1/[S]$ represents a linear transformation of the hyperbolic Michaelis-Menten equation.

The catalytic constant (k_{cat}), or turnover number for an enzyme, is the maximum number of molecules of substrate that can be transformed into product per molecule of enzyme (or per active site) per second and thus is equal to V_{max} divided by enzyme concentration. The unit for k_{cat} is s^{-1}. The ratio k_{cat}/K_m is an apparent second-order rate constant that governs the reaction of an enzyme with a substrate in dilute, nonsaturating solutions. Its value can provide a measure of the catalytic efficiency and substrate specificity of an enzyme.

The rates of enzyme-catalyzed reactions are also affected by the presence of inhibitors. Knowledge of enzyme-inhibition mechanisms aids in understanding the control of metabolism and the design of clinically useful drugs. Enzyme inhibitors may be classified as reversible or irreversible. Reversible inhibitors are bound noncovalently by the enzyme. They can be competitive (those that increase the apparent value of K_m, with no change in V_{max}), uncompetitive (those that decrease K_m and V_{max} proportionally), or noncompetitive (those that decrease V_{max}, with either no change in K_m—pure noncompetitive—or an increase or decrease in K_m—mixed noncompetitive). Inhibition studies can be used to determine the kinetic scheme of multisubstrate reactions (whether they are ordered, random, or ping-pong) and to determine the order of binding of substrates to an enzyme and the order of release of products.

Irreversible inhibitors form covalent bonds with the enzyme. By treating an enzyme with an irreversible inhibitor and then sequencing a segment of the protein, it is often possible to determine the identity of reactive amino acid residues, which can contribute to deciphering catalytic mechanisms. Site-directed mutagenesis represents another approach to investigating enzymes, including their catalytic mechanisms. Site-directed mutagenesis can be used to change the identity of a single amino acid residue within a protein, generating a new protein whose properties may reveal the role of the original amino acid residue.

Many proteases are synthesized as inactive zymogens that are activated extracellularly under appropriate conditions by selective proteolysis, the specific enzymatic cleavage of one or a few peptide bonds. Blood clotting depends on a cascade of zymogen activation involving over a dozen coagulation factors. Such cascades provide enormous signal amplification. The structures of proteins, as determined by X-ray crystallography, can reveal information about the active sites, including the binding of specific substrates.

Regulation of metabolism is often provided by allosteric modulators that act on certain key enzymes, almost all of which are oligomers. Allosteric modulators bind to a site other than the active site. They alter either the apparent K_m or V_{max} values of regulatory enzymes through control of the R:T ratio. Aspartate transcarbamoylase catalyzes the first committed step in the biosynthesis of pyrimidine nucleotides by *E. coli*. Its control properties have been elucidated at the molecular level.

Enzyme activity is also regulated by other mechanisms. For example, covalent modification of certain regulatory enzymes provides a control mechanism that is slower than allosteric interactions but more rapid than control achieved by changes in the concentration of enzyme, which is regulated at the level of gene expression. Some organisms or tissues contain isozymes, different proteins catalyzing the same chemical reaction. Isozymes can often be easily distinguished by kinetic or physical properties. The evolution of multienzyme complexes and multifunctional enzymes offers the advantages of metabolite channelling and coordinated regulation of multireaction processes.

Study Information

Please see Table 6·1 on Pages 30 and 31.

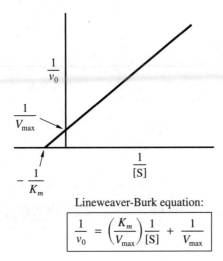

Lineweaver-Burk equation:

$$\frac{1}{v_0} = \left(\frac{K_m}{V_{max}}\right)\frac{1}{[S]} + \frac{1}{V_{max}}$$

Figure 6·6
Lineweaver-Burk (double-reciprocal) plot. This plot is derived from a linear transformation of the Michaelis-Menten equation. Values of $1/v_0$ are plotted as a function of $1/[S]$ values.

Table 6·4 Effects of reversible inhibitors on kinetic constants

Type of inhibitor	Effect	Rate law
Competitive (I binds to E only)	Raises K_m; V_{max} remains unchanged	$v_0 = \dfrac{V_{max}[S]}{K_m\left(1 + \dfrac{[I]}{K_i}\right) + [S]}$
Uncompetitive (I binds to ES only)	Lowers V_{max} and K_m; ratio of V_{max}/K_m remains unchanged	$v_0 = \dfrac{V_{max}[S]}{K_m + [S]\left(1 + \dfrac{[I]}{K_i}\right)}$
Noncompetitive (I binds to E and ES)		$v_0 = \dfrac{V_{max}[S]}{K_m\left(1 + \dfrac{[I]}{K_i}\right) + [S]\left(1 + \dfrac{[I]}{K_i'}\right)}$
Pure noncompetitive (I binds to E and ES equally)	Lowers V_{max}; K_m remains unchanged	$E + I \underset{}{\overset{K_i}{\rightleftarrows}} EI$
Mixed noncompetitive (I binds to E and ES unequally)	Lowers V_{max}; raises or lowers K_m	$ES + I \underset{}{\overset{K_i'}{\rightleftarrows}} ESI$

Table 6·1 Examples of enzymes from the six major classes

Major class	Enzyme	Reaction description
1. Oxidoreductase	Lactate dehydrogenase	Oxidation of the secondary alcohol L-lactate to pyruvate, a ketone
2. Transferase	Alanine transaminase (Alanine aminotransferase)	Transfer of an amino group
3. Hydrolase	Trypsin	Hydrolysis of Lys−Y (or Arg−Y) peptide bonds, where Y ≠ Pro
4. Lyase	Pyruvate decarboxylase	Decarboxylation of pyruvate
5. Isomerase	Alanine racemase	Interconversion of D and L isomers of alanine
6. Ligase	Glutamine synthetase	ATP-dependent synthesis of L-glutamine

Example of reaction catalyzed	Coenzyme involved
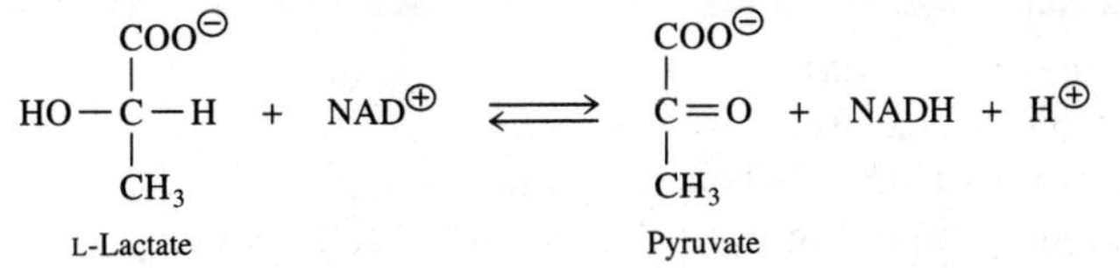 HO—C(H)(COO⁻)—CH₃ (L-Lactate) + NAD⁺ ⇌ C(=O)(COO⁻)—CH₃ (Pyruvate) + NADH + H⁺	NAD⁺ (Nicotinamide adenine dinucleotide)
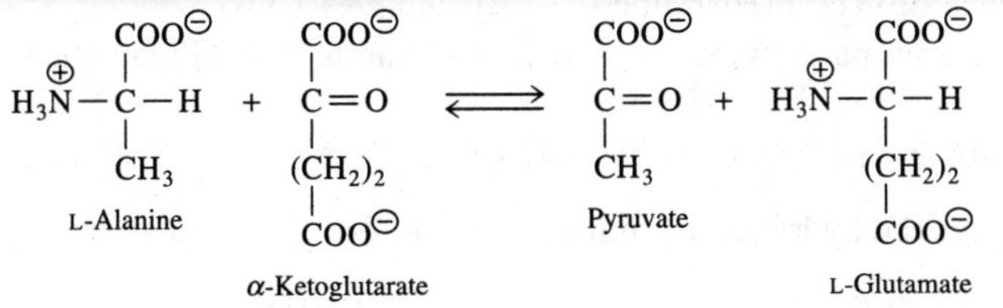 H₃N⁺—C(H)(COO⁻)—CH₃ (L-Alanine) + C(=O)(COO⁻)—(CH₂)₂—COO⁻ (α-Ketoglutarate) ⇌ C(=O)(COO⁻)—CH₃ (Pyruvate) + H₃N⁺—C(H)(COO⁻)—(CH₂)₂—COO⁻ (L-Glutamate)	Pyridoxal phosphate
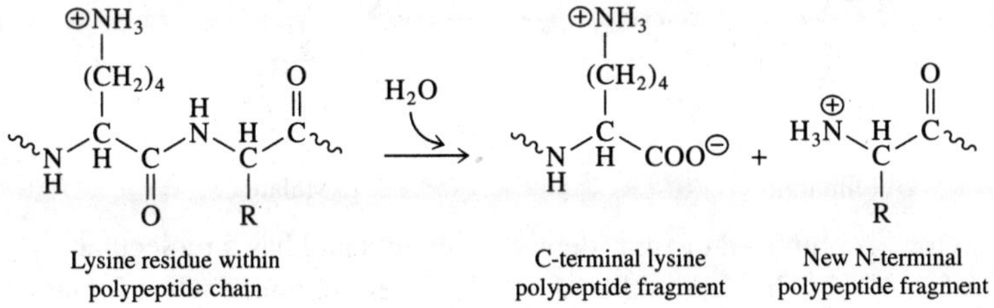 Lysine residue within polypeptide chain + H₂O ⟶ C-terminal lysine polypeptide fragment + New N-terminal polypeptide fragment	None
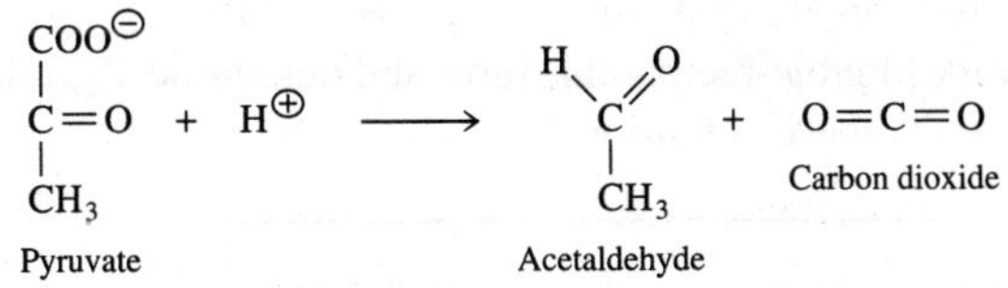 C(=O)(COO⁻)—CH₃ (Pyruvate) + H⁺ ⟶ CH₃—CHO (Acetaldehyde) + O=C=O (Carbon dioxide)	Thiamine pyrophosphate
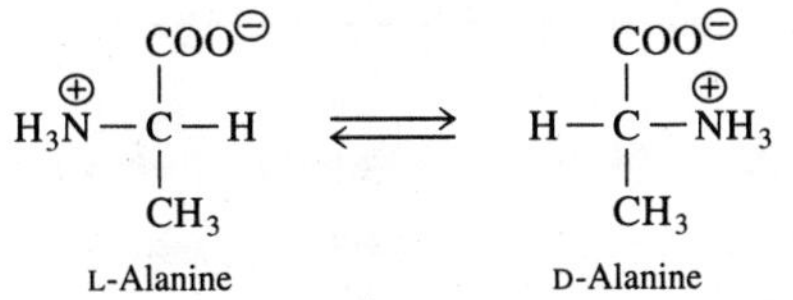 H₃N⁺—C(H)(COO⁻)—CH₃ (L-Alanine) ⇌ H—C(NH₃⁺)(COO⁻)—CH₃ (D-Alanine)	Pyridoxal phosphate
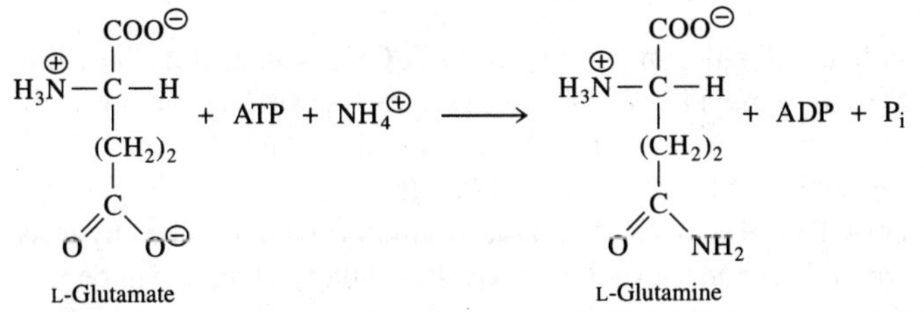 H₃N⁺—C(H)(COO⁻)—(CH₂)₂—COO⁻ (L-Glutamate) + ATP + NH₄⁺ ⟶ H₃N⁺—C(H)(COO⁻)—(CH₂)₂—C(=O)NH₂ (L-Glutamine) + ADP + Pᵢ	ATP

31

Problems

1. Determine the major enzyme class to which each of the following belongs.
   (a) Carboxypeptidase
   (b) Asparagine synthetase
   (c) Aspartate transcarbamoylase
   (d) Triose phosphate isomerase
   (e) Sucrase (Sucrose + $H_2O \longrightarrow$ Glucose + Fructose)
   (f) Alcohol dehydrogenase (Ethanol + $NAD^{\oplus} \rightleftharpoons$ Acetaldehyde + $NADH + H^{\oplus}$)
   (g) Histidine decarboxylase (Histidine $\longrightarrow$ Histamine + CO_2)
   (h) Aspartate transaminase (Aspartate + α-Ketoglutarate $\rightleftharpoons$ Glutamate + Oxaloacetate)
   (i) Carbonic anhydrase ($H_2O + CO_2 \rightleftharpoons H_2CO_3$)

2. Fumarase catalyzes the hydration of fumarate to L-malate.

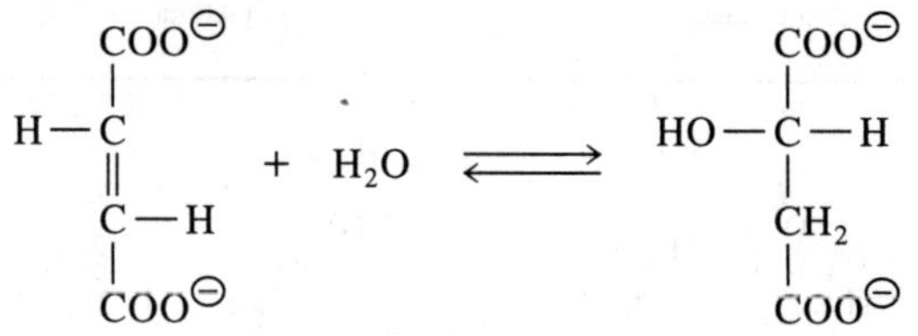

Fumarate   L-Malate

The enzyme is composed of four identical subunits and has a molecular weight of 194 000. The data in the table below were obtained when fumarate was used as the substrate, and the initial rates of hydration were measured at pH 5.7 and 25°C with an enzyme concentration of 2×10^{-6} M. Plot the data in Lineweaver-Burk (double-reciprocal) form and determine V_{max}, k_{cat}, and K_m for fumarase under these conditions.

Fumarate (mM)	Rate of product formation (mmol l^{-1} min^{-1})
2.0	2.5
3.3	3.1
5.0	3.6
10.0	4.2

3. Tosylamidophenylethyl chloromethyl ketone (TPCK) is an affinity-labelling reagent for chymotrypsin. TPCK inactivates chymotrypsin by reacting with His-57 at the active site (Figure 6·19). N-Tosylamidophenylethyl methyl ketone (TPMK) is a reversible inhibitor of chymotrypsin.
   (a) The effect of TPMK and TPCK on the chymotrypsin-catalyzed hydrolysis of a synthetic substrate such as tosylphenylalanylglycine can be measured. Sketch the Lineweaver-Burk (double-reciprocal) plots for TPMK and TPCK that would result if the Michaelis-Menten equation were applied to the data.
   (b) How could one modify the structure of TPCK to make an affinity label for trypsin?
   (c) How could one modify the structure to make an affinity label for elastase?
   (d) It is often necessary to inactivate or inhibit proteases during the isolation of proteins from tissue preparations. How could one inactivate trypsin, chymotrypsin, and elastase in a preparation that contained pancreatic proteases?

(e) How could one selectively inactivate trypsin in such a preparation without inactivating chymotrypsin or elastase?

4. Protein kinase A, a converter enzyme, contains two types of subunits, one of which (the catalytic subunit) can be inactivated by treatment with tosylamido-phenylethyl chloromethyl ketone (TPCK) since a specific cysteine residue of the catalytic subunit reacts with the chloromethyl ketone. Show the mechanism of this alkylation reaction and draw the structure of the modified amino acid that could be obtained from an acid hydrolysate of TPCK-treated protein kinase.

5. Subtilisin (M_r 27 600) is a bacterial protease that can catalyze hydrolysis of certain amino acid esters and amides. For the synthetic substrate N-acetyl-L-tyrosine ethyl ester (Ac-Tyr-OEt), subtilisin exhibits K_m and k_{cat} values of 0.15 M and 550 s^{-1}, respectively.
   (a) What is the V_{max} for the hydrolysis of Ac-Tyr-OEt when the subtilisin concentration is 0.4 mg ml^{-1}?
   (b) Indole is a competitive inhibitor of subtilisin with a K_i of 0.05 M. What is the V_{max} for Ac-Tyr-OEt hydrolysis by 0.40 mg ml^{-1} subtilisin in the presence of 6.25 mM indole?
   (c) What is the v_0 when 0.40 mg ml^{-1} subtilisin is incubated with 0.25 M Ac-Tyr-OEt and 1.0 M indole?

6. (a) Why does cleavage of the peptide bond between Arg-15 and Ile-16 in chymotrypsinogen result in activation of the zymogen?
   (b) Would either a chymotrypsin inhibitor or an elastase inhibitor be as effective as a trypsin inhibitor in protecting the pancreas from premature intracellular zymogen activation?
   (c) Thrombin, a key component in the blood clotting cascade of mammals, catalyzes hydrolysis of specific Arg—Gly bonds in the plasma protein fibrinogen, converting it to fibrin. Draw the structure of a chloromethyl ketone that could serve as an affinity label for thrombin.
   (d) Whereas 60% of the residues in the interior of trypsin are identical to those found in the interior of thrombin, only 10% of the residues on the surfaces of these homologous proteins are conserved. Explain why the interior residues are more conserved than the surface residues.

Solutions

1. (a) Hydrolase (carboxypeptidase catalyzes hydrolysis of C-terminal peptide bonds)
   (b) Ligase (compare with glutamine synthetase in Table 6·1)
   (c) Transferase (aspartate transcarbamoylase catalyzes transfer of the carbamoyl group of carbamoyl phosphate to the nitrogen of aspartate, as shown in Figure 6·31)
   (d) Isomerase
   (e) Hydrolase
   (f) Oxidoreductase (compare with lactate dehydrogenase in Table 6·1)
   (g) Lyase (compare with pyruvate decarboxylase in Table 6·1)
   (h) Transferase (compare with alanine transaminase in Table 6·1) This enzyme is sometimes called aspartate aminotransferase.
   (i) Lyase (note that water is added to a double bond in carbon dioxide, or in the reverse direction, water is cleaved from carbonic acid, creating a double bond)

2. To plot the kinetic data for fumarase (a lyase) in Lineweaver-Burk form, first calculate the reciprocals of substrate concentrations and initial rates of product formation. (Note the importance of including correct units as well as numerical values in calculating and plotting the data.)

Fumarate [S] (mM)	$\frac{1}{[S]}$ (mM^{-1})	Rate of product formation v_0 (mmol l^{-1} min^{-1})	$\frac{1}{v_0}$ (mmol^{-1} l min)
2.0	0.50	2.5	0.40
3.3	0.30	3.1	0.32
5.0	0.20	3.6	0.28
10.0	0.10	4.2	0.24

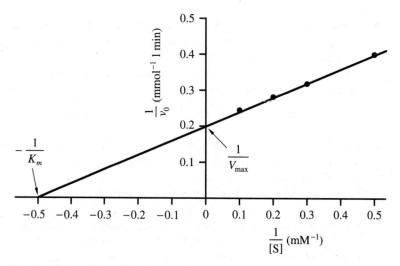

V_{max} is obtained by taking the reciprocal of $1/V_{max}$, whose value is read from the y intercept of the graph.

$$\frac{1}{V_{max}} = 0.20 \text{ mmol}^{-1} \text{ l min}$$

$$V_{max} = 5.0 \text{ mmol l}^{-1} \text{ min}^{-1}$$

The value of k_{cat} can be determined by dividing V_{max} by $[E]_{total}$ (Equation 6·9), where $[E]_{total}$ is the concentration of enzyme active sites. Since fumarase consists of four identical subunits, assume that there are four active sites per tetrameric enzyme molecule. A fumarase concentration of 2×10^{-6} M can be expressed as an active-site concentration of 8×10^{-6} M or 8×10^{-3} mmol l^{-1}. The units for k_{cat} are s^{-1}. Thus,

$$k_{cat} = \frac{(5.0 \text{ mmol l}^{-1} \text{ min}^{-1})}{(0.008 \text{ mmol l}^{-1})} \times \left(\frac{1 \text{ min}}{60 \text{ s}}\right) = 10.4 \text{ s}^{-1}$$

K_m is obtained by taking the reciprocal of $1/K_m$, whose value is obtained from the x intercept of the graph.

$$-\frac{1}{K_m} = -0.5 \text{ mM}^{-1}$$

$$K_m = 2.0 \text{ mM or } 2.0 \times 10^{-3} \text{ M}$$

3. (a)

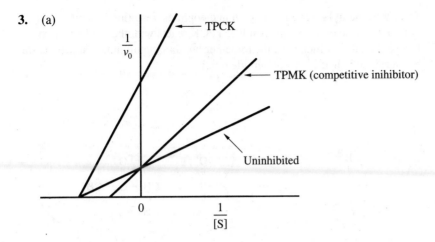

In this sketch of a double-reciprocal plot, TPMK shows kinetics characteristic of competitive inhibition: the V_{max} is unaffected by the presence of the inhibitor, but the apparent K_m is increased (compare with Figure 6·8).

In contrast, TPCK inactivates enzyme molecules with which it reacts, so that the apparent V_{max} decreases in direct proportion to the fraction of enzyme that has been covalently modified. Enzyme molecules that have not reacted with TPCK retain their original K_m. The plot gives the illusion of pure noncompetitive inhibition (Figure 6·11b), but the rate law for noncompetitive inhibition is not applicable to an irreversible inhibitor such as TPCK.

(b) The phenylalanyl side chain of TPCK (which could also be abbreviated Tos-Phe-CH$_2$Cl) fits in the hydrophobic pocket of the substrate-binding site of chymotrypsin, positioning the chloromethylene group so that it reacts with the imidazole group of His-57. Since trypsin specifically binds amino acid residues with positively charged side chains (Figure 6·26b), replacing the phenylalanine residue of the chloromethyl ketone with either a lysine residue or an arginine residue would result in an affinity-labelling reagent for trypsin (Tos-Lys-CH$_2$Cl or Tos-Arg-CH$_2$Cl, respectively).

(c) Replacing the phenylalanine residue of the chloromethyl ketone with glycine or alanine would result in an affinity-labelling reagent for elastase (Tos-Gly-CH$_2$Cl or Tos-Ala-CH$_2$Cl).

(d) Since trypsin, chymotrypsin, and elastase are homologous serine proteases, treatment of a preparation with diisopropyl fluorophosphate (DFP) could be used to inactivate all three. Because DFP is a nerve gas, a less toxic compound with similar reactivity is often used.

(e) Treatment of a preparation with Tos-Lys-CH$_2$Cl (or with Tos-Arg-CH$_2$Cl) would inactivate trypsin but not chymotrypsin or elastase.

Part 2 Structures and Functions of Biomolecules

4. The chloromethyl ketone reacts with nucleophiles, resulting in their alkylation. In the case of a cysteine residue, TPCK reacts with the thiol group to form a thioether, which (like the thioether bond in methionine residues) is stable during acid hydrolysis of peptide bonds.

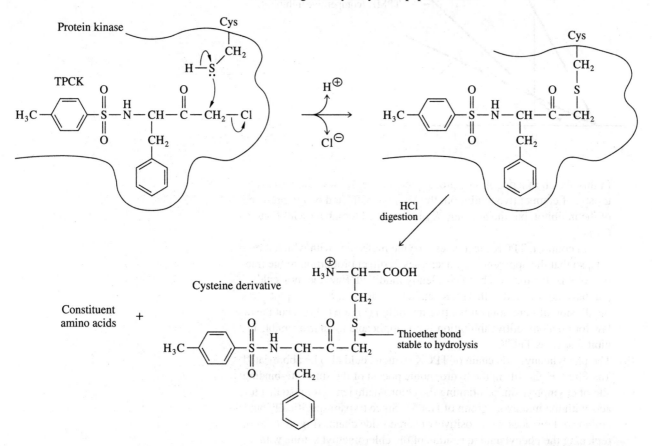

5. (a) First calculate the molar concentration of subtilisin. Then calculate V_{max}.

$$V_{max} = k_{cat}[E]_{total}$$

$$[E]_{total} = \frac{0.40 \text{ (mg/ml)}}{27\,600 \text{ (mg/mmol)}} = 1.45 \times 10^{-5} \text{ mmol ml}^{-1} = 1.45 \times 10^{-5} \text{M}$$

$$V_{max} = (550 \text{ s}^{-1})(1.45 \times 10^{-5} \text{M}) = 8.0 \times 10^{-3} \text{ M s}^{-1}$$
$$\text{or } 8.0 \times 10^{-3} \text{ mol l}^{-1} \text{ s}^{-1}$$

(b) Since indole is a *competitive* inhibitor, V_{max} is the same for both the inhibited and the uninhibited enzyme-catalyzed reactions. Note that V_{max} is not affected by [I] in the rate law for competitive inhibition (Table 6·4) nor in Lineweaver-Burk plots (Figure 6·8). Thus, V_{max} is $8.0 \times 10^{-3} \text{ M s}^{-1}$.

(c) Use the rate law that applies to competitive inhibition (Table 6·4).

$$v_0 = \frac{V_{max}[S]}{K_m\left(1 + \dfrac{[I]}{K_i}\right) + [S]}$$

$$v_0 = \frac{(8.0 \times 10^{-3} \text{ M s}^{-1})(0.25 \text{ M})}{(0.15 \text{ M})\left(1 + \dfrac{1.0 \text{ M}}{0.05 \text{ M}}\right) + 0.25 \text{ M}} = \frac{2.0 \times 10^{-3} \text{ M}^2 \text{ s}^{-1}}{3.4 \text{ M}}$$

$$v_0 = 5.9 \times 10^{-4} \text{ M s}^{-1} \text{ or } 5.9 \times 10^{-4} \text{ mol l}^{-1} \text{ s}^{-1}$$

6. (a) Cleavage of the peptide bond between Arg-15 and Ile-16 of chymo-trypsinogen allows protonation of the free α-amino group of isoleucine. The protonated α-amino group shifts so that it forms an ion pair with the β-carboxyl group of Asp-194. This local conformational change pulls the side chain of Met-192 away from some of its hydrophobic contacts, thereby generating the binding pocket that accepts the hydrophobic side chain of a substrate.

(b) Neither a chymotrypsin inhibitor nor an elastase inhibitor would be as effective as a trypsin inhibitor. Only the trypsin inhibitor could effectively prevent activation of other zymogens by blocking the action of trypsin, which triggers the pancreatic zymogen-activation cascade.

(c) The chloromethyl ketone that could act as an affinity-labelling reagent for thrombin would be Tos-Arg-CH$_2$Cl, an analog of tosylamidophenyl-ethyl chloromethyl ketone (TPCK, or Tos-Phe-CH$_2$Cl) with an arginine side chain.

(d) The fact that residues are more highly conserved in the interiors of trypsin and thrombin than on their surfaces implies that interior residues are critical in determining the overall conformation of the protein and the relative positions of the active-site residues. The three-dimensional structures of trypsin and thrombin are similar; therefore, the secondary structures (helices and β sheets) in the interiors, or "cores," of these proteins must be similar. The correct positions of residues involved in substrate binding and catalysis require precise protein folding. If an interior residue is replaced by another, there is greater likelihood that it will significantly affect protein folding (and therefore protein conformation) than if a surface residue is replaced by another. Most surface residues are hydrophilic; substitution of a polar or ionic surface residue with another would likely still permit hydrogen bonding with water and would probably have little or no effect on the overall conformation of the globular protein.

7

Mechanisms of Enzymes

Summary

Reaction mechanisms describe in atomic detail how a reaction proceeds. Kinetic experiments—measurements of reaction rates under varying conditions—offer an indirect approach to examining reactions that can contribute to the elucidation of mechanisms. Studies of the structures of enzymes supply additional mechanistic information.

The rate of a reaction depends on the rate of effective collisions between reactants. For each step in a reaction, there is a transition state, or energized configuration, that the reactants must pass through. The amount of energy needed to form the transition state, called the activation energy, affects the rate of the reaction. Catalysis provides a faster reaction pathway by lowering the energy of activation.

Ionizable and reactive amino acid residues in the active site of an enzyme form its catalytic center. Two major chemical modes of enzymatic catalysis are acid-base catalysis and covalent catalysis. In acid-base catalysis, proton transfer contributes to the acceleration of the reaction, with protons either donated by a weak acid or accepted by a base in the active site. The effect of pH on the rate of an enzymatic reaction can suggest the identities of active-site components. In covalent catalysis, the substrate or a portion of it is attached to the enzyme covalently to form a reactive intermediate. Nonenzymatic model-reaction systems are useful for simulating the mechanistic features of enzyme-catalyzed reactions.

The rates of catalysis for a few enzymes are so high that they approach the upper limit set by the rate at which reactants approach each other by diffusion. For enzymatic catalysis to be so rapid, each step of the reaction must be rapid, and the activation energies for the various steps must be balanced. The Circe effect allows catalysis by superoxide dismutase to exceed the diffusion-controlled limit.

Although acid-base catalysis and covalent catalysis are important, the greatest part of the rate acceleration achieved by an enzyme generally arises from the binding of reacting ligands to the enzyme. The initial formation of a noncovalent enzyme-substrate complex (ES) collects and orients reactants, which alone produces an acceleration of the reaction, a phenomenon termed the proximity effect. The binding of reactants must be relatively weak yet strong enough that the entropy of the reactants is considerably decreased. The energy of activation is further lowered by the binding of transition states with greater affinity than the binding of substrates. Support for transition-state stabilization as a major catalytic mode comes from the potent inhibitory activity of transition-state analogs, synthetic compounds that structurally resemble transition states. Furthermore, antibodies with catalytic activity can be induced by using transition-state analogs as antigens.

Enzymes that rely on induced-fit mechanisms use some of the substrate-binding energy for activating the enzyme. Conformational changes in some enzymes help to stabilize the transition state. Enzymes, therefore, catalyze reactions by assisting first in the formation, and then in the stabilization, of transition states.

The role of binding in catalysis has been demonstrated with tyrosyl-tRNA synthetase using structural information, kinetic experiments, and site-directed mutagenesis. The results of these experiments suggest that the transition state is stabilized by at least seven hydrogen bonds that form only with the reactant in the transition state, not with the reactant in its ground state.

Lysozyme presumably binds an unstable carbonium ion, stabilizing it. In addition, it uses proximity, acid-base catalysis, and substrate distortion as catalytic modes.

The serine proteases, exemplified by chymotrypsin, use both chemical and binding modes of catalysis. All serine proteases possess a hydrogen-bonded Ser-His-Asp catalytic triad in their active sites. The serine residue serves as a covalent catalyst, and the histidine residue serves as an acid-base catalyst. The aspartate residue, which is essential for maximum activity, aligns the histidine residue and stabilizes its protonated form. Anionic tetrahedral intermediates form additional hydrogen bonds with the enzyme; these bonds contribute to catalysis by stabilizing the transition state.

Study Information

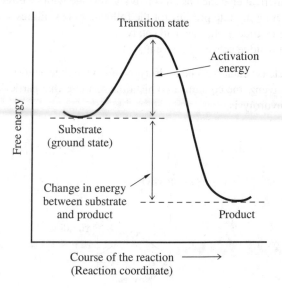

Course of the reaction ⟶
(Reaction coordinate)

Figure 7·1
Reaction diagram for a single-step reaction. This curve shows the lowest energy path between the substrate and the product. The upper arrow shows the energy of activation for the forward reaction. Molecules of substrate that have more free energy than the activation energy can pass over the activation barrier and become molecules of product. Because the substrate is at a higher free-energy level than the product, this is a spontaneous, or exergonic, reaction.

Table 7·1 Catalytic functions of ionizable amino acids

Amino acid	Reactive group	Net charge at pH 7	Principal functions
Aspartate	$-COO^{\ominus}$	−1	Cation binding; proton transfer
Glutamate	$-COO^{\ominus}$	−1	Cation binding; proton transfer
Histidine	Imidazole	Near 0	Proton transfer
Cysteine	$-S^{\ominus}$	Near 0	Covalent binding of acyl groups
Tyrosine	$-OH$	0	Hydrogen bonding to ligands
Lysine	$-NH_3^{\oplus}$	+1	Anion binding
Arginine	Guanidinium	+1	Anion binding
Serine	$-CH_2OH$	0	Covalent binding of acyl groups

Table 7·2 Typical pK_a values of ionizable groups of amino acids in proteins

Group	pK_a
Terminal α-carboxyl	3–4
Side-chain carboxyl	4–5
Imidazole	6–7
Terminal α-amino	7.5–9
Thiol	8–9.5
Phenol	9.5–10
ε-Amino	~10
Guanidine	~12
Hydroxymethyl	~16

Problems

1. A hypothetical enzyme has an active-site residue with an effective pK_a value of 5.0. Draw pH-rate profiles for the following possibilities:
   (a) The residue acts as an acid catalyst.
   (b) The residue acts as a base catalyst.

2. Ribonuclease catalyzes the hydrolysis of ribonucleic acid (RNA). The active site of the enzyme contains two histidine residues that participate in the first step of hydrolysis.

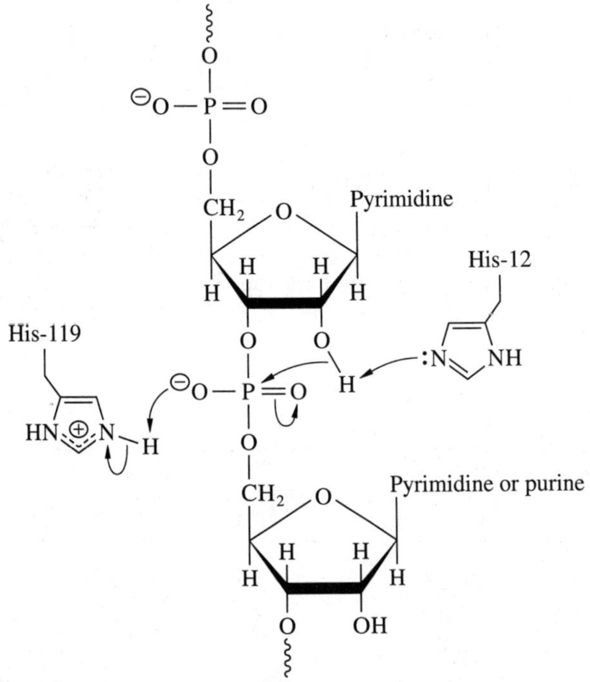

The pH-rate profile for the ribonuclease-catalyzed reaction is bell shaped, with inflection points at approximately pH 5.8 and pH 6.2. How is the pH-rate profile related to the mechanism for ribonuclease?

3. The rate of spontaneous hydrolysis of acetylsalicylate (aspirin) is about 100 times faster than that of phenyl acetate (both shown below). Write a mechanism that could account for this observation and explain how this observation is related to enzyme-catalyzed reactions.

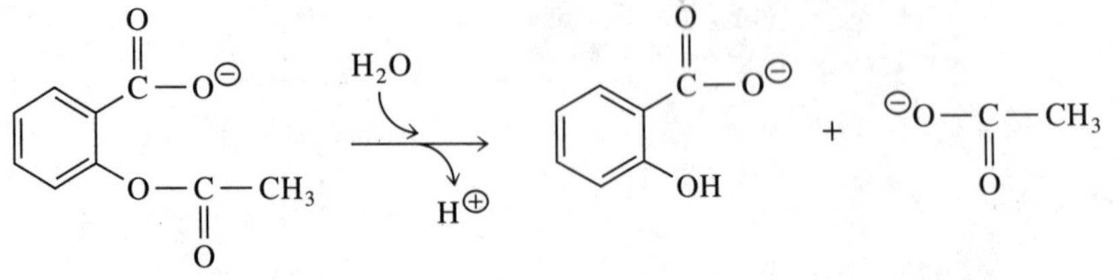

Acetylsalicylate

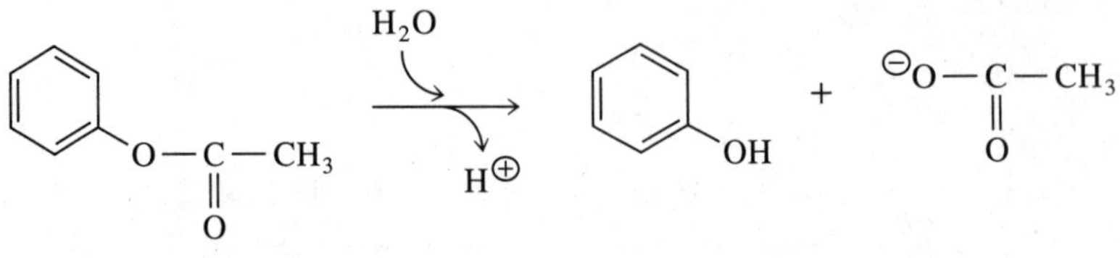

Phenyl acetate

4. Acetylcholinesterase catalyzes the hydrolysis of acetylcholine.

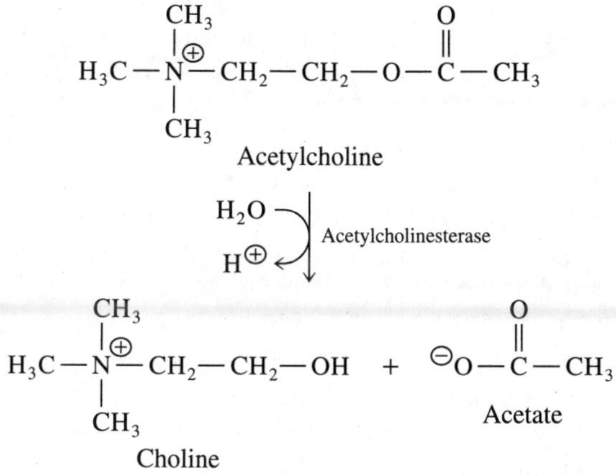

$$H_3C-\overset{\overset{\displaystyle CH_3}{|}}{\underset{\underset{\displaystyle CH_3}{|}}{N^{\oplus}}}-CH_2-CH_2-O-\overset{\overset{\displaystyle O}{\|}}{C}-CH_3$$

Acetylcholine

$$H_2O \qquad H^{\oplus} \qquad \text{Acetylcholinesterase}$$

$$H_3C-\overset{\overset{\displaystyle CH_3}{|}}{\underset{\underset{\displaystyle CH_3}{|}}{N^{\oplus}}}-CH_2-CH_2-OH \quad + \quad {}^{\ominus}O-\overset{\overset{\displaystyle O}{\|}}{C}-CH_3$$

Choline   Acetate

The enzyme is inactivated by treatment with diisopropyl fluorophosphate (DFP).
(a) Write a mechanism for the reaction of this residue with DFP, which leads to inactivation of acetylcholinesterase.
(b) Suggest a mechanism for the acetylcholinesterase-catalyzed hydrolysis of its substrate, acetylcholine.

5. The plant protease papain is inactivated by treatment with p-hydroxy-mercuribenzoate (PHMB) or by treatment with Tos-Lys-CH$_2$Cl. These and other experimental observations, together with the X-ray crystallographic analysis of the tertiary structure of the enzyme, have led to identification of a catalytic center consisting of Cys-25 and His-159. The bell-shaped pH-rate profile for papain-catalyzed hydrolysis has inflection points at pH 4.2 and pH 8.2.
(a) Depict the reaction of an active-site residue in papain with PHMB.
(b) Inactivation of papain by treatment with iodoacetate, chloroacetate, or Tos-Lys-CH$_2$Cl results in alkylation of Cys-25 but not His-159. What does this reveal about the apparent pK_a values exhibited in the pH-rate profile and the catalytic roles of Cys-25 and His-159?
(c) Propose a catalytic mechanism for the hydrolysis of a Lys–Phe peptide bond by the action of papain.

6. In the oligopeptide Ac-Leu–Leu–Phe-H, an acetyl group (Ac) is attached to the terminal amino group, and an aldehyde group has replaced the terminal carboxylate. This compound (shown below) is an extremely potent inhibitor of chymotrypsin, with a K_i resembling that of a transition-state analog. Propose a mechanism for inhibition by this aldehyde.

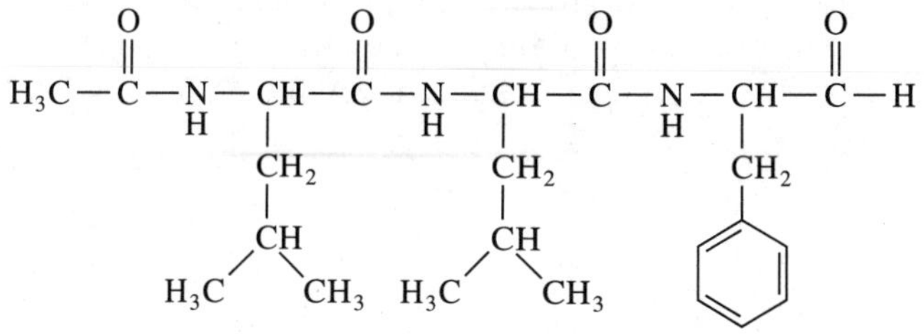

Solutions

1. (a) If the activity of the enzyme depends on acid catalysis, the enzyme would donate a proton in the rate-limiting step of the reaction. The protonated species (BH) would predominate over the conjugate base (B⊖) when the pH of the solution is *below* the pK_a of BH. Since the pK_a of BH in this example is 5.0, enzyme activity would increase below pH 5.0 and decrease above pH 5.0, coinciding with the titration of BH (BH ⇌ B⊖ + H⊕). The reaction would thus produce the pH-rate profile shown on the left, below.

    (b) If the activity of the enzyme depends on base catalysis, a proton would be abstracted by the enzyme during the rate-limiting step of the reaction. In this case, the unprotonated species (B⊖) would predominate at the active site, and enzymatic activity would increase above pH 5.0 and decrease below pH 5.0, giving the pH-rate profile shown on the right, below.

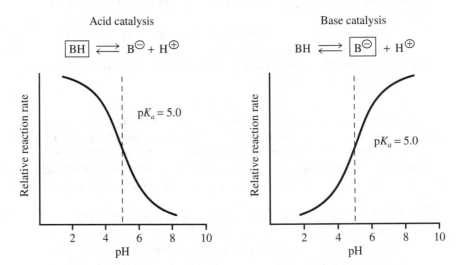

2. In order for ribonuclease to function according to the mechanism shown, His-12 must be in the unprotonated (imidazole) form to serve as a base catalyst and His-119 must be in the protonated (imidazolium ion) form to serve as an acid catalyst. The apparent pK_a values corresponding to the inflection points are approximately 5.8 (for His-12) and 6.2 (for His-119). The bell-shaped pH-rate profile can be related to protonation and deprotonation of the two histidine residues as follows: below pH 5.8, the majority of both histidine residues are protonated (imidazolium ions), precluding a concerted acid-base catalytic mechanism; above pH 6.2, the majority of both histidine residues are unprotonated (imidazole groups), also precluding concerted acid-base catalysis. At the pH optimum of the enzyme (pH 6, midway between pH 5.8 and pH 6.2), there is an optimal combination of protonated His-119 and unprotonated His-12.

3. Hydrolysis of phenyl acetate proceeds by the attack of water (or OH$^\ominus$) on the carbonyl carbon of the ester. The hydrolysis of the ester of aspirin is enhanced by the catalytic action of the carboxylate group *ortho* to the ester substituent. Two possible mechanisms of catalysis are shown below. In the first, the *o*-carboxylate serves as an attacking nucleophile that displaces the acetyl group from the phenolic oxygen to form an anhydride intermediate that is more rapidly hydrolyzed than the original ester. (Such an intermediate is analogous to an acyl-enzyme intermediate generated by nucleophilic attack during covalent catalysis.) A second possible mechanism involves base catalysis by the *o*-carboxylate, which accelerates the rate-limiting step (attack by OH$^\ominus$ on the carbonyl carbon of the ester bond). In fact, the first mechanism is followed by the 3,5-dinitro analog of aspirin but not by aspirin itself, which follows the second mechanism. Both mechanisms of ester hydrolysis provide an example of intramolecular catalysis, as observed with the unimolecular substrates that demonstrate the proximity effect.

(1)

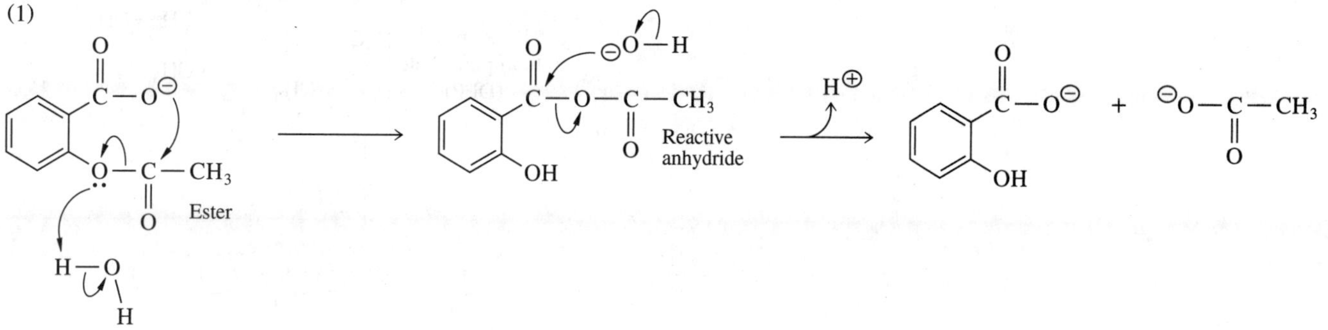

(2)

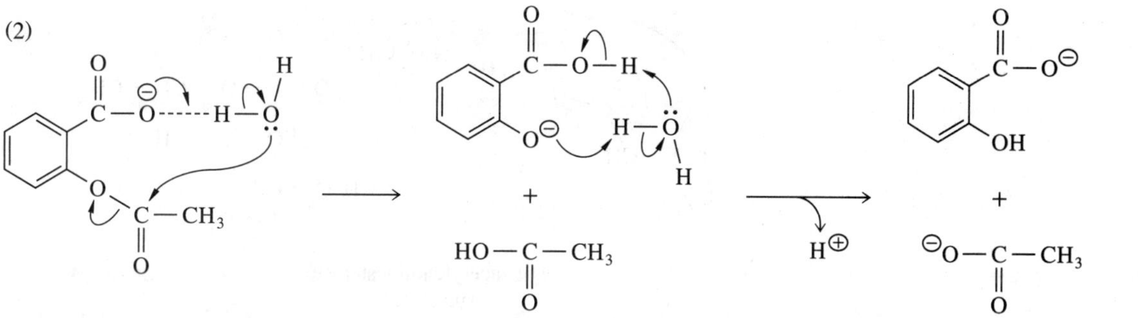

4. (a) Inactivation of acetylcholinesterase by DFP involves displacement of $F^{\ominus}$ from the phosphorus atom of the reagent by the oxygen of the side chain of a uniquely nucleophilic serine residue. The nucleophilicity of the hydroxymethyl group of the serine residue at the active site is strengthened by hydrogen bonding within the catalytic triad (serine to histidine to aspartate).

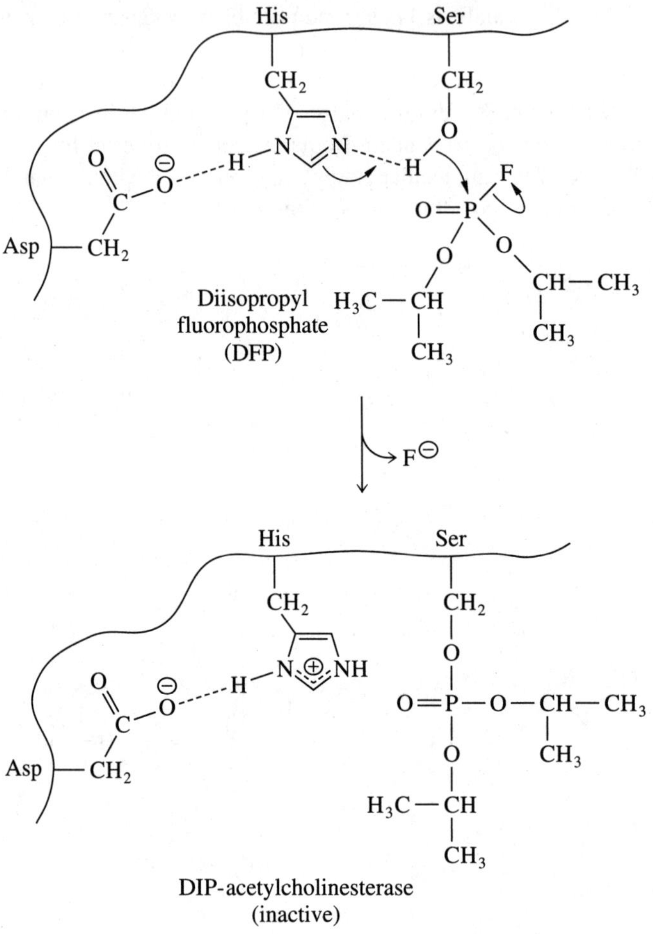

DIP-acetylcholinesterase
(inactive)

(b)

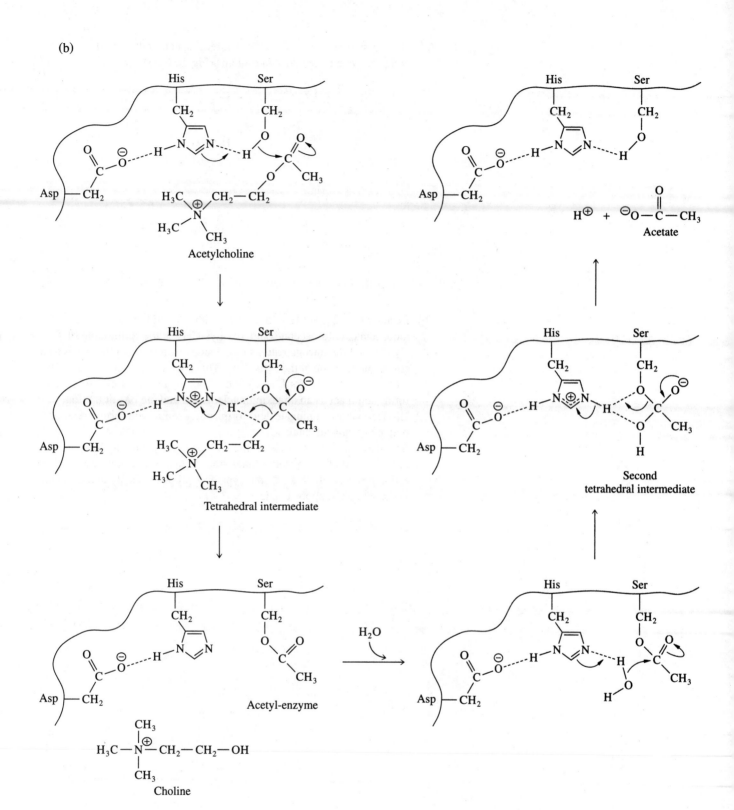

5. (a) Organic mercurials, such as PHMB, react specifically with the thiol group of cysteine residues in proteins (Figure 6·33).

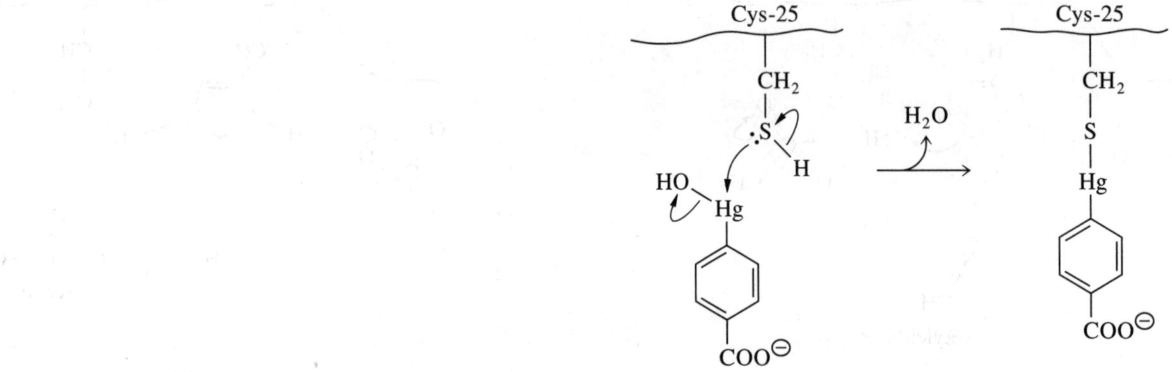

p-Hydroxymercuribenzoate
(PHMP)

(b) Because Cys-25, not His-159, reacts with alkylating reagents such as iodoacetate, chloroacetate, and Tos-Lys-CH$_2$Cl, the sulfur atom of Cys-25, not the nitrogen atoms of the side chain of His-159, serves as a strong nucleophile at the active site. This conclusion is consistent with the pK_a value of 4.2 for Cys-25, whose conjugate base form (thiolate ion form) could act as an attacking nucleophile, and the pK_a of 8.2 for His-159, whose conjugate acid form (imidazolium ion form) could participate as an acid catalyst.

(c) In papain, the role of nucleophilic Cys-25 in catalyzing hydrolysis of a peptide bond of a substrate is analogous to that of nucleophilic serine in serine proteases such as chymotrypsin. A proposed catalytic mechanism is shown on the opposite page.

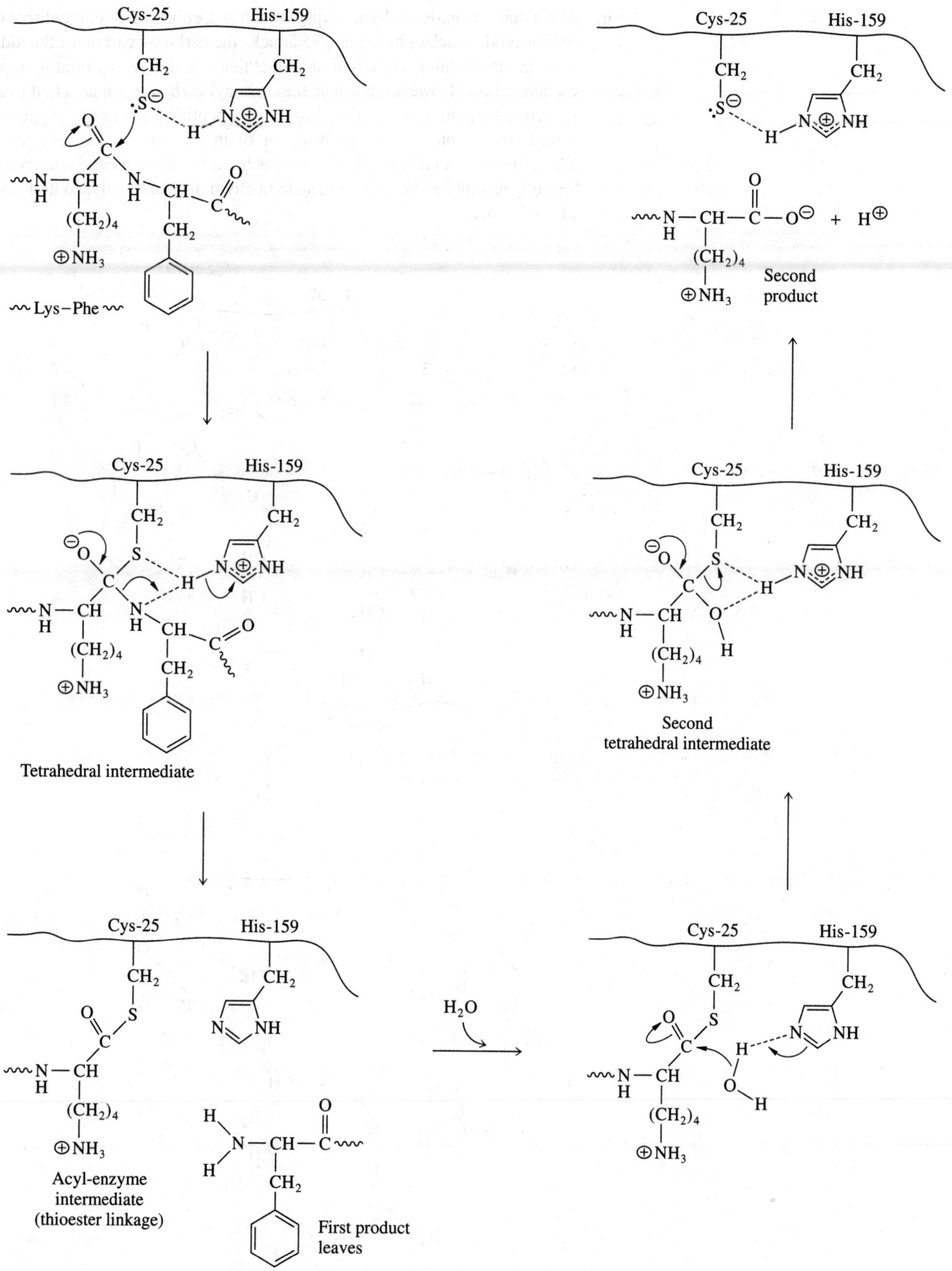
Cys-25
His-159
Lys–Phe
Tetrahedral intermediate
Acyl-enzyme
intermediate
(thioester linkage)
First product
leaves
H2O
Second
tetrahedral intermediate
Second
product

6. The aldehyde binds to chymotrypsin much as a substrate or competitive inhibitor binds: nucleophilic Ser-195 attacks the carbonyl carbon of the aldehyde group, forming a tetrahedral adduct that is stabilized by binding in the oxyanion hole. However, because the carbonyl carbon is not attached to a nucleophilic atom (such as the nitrogen of a peptide bond or the oxygen of an ester bond), displacement and formation of an acyl-enzyme cannot occur. Thus, the oligopeptide aldehyde remains bound as a tetrahedral adduct that strongly resembles the transition state in chymotrypsin-catalyzed hydrolysis of a substrate.

8

Coenzymes

Summary

Many enzyme-catalyzed reactions depend on the presence of cofactors. Cofactors include essential ions and group-transfer reagents called coenzymes. Inorganic ions, such as $K^{\oplus}$, $Mg^{2\oplus}$, $Ca^{2\oplus}$, $Fe^{2\oplus}$, and $Zn^{2\oplus}$, participate in the binding of substrates, in stabilization of enzymes, and as active catalytic components. Some coenzymes are synthesized from common metabolites, and a number of others are derived from B vitamins. Vitamins are organic compounds that must be supplied in small amounts in the diets of humans and other animals to avoid nutritional-deficiency diseases. Coenzymes that carry mobile metabolic groups may function as cosubstrates of enzymes, or they may be firmly attached to enzymes as prosthetic groups.

ATP is the most common metabolite coenzyme. It can donate a phosphoryl, pyrophosphoryl, adenylyl, or adenosyl group. Other metabolite coenzymes include S-adenosylmethionine, phosphoadenosine phosphosulfate, and the sugar nucleotides. Cytidine nucleotides participate in the biosynthesis of lipids.

The pyridine nucleotide coenzymes, $NAD^{\oplus}$ and $NADP^{\oplus}$, are derived from nicotinic acid (niacin). Pyridine nucleotide–dependent dehydrogenases catalyze transfer of a hydride ion ($H^{\ominus}$) from a specific substrate to position 4 of the pyridine ring of $NAD^{\oplus}$ or $NADP^{\oplus}$, reducing the coenzyme to NADH or NADPH, respectively, with release of a proton. The pyridine nucleotides accept or donate two electrons at a time.

Riboflavin (vitamin B_2) consists of an isoalloxazine ring system and a ribitol residue. Characteristically, the coenzyme forms of riboflavin—FAD and FMN—are tightly bound as prosthetic groups in flavoproteins. FAD and FMN are reduced by hydride (two-electron) transfers to form $FADH_2$ and $FMNH_2$. The reduced flavin coenzymes donate electrons one at a time, passing through the stable free-radical forms $FADH\cdot$ and $FMNH\cdot$.

Coenzyme A is derived from pantothenic acid. It is the principal coenzyme involved in acyl-group–transfer reactions. The energy of the thioester bond of acetyl CoA is about the same as the energy of the phosphoanhydrides of ATP. The prosthetic group of acyl carrier protein (ACP), 4'-phosphopantetheine, is also derived from pantothenic acid.

The first vitamin isolated, thiamine (vitamin B_1), was found to prevent or cure beriberi in humans subsisting largely on polished rice. Its coenzyme form is thiamine pyrophosphate (TPP), whose thiazole ring binds the aldehyde generated upon decarboxylation of an α-keto acid substrate. A number of α-keto acid dehydrogenases require TPP as a coenzyme or prosthetic group.

Pyridoxal 5'-phosphate (PLP) serves as a prosthetic group for many enzymes involved in amino acid metabolism. It is derived from vitamin B_6, pyridoxine. The aldehyde group at C-4 of the pyridine ring of PLP reversibly reacts with a free amino group to form an aldimine (Schiff base). In many cases, the coenzyme is linked to a lysine residue at the active site of a PLP-dependent enzyme. PLP participates in the catalytic mechanism of such enzymes by forming a Schiff base with an amino acid substrate, through which it stabilizes the carbanion that is generated upon cleavage of a bond to the α-carbon atom of the amino acid.

Biotin serves as a prosthetic group for several ATP-dependent carboxylases and carboxyltransferases. It is covalently linked by an amide bond to the ε-amino group of a lysine residue at the enzyme active site; the biotinyl lysine residue is sometimes referred to as biocytin.

Tetrahydrofolate, the reduced form of folic acid, is involved in the transfer of one-carbon units at the oxidation levels of methanol, formaldehyde, and formic acid. Such transfers are especially important in protein synthesis and in the biosynthesis of purine nucleotides and dTMP. The tetrahydropterin moiety of tetrahydrofolate is also found in tetrahydrobiopterin, a coenzyme involved in O_2-dependent hydroxylation reactions, and as a component of a molybdenum-containing cofactor.

Vitamin B_{12} and its coenzyme forms contain a corrin ring system. The adenosylcoenzyme derivative of vitamin B_{12}, adenosylcobalamin, is involved in some intramolecular rearrangement reactions. The methyl form, methylcobalamin, is an intermediate in the biosynthesis of methionine from 5-methyltetrahydrofolate and homocysteine.

Lipoic acid, a prosthetic group for α-keto acid dehydrogenase multienzyme complexes, can be synthesized by most organisms. It accepts acyl groups, forming a thioester that passes the acyl group to a second acceptor such as coenzyme A. In this process, dihydrolipoic acid is formed.

The four fat-soluble, or lipid, vitamins are A, D, E, and K. Vitamin A functions as a light-sensitive compound in vision. A derivative of vitamin D regulates Ca^{2+} utilization. Vitamin E helps prevent oxidative damage to membrane lipids. Vitamin K is essential for the conversion of glutamate to γ-carboxyglutamate residues in certain blood-coagulation proteins.

Ubiquinone (coenzyme Q) is a lipid-soluble electron carrier present in the inner membrane of mitochondria. It transfers electrons between enzyme complexes that catalyze the oxidation of NADH and $FADH_2$. Ubiquinone can accept or donate either one or two electrons. Plastoquinone in chloroplasts is similar in structure and function to ubiquinone.

Certain enzymes undergo modifications in which amino acid residues are converted into active-site prosthetic groups. The best-known example is the pyruvamide group of bacterial histidine decarboxylase, which arises from modification of a serine residue. This pyruvate residue plays a mechanistic role similar to that of the aldehyde group of PLP.

Some proteins are not catalysts themselves but function as group-transfer agents, very much like coenzymes. These proteins usually have relatively low molecular weights and are heat stable. Examples are cytochrome *c,* ferredoxin, and acyl carrier protein (ACP).

Study Information

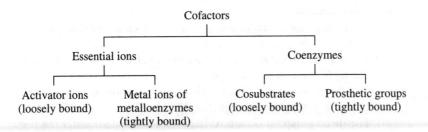

Figure 8·1
Types of cofactors. Cofactors can be divided into two types—essential ions and coenzymes—which can be further distinguished by the strength of interaction with their apoenzymes. An essential ion may be an activator, which binds loosely to an enzyme, or a metal ion that is bound tightly within a metalloenzyme. Similarly, a coenzyme may be a loosely bound cosubstrate or a tightly bound prosthetic group.

Table 8·1 Major coenzymes

Coenzyme	Vitamin source	Major metabolic roles	Mechanistic role
Adenosine triphosphate (ATP)	No	Transfer of phosphoryl or nucleotidyl groups	Cosubstrate
S-Adenosylmethionine	No	Transfer of methyl groups	Cosubstrate
Phosphoadenosine phosphosulfate (PAPS)	No	Transfer of sulfuryl groups	Cosubstrate
Nucleotide sugars	No	Transfer of carbohydrate groups	Cosubstrate
Cytidine diphosphate (CDP) alcohols	No	Transfer of alcohols in lipid synthesis	Cosubstrate
Nicotinamide adenine dinucleotide (NAD$^{\oplus}$) and nicotinamide adenine dinucleotide phosphate (NADP$^{\oplus}$)	Niacin	Oxidation-reduction reactions involving two-electron transfers	Cosubstrate
Flavin mononucleotide (FMN) and flavin adenine dinucleotide (FAD)	Riboflavin (B$_2$)	Oxidation-reduction reactions involving one- and two-electron transfers	Prosthetic group
Coenzyme A (CoA)	Pantothenic acid (B$_3$)	Transfer of acyl groups	Cosubstrate
Thiamine pyrophosphate (TPP)	Thiamine (B$_1$)	Transfer of aldehyde groups	Prosthetic group
Pyridoxal phosphate (PLP)	Pyridoxine (B$_6$)	Transfer of groups to and from amino acids	Prosthetic group
Biocytin (biotin bound to ε-amino group in a biotinylated enzyme)	Biotin (H)	ATP-dependent carboxylation of substrates or carboxyl-group transfer between substrates	Prosthetic group
Tetrahydrofolate	Folic acid (B$_c$)	Transfer of one-carbon substituents, especially formyl and hydroxymethyl groups; provides the methyl group for thymine in DNA	Cosubstrate
Adenosylcobalamin and methylcobalamin	Cobalamin (B$_{12}$)	Intramolecular rearrangements and transfer of methyl groups	Prosthetic group
Lipoamide residue (lipoyl group bound to ε-amino group in a protein)	No	Oxidation of a hydroxyalkyl group from TPP and subsequent transfer as an acyl group	Prosthetic group
cis-Retinal	Vitamin A	Vision	Prosthetic group
Vitamin K	Vitamin K	Carboxylation of some glutamate residues	Prosthetic group
Ubiquinone	No	Lipid-soluble electron carrier	Cosubstrate
Plastoquinone	No	Lipid-soluble electron carrier	Cosubstrate

Problems

1. Identify each of the following coenzymes and indicate the vitamin from which it is derived.
   (a) Cosubstrate for the reduction of a ketone (such as pyruvate) to a secondary alcohol (such as lactate).
   (b) Cosubstrate for the oxidation of a primary alcohol (such as ethanol) to an aldehyde (such as acetaldehyde).
   (c) Prosthetic group for ATP-dependent carboxylation reactions, such as carboxylation of pyruvate to form oxaloacetate.
   (d) Prosthetic group for decarboxylation and aldehyde-transfer reactions, such as the decarboxylation of pyruvate to form acetaldehyde.
   (e) Cosubstrate for transfer of formyl or methylene (hydroxymethyl) groups.
   (f) Cosubstrate for transfer of acetyl (two-carbon) or longer acyl groups.
   (g) Prosthetic group for the removal and replacement of groups from the α-carbon of amino acids.

2. Explain why a particular enzyme preparation may show full activity in a potassium phosphate buffer but only partial activity in a Tris buffer.

3. Alcohol dehydrogenase from yeast catalyzes the $NAD^{\oplus}$-dependent oxidation of ethanol to acetaldehyde. This metalloenzyme contains a zinc atom at the reactive site.
   (a) Draw a mechanism for the alcohol dehydrogenase reaction that is analogous to the mechanism for lactate dehydrogenase (Figure 8·19).
   (b) Does alcohol dehydrogenase require a residue analogous to His-195 in lactate dehydrogenase?
   (c) Does alcohol dehydrogenase require a residue analogous to Arg-171 in lactate dehydrogenase?

4. A sample of mammalian liver is homogenized in a mixture of chloroform and water. In which phase would you find each of the following: vitamin A, vitamin B_6, vitamin C, and vitamin D?

5. Enzyme-catalyzed transfer of a hydride ion from NADH to a substrate and the reverse transfer from the reduced form of the substrate to $NAD^{\oplus}$ are strictly stereospecific. Propose an experiment using [^{3}H]-labelled lactate to determine whether yeast alcohol dehydrogenase (which catalyzes the conversion of ethanol to acetaldehyde) and muscle lactate dehydrogenase (which catalyzes the conversion of lactate to pyruvate; Figure 8·19) catalyze transfer of a hydride ion to and from the same face of the coenzyme (NADH). Specify which hydrogen atom in lactate should be labelled.

6. Explain why mammalian cells cannot survive more than a few hours in a buffered salt solution.

7. UDP-glucose 4'-epimerase is an isomerase that catalyzes inversion of configuration at C-4' of uridine diphosphate glucose to produce UDP-galactose.

   UDP-glucose   UDP-galactose

   The epimerase uses $NAD^{\oplus}$ as a prosthetic group rather than as a cosubstrate. Provide a mechanism that shows the role of $NAD^{\oplus}$ in the conversion of UDP-glucose to UDP-galactose.

8. Virtually all organisms can synthesize purine nucleotides from smaller precursor molecules. Some parasitic protozoa, such as the one that causes
malaria, rely on their hosts to supply purines. Is ATP a metabolite coenzyme
for these microorganisms?

Solutions

1. (a) NADH (reduced nicotinamide adenine dinucleotide) or, in some cases,
NADPH (reduced nicotinamide adenine dinucleotide phosphate); the
vitamin precursor is niacin in either case.
   (b) $NAD^{\oplus}$ (nicotinamide adenine dinucleotide); the vitamin precursor is
niacin.
   (c) Biocytin (a biotinyl-lysine residue); the vitamin precursor is biotin.
   (d) TPP (thiamine pyrophosphate); the vitamin precursor is thiamine (vitamin B_1).
   (e) Tetrahydrofolate; the vitamin precursor is folate.
   (f) CoA (coenzyme A); the vitamin precursor is pantothenate.
   (g) PLP (pyridoxal 5'-phosphate); the vitamin precursor is pyridoxine (vitamin B_6).

2. The enzyme may be stabilized or activated by potassium ions, which are lacking in the Tris buffer, or the enzyme may be inhibited by Tris.

3. (a) In one proposed mechanism, a water molecule bound to the zinc ion of
alcohol dehydrogenase forms $OH^{\ominus}$, in the same manner as the water
bound to carbonic anhydrase (Section 8·1B). The basic hydroxide ion ab-
stracts the proton from the hydroxyl group of ethanol to form H_2O. (Another mechanism proposes that the zinc also binds to the alcoholic oxygen of the ethanol, polarizing it.)

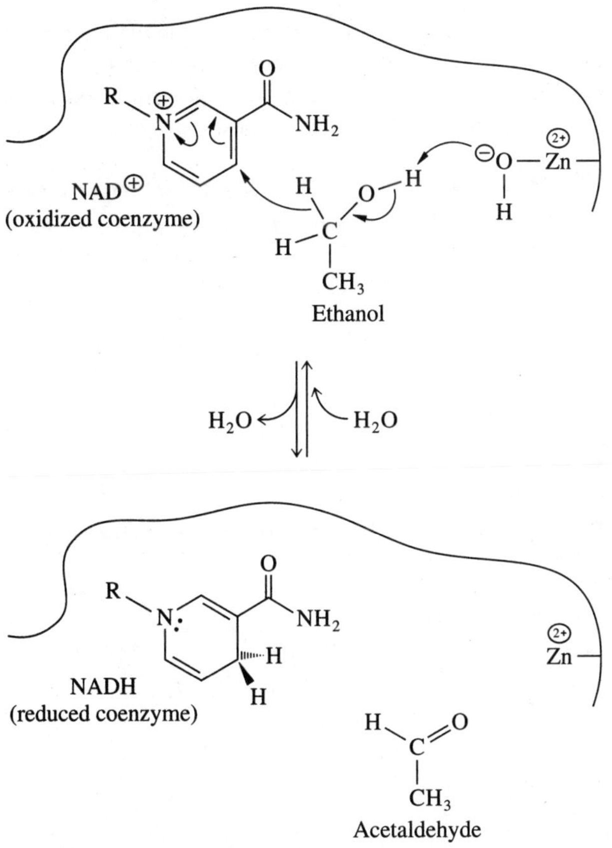

55

(b) No, a basic residue is not required. The zinc ion serves the same purpose as His-195 of lactate dehydrogenase, promoting ionization of the alcoholic group and facilitating transfer of the hydride ion from C-1 of ethanol to C-4 of $NAD^{\oplus}$.

(c) No, a residue such as arginine is not required. Ethanol, unlike lactate, lacks a carboxylate group that can bind electrostatically to the arginine side chain.

4. Vitamin B_6 and vitamin C would be found in the aqueous phase; vitamins A and D are lipid vitamins and would be found in the organic phase.

5. Incubate 2-[^{3}H]-lactate and $NAD^{\oplus}$ with lactate dehydrogenase to generate tritium-labelled NADH (NAD[^{3}H]).

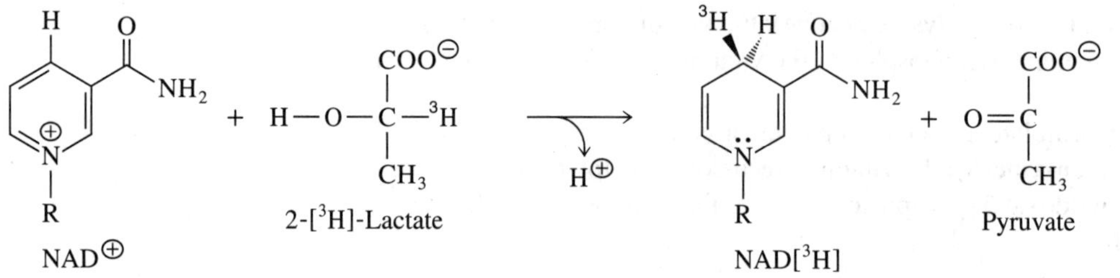

Isolate NAD[^{3}H] by a procedure such as gel-filtration or ion-exchange chromatography and then incubate the NAD[^{3}H] with acetaldehyde and alcohol dehydrogenase. Isolate the products of this reaction and determine whether ethanol or $NAD^{\oplus}$ contains tritium. If 1-[^{3}H]-ethanol is formed, the stereospecificity of alcohol dehydrogenase and lactate dehydrogenase is the same, but if the $NAD^{\oplus}$ retains the tritium label, the enzymes have opposite stereospecificity. (In fact, these two dehydrogenases do have the same stereospecificity, which is shown above.)

6. Mammalian cells survive for several hours when maintained in a solution of the appropriate pH and salt concentration. However, the cells gradually lose essential ions and small metabolites. For long-term viability, cells must be maintained in a medium that includes these substances as well as a source of energy.

7. UDP-glucose is first oxidized by the action of the epimerase, resulting in a reduced prosthetic group (NADH) and a keto-sugar intermediate. The keto-
   sugar intermediate then rotates within the active site of the enzyme, allowing
   NADH to reduce the keto group from the opposite side. The product of this
   reduction is UDP-galactose.

8. ATP is not a metabolite coenzyme in these organisms. Exogenous adenine is
   required for the growth of these organisms, just as vitamins are required by
   some animals. ATP is therefore analogous to a vitamin-derived cofactor.

9

Carbohydrates and Glycoconjugates

Summary

Carbohydrates consist of hydroxyaldehydes (aldoses) and hydroxyketones (ketoses) and their derivatives. They include monosaccharides and disaccharides (simple sugars), oligosaccharides, and polysaccharides. Except for the simplest ketose, dihydroxyacetone, carbohydrates are chiral and therefore exhibit optical activity. For a given monosaccharide, there are 2^n possible stereoisomers, where n is the number of chiral carbon atoms. Among these stereoisomers, any two that are non-superimposable mirror images of each other are referred to as enantiomers. Two that differ in configuration at only one of several chiral centers are known as epimers. A monosaccharide is designated D or L, depending on the configuration of the chiral carbon farthest from the aldehydic (C-1) or ketonic (usually C-2) carbon atom.

Aldopentoses, aldohexoses, and ketohexoses exist principally as cyclic hemiacetals known as furanoses and pyranoses. In sugar hemiacetals, the anomeric carbon (the carbonyl carbon in the open-chain form) has four substituents, giving these structures an additional asymmetric center. The chirality of the anomeric carbon is designated either α or β. Two optical isomers that differ in configuration only at their anomeric carbon atoms are referred to as anomers. Although furanoses adopt envelope or twist conformations and pyranoses exist primarily in chair conformations, they are often depicted in Haworth projections, which reveal the chirality of each asymmetric carbon atom including the anomeric carbon.

There are several classes of biologically important nonpolymerized derivatives of monosaccharides. These include sugar phosphates, deoxy sugars, amino sugars, sugar alcohols, and sugar acids. Ascorbic acid is also an important monosaccharide derivative.

Monosaccharide residues can be linked via glycosidic bonds to form oligosaccharides and polysaccharides. Four important disaccharides are maltose, cellobiose, lactose, and sucrose. Lactose, an epimer of cellobiose, is the major carbohydrate in milk. Sucrose is synthesized in many plants and is the most abundant disaccharide found in nature. The anomeric carbons of both monosaccharide residues in sucrose are involved in the glycosidic linkage of the disaccharide.

Glycosides are compounds formed when the anomeric carbons of sugars form glycosidic linkages with hydroxyl groups of other sugar molecules or with organic nonsugar molecules. Nucleotides are commonly encountered glycosides.

Glucose is the repeating monomeric unit of the storage polysaccharides amylose, amylopectin, and glycogen and of the structural polysaccharide cellulose, which is the most abundant organic substance in the biosphere. Chitin, the second most abundant organic compound on earth, is another example of a storage homopolysaccharide. Its monomeric unit is β-$(1\rightarrow4)$-linked N-acetylglucosamine.

Important glycoconjugates include peptidoglycans, glycoproteins, and proteoglycans. A peptidoglycan is a major component of bacterial cell walls, which maintain the shape and functional integrity of bacteria. The glycan moiety of this peptidoglycan is a polymer of a repeating disaccharide, composed of N-acetylglucosamine and N-acetylmuramic acid joined by β-$(1\rightarrow4)$ linkages. The action of lysozyme cleaves the linkage between N-acetylmuramic acid and N-acetylglucosamine, depolymerizing the glycan moiety. The carboxyl group of N-acetylmuramic acid is linked to a tetrapeptide containing a mixture of L and D amino acids. The C-terminal D-alanine of this tetrapeptide is in turn cross-linked to the penultimate residue of a tetrapeptide on a neighboring peptidoglycan molecule via a linker peptide consisting of five glycine residues. This cross-linking of peptidoglycans results in a giant macromolecule that provides great overall rigidity to the cell wall.

The pathway of biosynthesis of peptidoglycans features stepwise addition of amino acids to MurNAc and involvement of glycyl–transfer RNA. The final step involves linkage of a D-alanine residue to a pentaglycine cross-link of a neighboring peptidoglycan molecule, catalyzed by a transpeptidase. This step is inhibited by penicillin. Certain other antibiotics also inhibit various steps in peptidoglycan synthesis.

Other important carbohydrate-containing molecules are components of cell walls and outer membranes of certain bacteria. The teichoic acids (polymers of glycerol phosphate or ribitol phosphate) are found in the cell walls of certain Gram-positive bacteria, and lipopolysaccharides are found in the outer membranes of Gram-negative bacteria.

Many proteins with a wide variety of functions are glycosylated. The glycan chains exhibit great diversity in structure due to the variations in the monosaccharides added, the atoms involved in the glycosidic bond, and the potential for branched structures. Most glycoproteins can be classified according to the nature of the linkage joining the protein to its carbohydrate moiety. The three major classes exhibit either an *O*-glycosidic linkage, an *N*-glycosidic linkage, or a linkage to a phosphatidylinositol-glycan structure. The major linkage in *O*-linked glycoproteins is between *N*-acetylgalactosamine and the hydroxyl group of serine or threonine. In *N*-linked glycoproteins, the linkage involves *N*-acetylglucosamine and the nitrogen of the amide group of asparagine. A number of proteins that are present on the extracellular surface of plasma membranes are anchored to its outer leaflet by phosphatidylinositol-glycan structures.

The biosynthesis of *O*-linked glycoproteins uses nucleotide sugars as glycosyl-group donors. Glycosyltransferases sequentially catalyze the addition of individual sugars. In the biosynthesis of *N*-linked glycoproteins, the initial saccharide chain donated by dolichol pyrophosphate–oligosaccharide is processed by specific glycosidases and glycosyltransferases to produce high-mannose chains or complex chains. In certain cases, the result is a hybrid oligosaccharide chain, a branched chain in which one branch is of the high-mannose type and the other is of the complex type.

A variety of methods are available to detect and purify glycoproteins. The principal methods used to characterize their oligosaccharide chains are binding to lectins, compositional analysis, the use of specific glycosidases, mass spectrometry, NMR spectroscopy, and methylation analysis.

The oligosaccharide chains of glycoproteins play roles in many biological processes, including the clearance of certain proteins from the plasma, the spread of cancer cells, and certain aspects of inflammation.

Proteoglycans are composed of specific proteins (core and link) and glycosaminoglycans. Glycosaminoglycans are usually composed of repeating disaccharides of amino sugars (e.g., GlcNAc and GalNAc) and uronic acids (glucuronic and iduronic acids). At least six glycosaminoglycans have been isolated and characterized: chondroitin sulfate, dermatan sulfate, heparan sulfate, heparin, hyaluronic acid, and keratan sulfate.

The principal proteoglycan of cartilage is an enormous molecular aggregate, composed of hyaluronic acid, chondroitin sulfate, keratan sulfate, core proteins, and link proteins, all of which are organized in a specific manner. Like other polysaccharides, the glycan chains are synthesized by the addition of sugar residues donated by nucleotide sugars. Sulfation and epimerization of specific residues may also occur. Certain diseases are due to defects in the degradation of glycosaminoglycans. Proteoglycans have a variety of functions including interactions with other omponents of the extracellular matrix.

Study Information

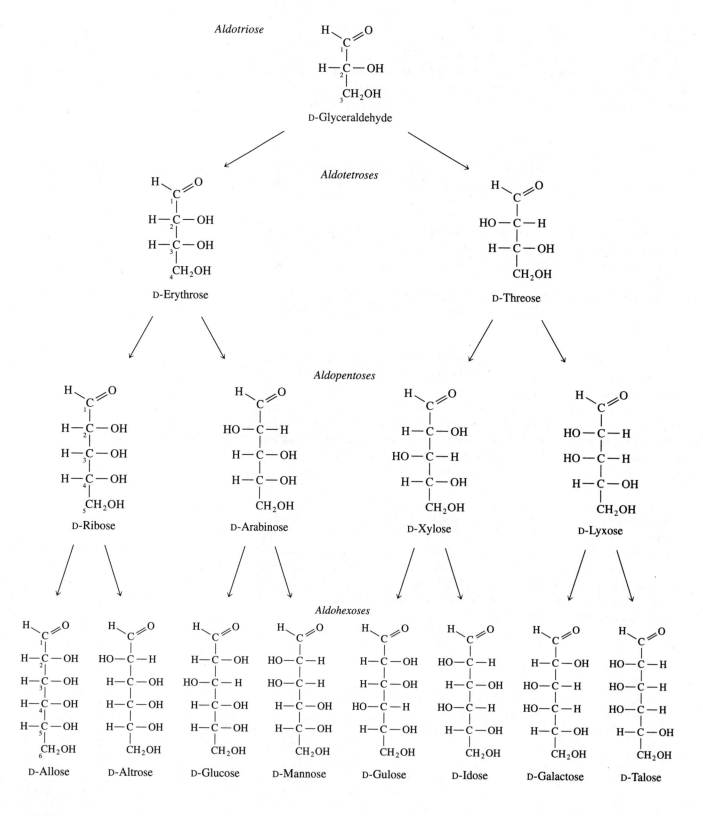

Figure 9·4
Structures of the three- to six-carbon
D-aldoses, shown as Fischer projections.

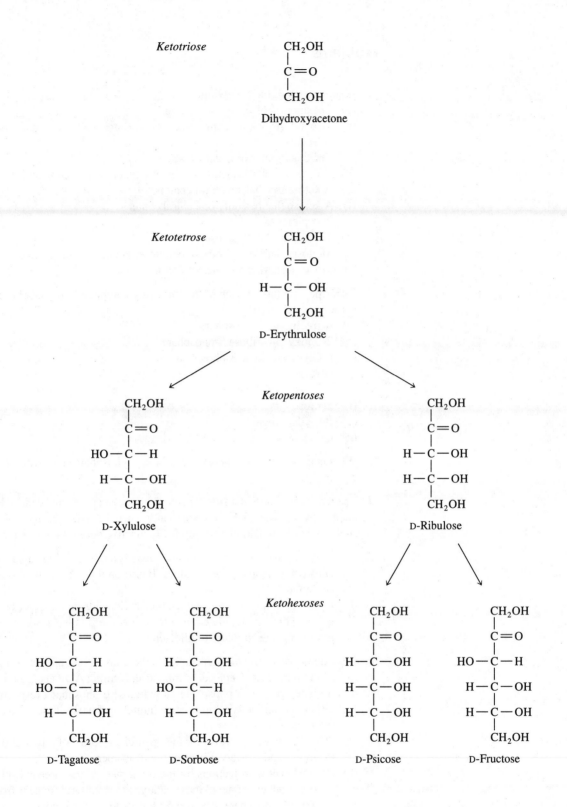

Ketotriose

CH₂OH
|
C=O
|
CH₂OH
Dihydroxyacetone

Ketotetrose

CH₂OH
|
C=O
|
H—C—OH
|
CH₂OH
D-Erythrulose

Ketopentoses

CH₂OH
|
C=O
|
HO—C—H
|
H—C—OH
|
CH₂OH
D-Xylulose

CH₂OH
|
C=O
|
H—C—OH
|
H—C—OH
|
CH₂OH
D-Ribulose

Ketohexoses

CH₂OH
|
C=O
|
HO—C—H
|
HO—C—H
|
H—C—OH
|
CH₂OH
D-Tagatose

CH₂OH
|
C=O
|
H—C—OH
|
HO—C—H
|
H—C—OH
|
CH₂OH
D-Sorbose

CH₂OH
|
C=O
|
H—C—OH
|
H—C—OH
|
H—C—OH
|
CH₂OH
D-Psicose

CH₂OH
|
C=O
|
HO—C—H
|
H—C—OH
|
H—C—OH
|
CH₂OH
D-Fructose

Figure 9·7
Structures of the three- to six-carbon D-ketoses.

Problems

1. Identify each of the following.
   (a) The anomer of β-D-glucopyranose
   (b) Two aldoses whose configuration at carbons 3, 4, and 5 matches that of D-fructose
   (c) The enantiomer of D-galactose
   (d) An epimer of D-galactose that is also an epimer of D-mannose
   (e) A ketose that has no chiral centers
   (f) A ketose that has only one chiral center
   (g) An epimer of α-lactose
   (h) The anomer of α-lactose
   (i) Monosaccharide residues of cellulose, amylose, and glycogen
   (j) Monosaccharide residues of chitin

2. Draw the structure of each of the following compounds and label each chiral carbon with an asterisk.
   (a) α-D-Glucose 1-phosphate
   (b) 2-Deoxy-β-D-ribose 5-phosphate
   (c) D-Glyceraldehyde 3-phosphate
   (d) Glycerol
   (e) Glycerol 3-phosphate
   (f) Sucrose
   (g) β-D-Glucuronate
   (h) L-Gluconate

3. What is the relationship between L-ribitol 1-phosphate and D-ribitol 5-phosphate?

4. The aldehyde group of a D-aldopentose has been chemically reduced to an alcohol, which was found to be optically inactive. Which of the four D-aldopentoses would yield an optically inactive pentitol upon reduction?

5. (a) A disaccharide known as α,β-trehalose is a nonreducing sugar that contains two glucopyranose residues. Based on this information, draw its structure.
   (b) In *Escherichia coli,* lactose metabolism depends on the expression of *lac* genes, which are regulated by allolactose, the β-($1\rightarrow6$) isomer of lactose. Draw the structure of allolactose.

6. Hexokinase is an enzyme that catalyzes the transfer of a phosphoryl group from ATP to glucose, fructose, or mannose, forming ADP and glucose 6-phosphate, fructose 6-phosphate, or mannose 6-phosphate, respectively. *N*-Acetylglucosamine is a competitive inhibitor of hexokinase-catalyzed phosphorylation of these sugars.
   (a) Draw Fischer projections for the open-chain forms of glucose, fructose, mannose, and *N*-acetylglucosamine (D isomers).
   (b) Draw Haworth projections for the α anomers of the three phosphorylated hexose products of hexokinase-catalyzed phosphoryl transfer from ATP (glucose 6-phosphate, fructose 6-phosphate, and mannose 6-phosphate).
   (c) Based on the four hexoses in (a), what conclusions can be drawn regarding the substrate specificity of hexokinase and inhibition of the enzyme by *N*-acetylglucosamine?

7. Describe how (a) penicillin, (b) bacitracin, and (c) cycloserine exert their antibacterial actions.

8. Cite four factors that contribute to the structural diversity of the oligosaccharide chains of glycoproteins.

9. You wish to determine whether a purified glycoprotein contains *N*- or *O*-linked oligosaccharide chains, or both. How would you proceed?

10. You detect, by SDS-PAGE, a protein present in the plasma membrane of red blood cells. How would you determine whether the protein contains a GPI membrane anchor?

11. What is the rationale for the use of swainsonine as an anticancer agent?

Solutions

1. (a) α-D-Glucopyranose
   (b) D-Glucose and D-mannose
   (c) L-Galactose
   (d) D-Glucose or D-talose
   (e) Dihydroxyacetone
   (f) Erythrulose (either D or L)
   (g) α-Cellobiose
   (h) β-Lactose
   (i) D-Glucose
   (j) *N*-Acetylglucosamine

2.

(a)

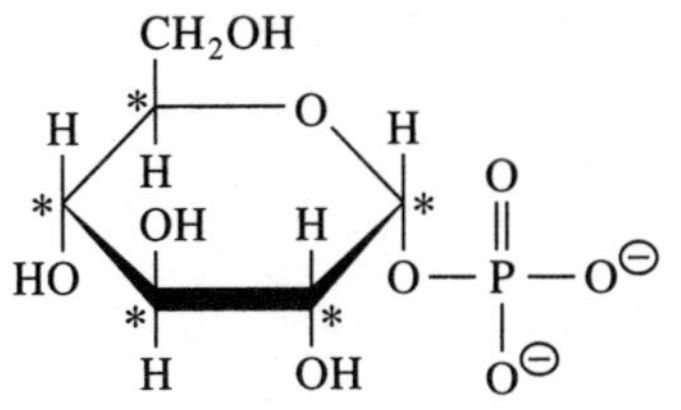

(b)

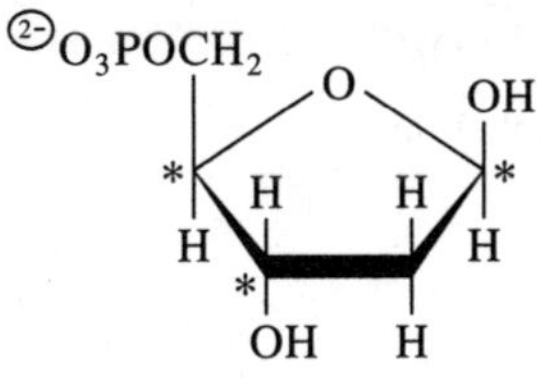

(c)

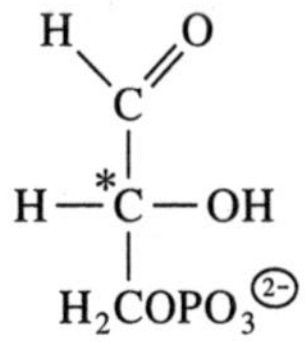

(d)

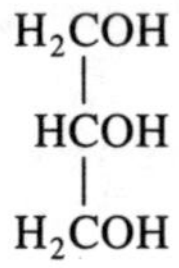

(e)

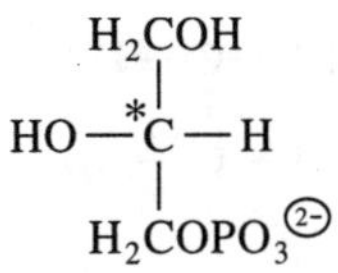

(f)

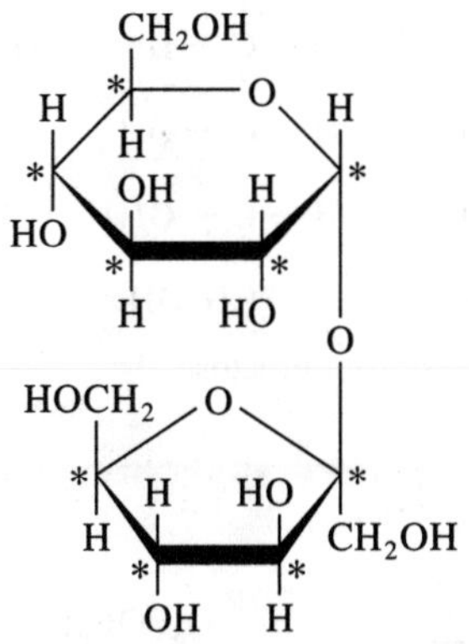

(g)

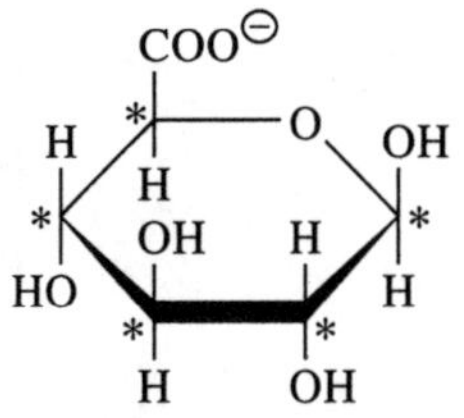

(h)

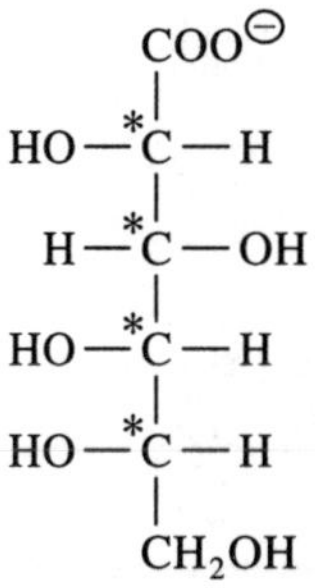

3. They are the same compound. (The enantiomer is D-ribitol 1-phosphate or L-ribitol 5-phosphate.)

4. Two of the four D-aldopentoses, D-ribose and D-xylose, would yield optically inactive alditols upon reduction (ribitol and xylitol, respectively). Although both C-2 and C-4 are chiral, a plane of symmetry can be drawn through C-3

(including the —H and —OH groups bound to it); the optical rotation contributed by one half is cancelled by equal but opposite rotation contributed by the other half.

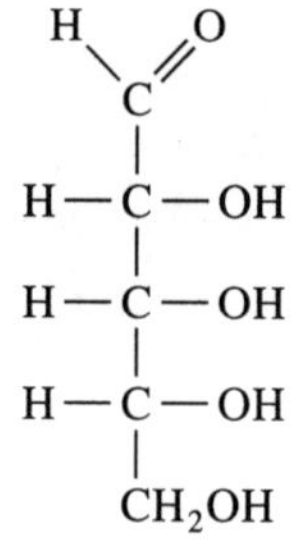

D-Ribose 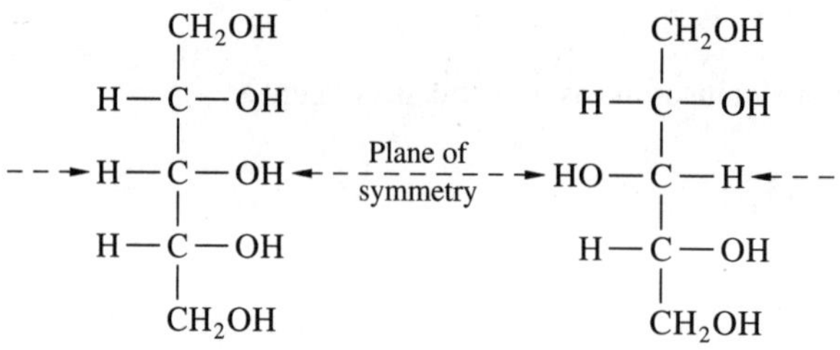   Ribitol   Xylitol   D-Xylose

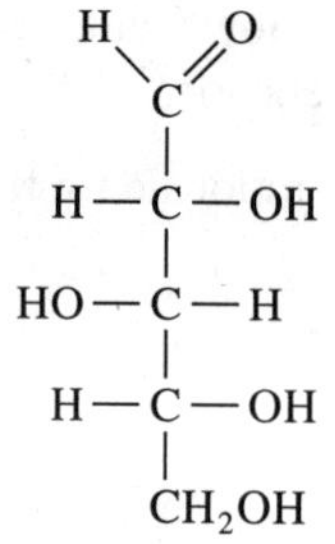

5.

(a)

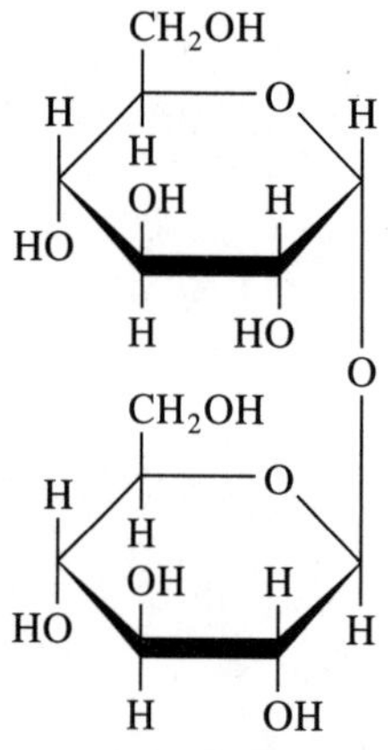

α,β-Trehalose

(b)

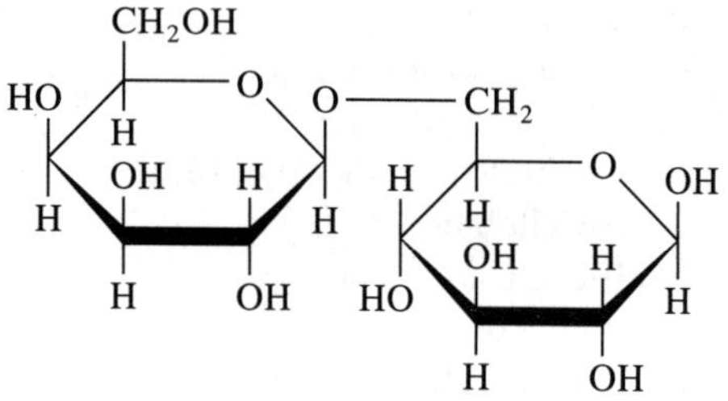

Allolactose (β anomer)
(β-D-Galactopyranosyl-(1→6)-β-D-glucopyranose)

6.

(a)

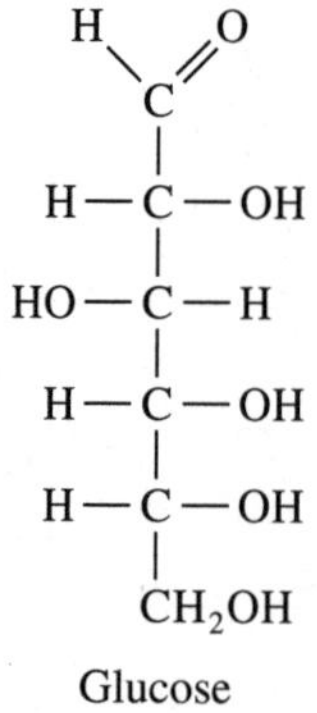

Glucose 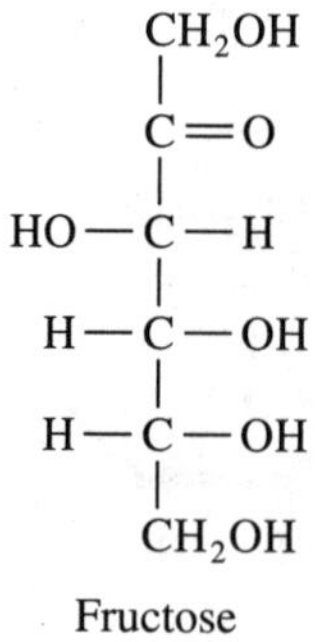   Fructose 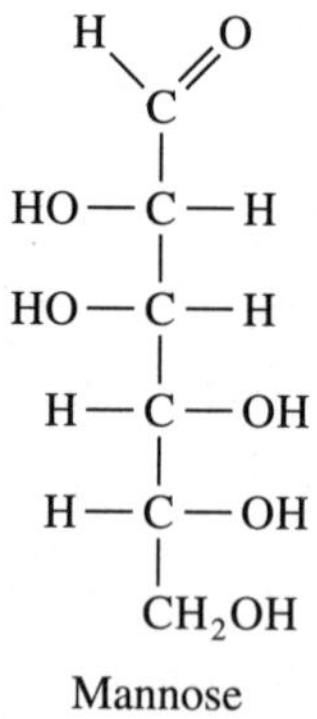   Mannose 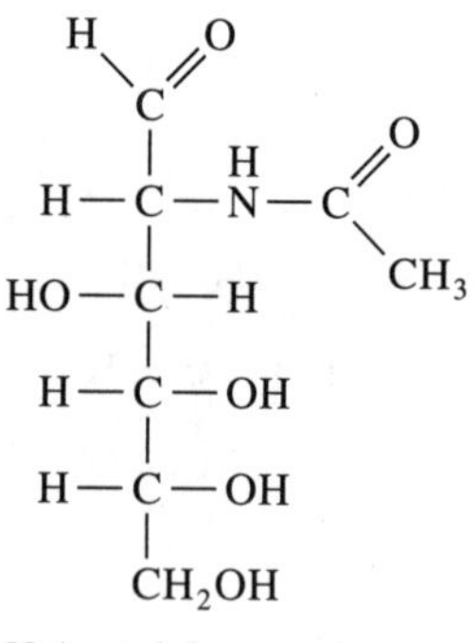   N-Acetylglucosamine

(b)

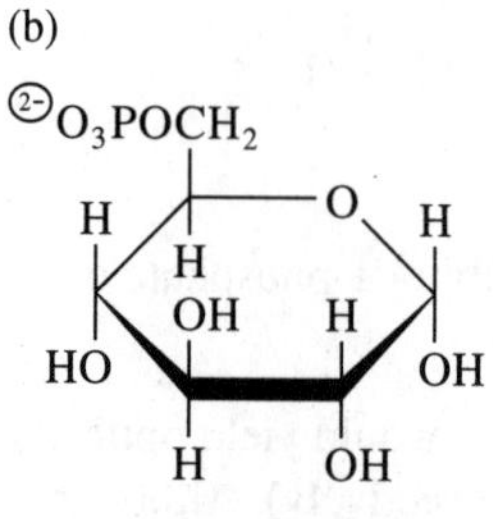

α-D-Glucopyranose 6-phosphate α-D-Fructofuranose 6-phosphate α-D-Mannopyranose 6-phosphate

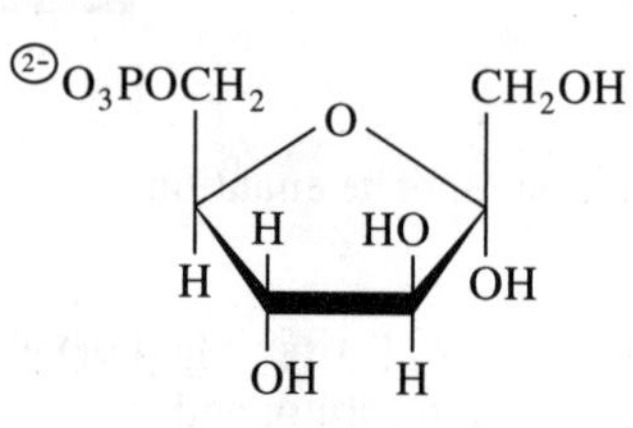

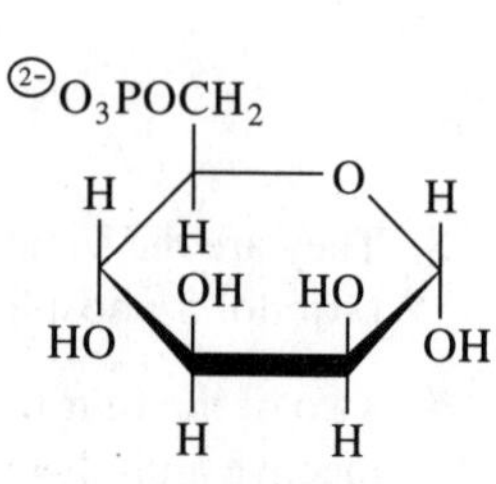

66

(c) The configuration of chiral carbons 3, 4, and 5 is the same for all three substrates; thus, hexokinase may interact with positions 3, 4, 5, and 6 of these hexoses. In contrast, C-2 may be in either the D or L configuration (glucose and mannose) or may even be ketonic (fructose); thus C-2 does not appear to play a major role in hexokinase substrate specificity.

The substituent of C-2 in *N*-acetylglucosamine, however, plays a role in inhibiting the phosphorylase activity of hexokinase. When *N*-acetylglucosamine binds to hexokinase, the bulky acetylamino group on C-2 sterically prevents the transfer of the phosphoryl group from ATP to the oxygen atom on C-6.

7. All three compounds act by inhibiting specific steps in the synthesis of peptidoglycan, a major component of the cell wall of Gram-positive bacteria (Figure 9·36).
   (a) Penicillin inhibits the transpeptidase that catalyzes the final step in the synthesis of peptidoglycan, in which the penultimate D-alanine residue of one strand is cross-linked to a terminal glycine of a neighboring strand, with the release of the terminal alanine of the first strand. The structure of penicillin resembles that of the terminal D-Ala–D-Ala dipeptide.
   (b) Bacitracin inhibits the dephosphorylation of bactoprenol pyrophosphate, thus diminishing the supply of bactoprenol phosphate, an essential component in peptidoglycan synthesis.
   (c) Cycloserine inhibits the conversion of L-alanine to D-alanine and also inhibits the formation of the D-Ala–D-Ala dipeptide.

8. (a) A number of different sugars—up to eight in eukaryotic glycoproteins—can occur in an oligosaccharide chain.
   (b) The glycosidic linkages joining sugars may be either α or β.
   (c) Various carbon atoms may be involved in the glycosidic linkages, for example, C-1, -2, -3, -4, and -6 of a hexose.
   (d) Oligosaccharide chains may contain a number of branches.

9. To test for the presence of *N*-linked chains, the protein could be treated with peptide *N*-glycosidase F. This enzyme catalyzes cleavage of the bond between the polypeptide chain and the innermost GlcNAc residue in an *N*-linked glycoprotein, reducing the molecular weight of the protein and increasing its migration in SDS-PAGE.

   To test for the presence of *O*-linked chains, the protein could be treated with *O*-glycanase. This enzyme catalyzes cleavage of the bond between the polypeptide chain and the innermost GalNAc residue in an *O*-linked glycoprotein. Alternatively, the protein could be treated with mild alkali, which causes elimination at *O*-glycosidic linkages. In both cases, the removal of *O*-linked oligosaccharide chains causes a reduction in molecular weight, which can be detected by SDS-PAGE.

10. The red cell membrane could be treated with a GPI-specific phospholipase C. If the protein under consideration contains a GPI anchor, the anchor would be cleaved by the enzyme and the protein released into the supernatant solution. SDS-PAGE analysis would show that the protein was present in the supernatant solution and absent in the plasma membrane fraction.

11. Swainsonine inhibits a mannosidase involved in the processing of high-mannose chains present in *N*-linked glycoproteins. Mannosidase inhibition results in inhibition of synthesis of complex *N*-linked chains and in the accumulation of high-mannose chains. It is thought that the presence of certain complex chains in plasma membrane glycoproteins allows at least some types of cancer cells to metastasize. If these chains are not formed, due to the action of swainsonine, the ability to metastasize may be impaired.

10

Nucleotides

Summary

A nucleotide consists of a heterocyclic base linked to a phosphorylated pentose. The nitrogenous bases of nucleotides are derivatives of pyrimidine and purine. The most common pyrimidines are cytosine (C), thymine (T), and uracil (U); the most common purines are adenine (A) and guanine (G). The amino and lactam tautomers of the bases predominate.

In a nucleoside, the base is linked to a carbohydrate via a β-N-glycosidic bond. In ribonucleosides, the sugar is ribose; in deoxyribonucleosides, it is deoxyribose. The glycosidic bond is more often found in the *anti* conformation than in the *syn* conformation in nucleic acids.

Nucleosides can be phosphorylated, usually at the 5′-hydroxyl group, to form mono-, di-, and triphosphates, called nucleotides. Magnesium ions form complexes with the phosphate groups of nucleoside di- and triphosphates.

Transfer of phosphoryl groups among nucleotides is catalyzed by two types of kinases. Phosphoryl groups originating in ATP are ultimately transferred to mono- and diphosphates to convert them to the triphosphates needed for synthetic reactions. Nucleotides are linked by 3′–5′-phosphodiester linkages in nucleic acids.

Nucleosides and nucleotides have additional roles as regulatory molecules. Cyclic AMP and cyclic GMP are second messengers, molecules that play a role in the transmission of information from extracellular hormones to intracellular enzymes. Adenosine, formed from the catabolism of adenine nucleotides, has some properties of a hormone. GTP binds to intracellular G proteins that are involved in many essential cell processes. Several alarmones, or intracellular signal nucleotides, have been isolated, including ppGpp and ApppppA.

Study Information

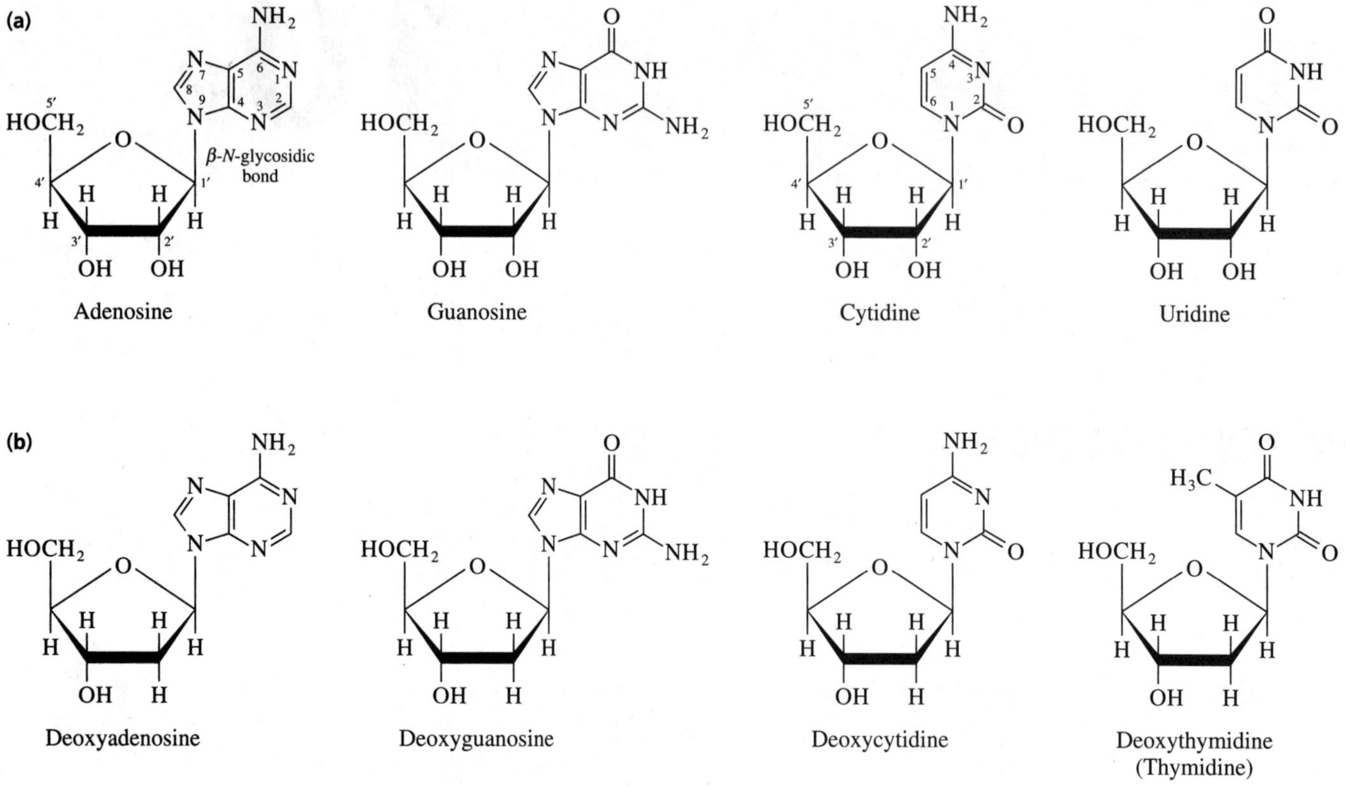

Figure 10·8
Structures of some important nucleosides.
(a) Ribonucleosides. **(b)** Deoxyribonucleosides.

Problems

1. Draw the structures of the following compounds.
   (a) N^6,N^6-Dimethyladenine
   (b) 3′-O-Methyldeoxycytidine 5′-monophosphate
   (c) 5-Hydroxymethyldeoxycytidine 5′-monophosphate

2. Purines and pyrimidines are weak bases. The pK_a values of their ring nitrogens are about 9.5. Titration shows that uridine has a pK_a of 9.2. Draw the predominant ionic forms of uridine at pH 7 and pH 10.

3. Indicate all hydrogen-bond donors and acceptors in the two tautomers of cytosine.

4. The adenosine receptor system of heart tissue is an active area of research. Explain why.

Solutions

1.

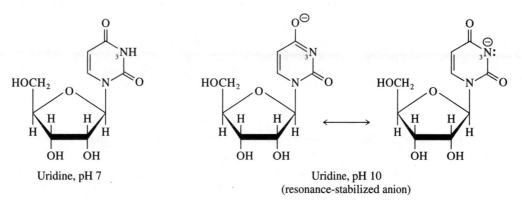

(a)

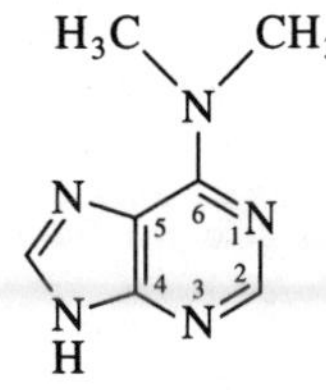

N^6,N^6-Dimethyladenine

(b)

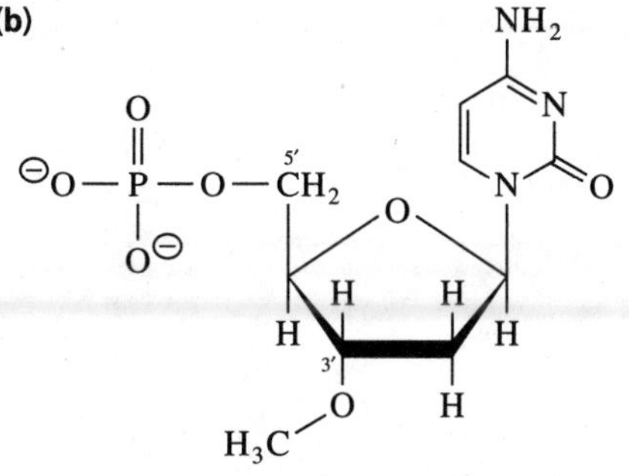

3'-O-Methyldeoxycytidine 5'-monophosphate

(c)

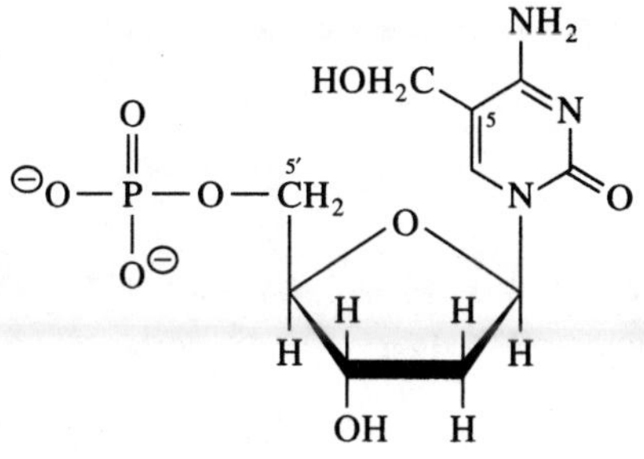

5-Hydroxymethyldeoxycytidine 5'-monophosphate

2. The pK_a is for loss of the proton at N-3. Therefore, draw the protonated form for pH 7 and two forms (resonance-stabilized anion and N$^{\ominus}$) with the N-3 proton removed for pH 10.

Uridine, pH 7

Uridine, pH 10
(resonance-stabilized anion)

3. Refer to Figure 10·7 for the structures of cytosine tautomers. In the amino tautomer, the nitrogen atom attached to C-4 can potentially donate two hydrogen atoms for hydrogen bonds. N-1 and N-3 are hydrogen acceptors, and the C-2 carbonyl oxygen atom is also a hydrogen acceptor. In the imino tautomer, N-3 and the nitrogen atom attached to C-4 are hydrogen donors; the latter can also be a hydrogen acceptor. N-1 and the C-2 carbonyl oxygen atom are hydrogen acceptors.

4. Adenosine affects the function of both the heart and blood vessels by increasing blood flow when the cellular ATP concentration is low. Therefore, an understanding of normal cardiovascular function or cardiovascular disease requires an understanding of the roles of adenosine and other purine derivatives (including ATP). Evidence for a large number of adenosine and ATP receptors is accumulating, and the role of each receptor must be examined.

11

Lipids

Summary

Lipids are water-insoluble organic compounds that can be extracted from biological samples with nonpolar organic solvents. Lipids are quite diverse, both structurally and functionally.

Fatty acids are relatively long-chain monocarboxylic acids. The majority of naturally occurring fatty acids contain an even number of carbon atoms, ranging from 12 to 20. Fatty acids that contain no carbon-carbon double bonds are classified as saturated fatty acids; those that contain one carbon-carbon double bond are classified as monounsaturated, and those that contain more than one carbon-carbon double bond are classified as polyunsaturated. Most of the double bonds found in unsaturated fatty acids have the *cis* configuration. As esters, saturated and unsaturated fatty acids are constituents of a wide variety of lipids.

Fatty acids are generally stored as complex lipids called triacylglycerols (fats and oils). Triacylglycerols are neutral and nonpolar. Waxes, also neutral, nonpolar lipids, are esters of long-chain aliphatic alcohols and fatty acids. Eicosanoids are physiologically important derivatives of C_{20} fatty acids such as arachidonate.

Glycerophospholipids are among the major amphipathic lipid components of biological membranes. They include phosphatidylcholine, phosphatidylethanolamine, phosphatidylserine, and phosphatidylinositol. Their polar heads include an anionic phosphodiester group that links C-3 of the glycerol backbone to another water-soluble component, whereas their nonpolar tails are made up of fatty acyl residues esterified to C-1 and C-2 of the glycerol moiety. Plasmalogens are glycerophospholipids in which the C-1 oxygen of the glycerol 3-phosphate moiety is bound to a hydrocarbon chain as a vinyl ether. Platelet activating factor is a biologically active lipid that has an alkyl ether group at C-1 and an acetyl group at C-2. Phosphatidylinositol 4,5-*bis*phosphate, a polar derivative of phosphatidylinositol, is involved in transmembrane signalling.

Other major classes of lipids include the sphingolipids, steroids, and fat-soluble, or lipid, vitamins. The long-chain amino alcohol sphingosine provides the backbone for sphingolipids. Three major classes of sphingolipids are sphingo-myelins, cerebrosides, and gangliosides. The lipid vitamins are examples of poly-prenyl compounds, or isoprenoids, lipids synthesized from a five-carbon compound related to isoprene. Cholesterol, a steroid, is an important component of animal membranes and serves as the precursor to a variety of hormones. Other steroids are found in plants and other eukaryotes.

Because they are poorly soluble in water, lipids are extracted and purified in organic solvents.

Figure 11·1
Organization of the major types of lipids based on structural relationships. Fatty acids are the simplest lipids in terms of structure. A number of other types of lipids either contain or are derived from fatty acids. These include the triacylglycerols, the glycerophospholipids, and the sphingolipids, as well as the eicosanoids and the waxes. The glycerophospholipids and the sphingomyelins all contain phosphate and are thus classified as phospholipids. Cerebrosides and gangliosides contain monosaccharides or their derivatives and are thus classified as glycosphingolipids. Steroids, lipid vitamins, and terpenes are derived from the five-carbon molecule isoprene and are classified as polyprenyl compounds, or isoprenoids.

Study Information

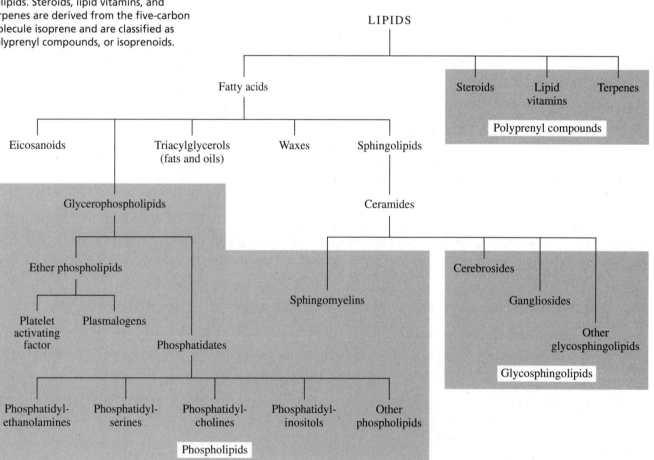

74

Table 11·2 Some common substituents attached to the phosphate group of glycerophospholipids

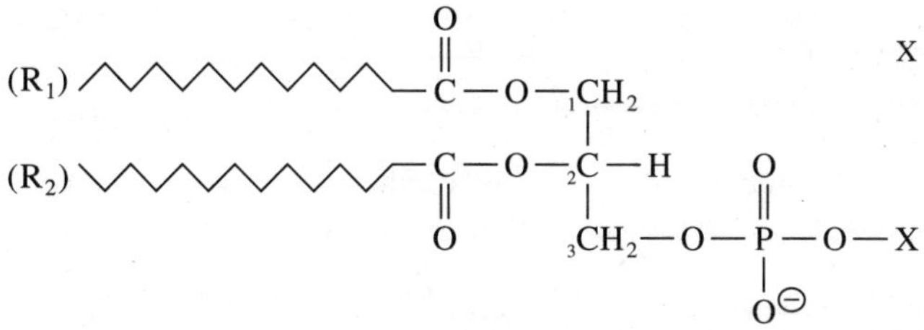

X = rest of polar head

Precursor of X (HO — X)	Formula of X	Name of resulting glycerophospholipid family
Water	$-H$	Phosphatidate
Choline	$-CH_2CH_2\overset{\oplus}{N}(CH_3)_3$	Phosphatidylcholine
Ethanolamine	$-CH_2CH_2\overset{\oplus}{N}H_3$	Phosphatidylethanolamine
Serine	$-CH_2-CH\begin{smallmatrix}\overset{\oplus}{N}H_3\\ COO^{\ominus}\end{smallmatrix}$	Phosphatidylserine
Glycerol	$-CH_2CH(OH)-CH_2OH$	Phosphatidylglycerol
Phosphatidyl-glycerol	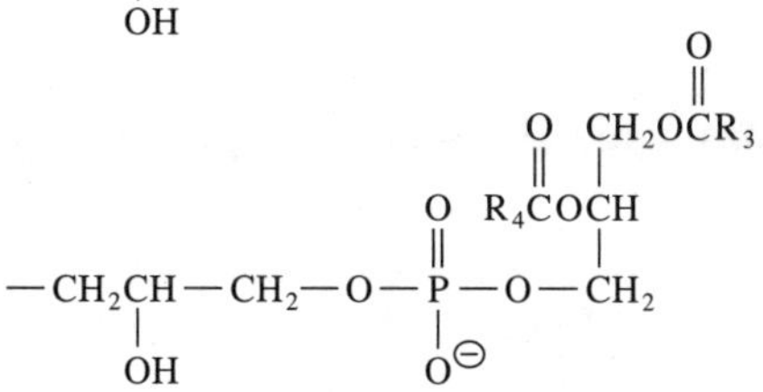 $-CH_2CH(OH)-CH_2-O-P(O^{\ominus})(=O)-O-CH_2$ ...	Diphosphatidylglycerol (Cardiolipin)
myo-Inositol	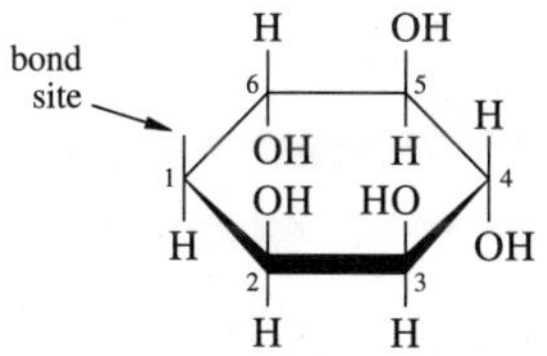 (inositol ring structure)	Phosphatidylinositol

Problems

1. Rank the following lipids in order of increasing polarity at pH 7: cholesterol, PIP_2, phosphatidylserine, phosphatidylcholine, triacylglycerol.

2. How many moles of hydrogen gas are required to completely saturate 10 ml of a solution of 0.1 mM linoleic acid ($\Delta^{9,12}$) in methanol?

3. Draw the structure of the 1-stearoyl-2-arachidonoyl species of phosphatidyl-inositol.

4. How would you determine the C-2 fatty acid composition of a natural mixture of phosphatidylethanolamines?

5. Draw the structure of oleate with the double bond in the *cis* configuration. Also draw the structure with a *trans* rather than a *cis* double bond. Why are the properties of the *trans* isomer similar to those of saturated fatty acids?

6. What are the products of phospholipase C attack on sphingomyelin?

7. The information given below was obtained from a sample of phosphatidyl-serine (PS) purified from human platelets. Identify the two major molecular species of PS in the sample.

Total fatty acid methyl esters from the PS sample	Moles (%)	Fatty acid methyl esters derived from fatty acids released during phospholipase A_2 attack on PS	Moles (%)
16:0	1.5	16:0	0.2
16:1	0.7	16:1	1.5
18:0	44.9	18:0	2.1
18:1	26.7	18:1	50.0
20:0	1.6	20:0	0.1
20:4	22.6	20:4	45.2

8. Indicate the isoprene units in the aliphatic terpene derivative geraniol (from roses) and in limonene (Figure 11·26).

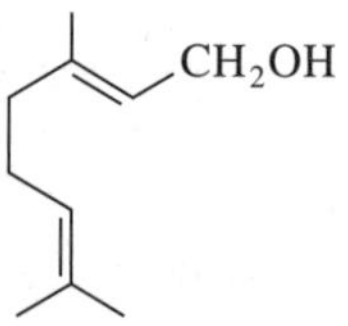

Geraniol

Solutions

1. The order of increasing polarity at pH 7 is triacylglycerol, cholesterol, phosphatidylcholine, phosphatidylserine, PIP_2.

2. Linoleic acid contains two double bonds; thus, four moles of hydrogen atoms, or two moles of hydrogen gas, are required to saturate each mole of fatty acid. The number of moles of linoleic acid is

$$10^{-2}1 \times 10^{-4}\,\text{mol l}^{-1} = 10^{-6}\,\text{mol}$$

Therefore, $2 \times 10^{-6}\,\text{mol}$ hydrogen gas is required.

3.

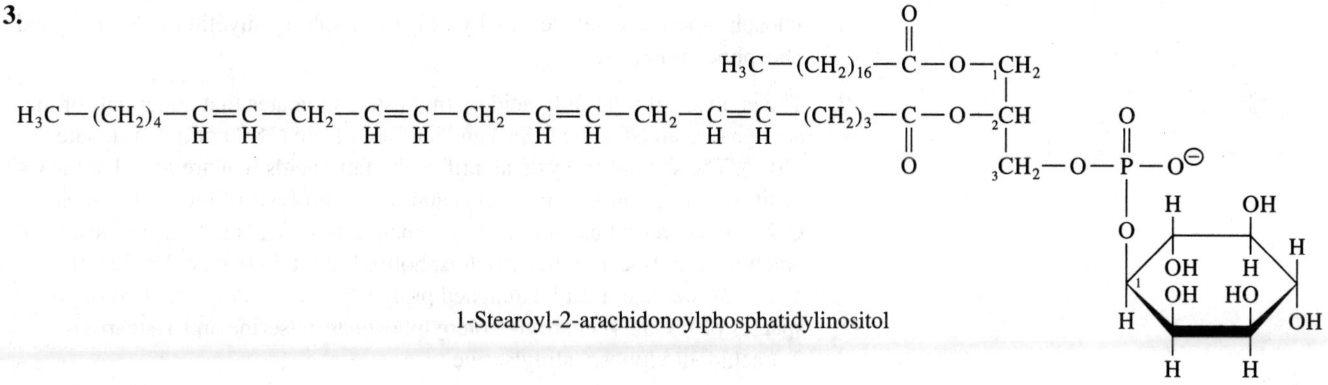

$$H_3C-(CH_2)_4-\underset{H}{\overset{}{C}}=\underset{H}{\overset{}{C}}-CH_2-\underset{H}{\overset{}{C}}=\underset{H}{\overset{}{C}}-CH_2-\underset{H}{\overset{}{C}}=\underset{H}{\overset{}{C}}-CH_2-\underset{H}{\overset{}{C}}=\underset{H}{\overset{}{C}}-(CH_2)_3-\overset{O}{\overset{\|}{C}}-O-_2CH$$

with the glycerol backbone bearing:

$$H_3C-(CH_2)_{16}-\overset{O}{\overset{\|}{C}}-O-_1CH_2$$

1-Stearoyl-2-arachidonoylphosphatidylinositol

4. First, treat the phosphatidylethanolamine with phospholipase A_2 to release the fatty acyl groups attached to C-2. Isolate the free fatty acids by chromatography. Next, treat the free fatty acids with acidified methanol to produce fatty acyl methyl esters. Finally, identify these compounds by gas chromatography.

5.

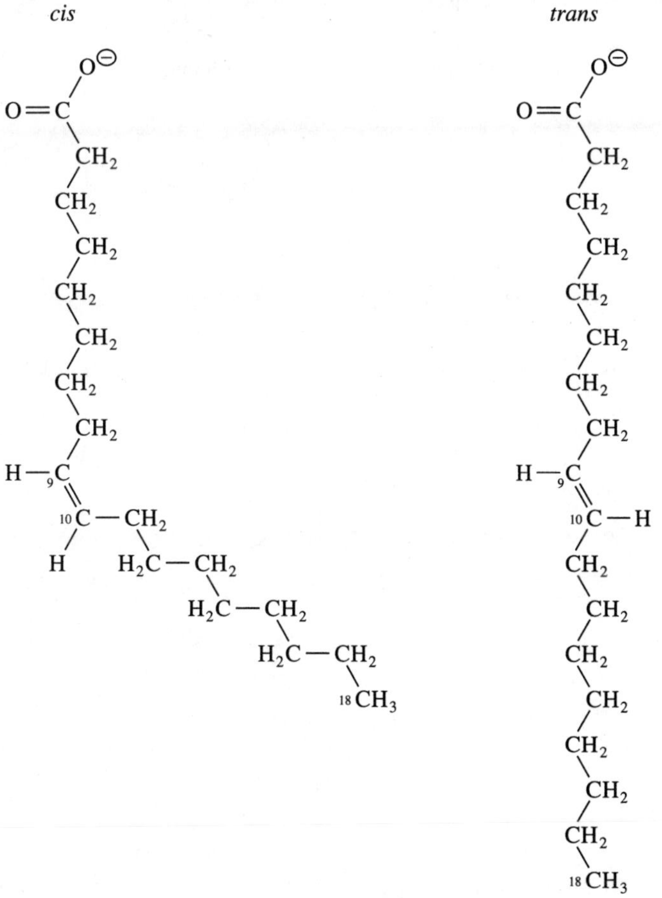

cis   *trans*

The hydrocarbon chain of the *cis* isomer of oleate contains a kink where the double bond occurs. The kink prevents close packing of the fatty acid molecules, leading to weak van der Waals interactions. The hydrocarbon chain of the *trans* isomer, in contrast, is nearly straight. The geometry of the *trans* isomer resembles that of a saturated fatty acid. Since there are no kinks, the hydrocarbon chains pack almost as closely as do the chains of saturated fatty acids. The *trans* isomers, like saturated fatty acids, have more extensive intermolecular van der Waals contact than the *cis* isomers.

6. Phospholipase C catalyzes the hydrolysis of sphingomyelin to ceramide and phosphocholine.

7. The analysis of total fatty acid methyl esters indicates that about half of the acyl groups are stearate (18:0) and half are oleate (18:1) or arachidonate (20:4). The second analysis identifies the fatty acids that are found at the C-2 position, since phospholipase A_2 catalyzes hydrolysis of the acyl groups at C-2 of the glycerol backbone. The principal fatty acids at C-2 are oleate and arachidonate. Because glycerophospholipids contain two acyl chains (at C-1 and C-2), stearate must be attached primarily at C-1. Thus, the two major species of PS are 1-stearoyl-2-oleoylphosphatidylserine and 1-stearoyl-2-arachidonoylphosphatidylserine.

8. Geraniol is a metabolic precursor of limonene. Both geraniol and limonene are derived from two isoprene units (separated by dashed lines in the figure).

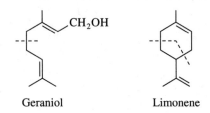

Geraniol   Limonene

12

Biological Membranes

Summary

Biological membranes define the external boundary of the cell and separate compartments within the cell. A typical membrane consists of lipids and proteins, with small amounts of carbohydrate on glycosphingolipids and glycoproteins. Membranes are a fluid mosaic of proteins in a lipid bilayer matrix. Amphipathic lipids, such as glycerophospholipids and sphingolipids, assemble into bilayers spontaneously. Lateral diffusion of lipids in the bilayer is rapid, whereas transverse diffusion from one leaflet to the other takes place very slowly. Specific lipids are asymmetrically distributed between the inner and outer leaflets of biological membranes. At low temperatures, a lipid bilayer exists in an ordered gel state in which the acyl chains are extended. The bilayer undergoes a phase transition on warming and adopts the fluid liquid-crystalline phase in which the acyl chains constantly bend and flex. *Cis* double bonds create kinks in the acyl chain, thus decreasing the phase-transition temperature and increasing fluidity. Cholesterol modulates membrane fluidity in animal cell membranes by disrupting the packing of gel-phase lipids and restricting the motion of acyl chains of liquid-crystalline–phase lipids.

Most integral membrane proteins span the hydrophobic interior of the bilayer, whereas peripheral membrane proteins are more loosely associated with the membrane surface. Lipid-anchored proteins are tethered to the bilayer by covalent linkage to either a fatty acyl chain, a molecule of glycosylphosphatidylinositol, or an isoprenoid group. Nearly all integral membrane proteins contain α-helical segments that span the lipid bilayer. Receptor proteins often possess only a single α-helical region, whereas transport proteins always have multiple membrane-spanning segments connected by loops at the membrane surface. A stretch of about 20 amino acid residues is sufficient to completely span the bilayer. Although these amino acids are usually predominantly hydrophobic, they may include some charged residues. The topology of a membrane protein can be determined by experimental methods such as X-ray crystallography, electron microscopy, and vectorial labelling and may be predicted using hydropathy plots.

Although many proteins are free to diffuse laterally in the membrane, others are anchored to the cytoskeletal network, which is attached at the inner face of the plasma membrane and provides mechanical strength. In the human erythrocyte, a meshwork of spectrin and actin is attached to the cytoplasmic domains of integral membrane proteins by peripheral linker proteins. In other cell types, the cytoskeleton consists of actin microfilaments, microtubules, and intermediate filaments. All three networks are highly dynamic structures that are constantly assembled and disassembled, depending on the needs of the cell.

The oligosaccharide chains of glycolipids and glycoproteins appear to be located exclusively on the external surface of the cell. Glycosphingolipid sugar chains and the *N*- and *O*-linked oligosaccharides of membrane glycoproteins form a sugar coat, called the glycocalyx, which displays a "fingerprint" of the cell to the exterior. Oligosaccharides make up blood group antigens and act as receptors for viruses, parasites, and protein toxins.

A lipid bilayer is a selectively permeable barrier that is impenetrable to most charged species but allows water and hydrophobic molecules to diffuse freely across it. The rate of diffusion of ions across a membrane may be greatly enhanced by certain ionophores. Specific transport, channel, or pore proteins mediate the movement of ions and polar molecules across membranes. Channel proteins allow rapid diffusion of large numbers of specific ions or small molecules through a central pore, down a concentration gradient. Transport proteins bind a substrate and move it across the membrane by alternating between an outward-facing and an inward-facing conformation. Passive transporters move molecules down a concentration gradient and do not require energy. Active transporters, which move substrates up a concentration gradient, require energy input. In primary active transport, energy is supplied directly from ATP hydrolysis, light, or electron transport. Secondary active transport is driven by an ion gradient; "uphill" transport of the substrate is coupled to "downhill" transport of the ion. Large protein molecules can be moved into and out of the cell by the processes of endocytosis and exocytosis, respectively, which involve creation and fusion of lipid vesicles.

Hormones and growth factors bind to specific receptors in the plasma membrane, which convert the external stimulus to an intracellular signal. After ligand binding to the receptor, a G-protein transducer is often activated by exchange of bound GDP for GTP. The activated G protein passes the signal to a membrane effector enzyme, which produces one or more second messengers. These small molecules or ions then carry the signal to the cell interior. These second messengers often activate protein kinases. Phosphorylation of amino acid side chains by these kinases alters the function of target proteins and produces a cellular response. Only a few common routes are used for signal transduction. One signalling pathway results in the G protein–mediated activation of membrane-bound adenylate cyclase, which produces the second messenger cAMP. This in turn activates protein kinase A, a serine-threonine protein kinase. In another major pathway, a G protein activates phospholipase C, which catalyzes the hydrolysis of the lipid phosphatidylinositol 4,5-*bis*phosphate. One product (inositol 1,4,5-*tris*phosphate) increases the cytoplasmic Ca^{2+} concentration, whereas the other product (diacylglycerol) activates protein kinase C. Many growth factors activate a protein tyrosine-kinase domain on the cytoplasmic side of their membrane receptors. This activation results in phosphorylation of tyrosine residues of the receptor as well as other target proteins. Oncogenes, genes that transform normal cells to cancer cells, often encode altered versions of proteins involved in the signalling pathways leading to cell proliferation.

Study Information

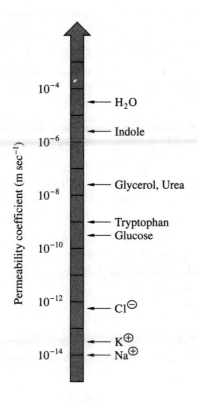

Figure 12·22
Permeability coefficients for diffusion of various species across a lipid bilayer.

Table 12·3 Characteristics of different types of membrane transport

	Protein carrier	Saturable with substrate	Movement relative to concentration gradient	Energy input required
Simple diffusion	No	No	Down	No
Channels and pores	Yes	No	Down	No
Passive transport	Yes	Yes	Down	No
Active transport				
Primary	Yes	Yes	Up	Yes (direct source)
Secondary	Yes	Yes	Up	Yes (ion gradient)

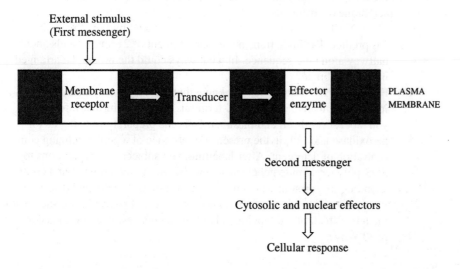

Figure 12·39
Signal transduction across the plasma membrane of a cell.

Problems

1. In what ways is the plasma membrane of a eukaryotic cell asymmetric?

2. Some pharmaceutical compounds can exert their effects only after they gain entry to the inside of a living cell. Yet many of these drugs are charged or polar and cannot passively diffuse across the plasma membrane. Liposomes have been investigated as potential agents for delivery of certain drugs to the interior of cells. Explain how liposomes might carry out this function.

3. What is the driving force for the formation of phospholipid bilayers?

4. Rank from lowest to highest the phase-transition temperatures for the following phosphatidylcholine species: dioleoylphosphatidylcholine (18:1, *cis* double bond), dielaidoylphosphatidylcholine (18:1, *trans* double bond), dilinoleoylphosphatidylcholine (18:2, *cis* double bonds), distearoylphosphatidylcholine (18:0).

5. Which of the following strategies can be used to restore membrane fluidity to a bacterium transferred from a growth temperature of 37°C to 25°C?
   (a) Production of longer fatty acid chains
   (b) Production of shorter fatty acid chains
   (c) Decreased amount of cholesterol
   (d) Production of more unsaturated fatty acid chains
   (e) Production of more saturated fatty acid chains

6. Bacterial membrane lipids contain cyclopropane fatty acid residues such as lactobacillic acid (below).

$$H_3C-(CH_2)_5-\overset{\displaystyle CH_2}{\overset{\diagup \diagdown}{CH-CH}}-(CH_2)_9-COO^{\ominus}$$

   What effect do these fatty acid chains have on fluidity of the membrane?

7. The peptide melittin is the main toxin of bee venom. Melittin damages cell membranes, causing leakage of internal enzymes and proteins. Given the amino acid sequence below, what can you say about the molecular properties of melittin, and how can these explain the effects of the peptide on cell membranes?

   $$\overset{1}{Gly}-Ile-Gly-Ala-Val-Leu-Lys-Val-Leu-Thr-Thr-Gly-Leu-$$
   $$Pro-Ala-Leu-Ile-Ser-Trp-Ile-Lys-Arg-Lys-Arg-Gln-\underset{26}{Gln}$$

8. You are trying to determine the topology of a receptor protein (M_r 50 000) in the membrane of liver cells.
   (a) The DNA for the protein has been sequenced, and hydropathy analysis has predicted a single transmembrane segment of 21 amino acids about halfway along the sequence. In what ways could the protein be arranged in the membrane?
   (b) From the cDNA sequence, you know that the only two tyrosine residues in the protein are close to the C-terminus. To obtain further information, you label several cell and plasma membrane preparations using lactoperoxidase and ^{125}I, in the presence or absence of a permeabilizing concentration of detergent. After labelling, you subject the preparations to SDS-polyacrylamide gel electrophoresis, cut out the protein band corresponding to a protein with a molecular weight of 50 000, and determine the amount of ^{125}I incorporated into the protein. The results are shown at the left. Which of the topological arrangements described in (a) is correct? Why?

Preparation	^{125}I (counts per minute per μg of protein)
Intact liver cells	125
Intact liver cells plus detergent	15 724
Right-side-out sealed plasma membrane vesicles	118
Inside-out sealed plasma membrane vesicles	16 093

9. Rank the following species in order (highest to lowest) of their predicted rate of diffusion across a lipid bilayer: galactose, phenylalanine, chloride ion, toluene, water.

10. Explain why the rate of transport of potassium ions across a bilayer of dimyristoylphosphatidylcholine by valinomycin greatly decreases when the temperature is gradually reduced from 25°C to 15°C, whereas the rate of transport of potassium ions by gramicidin D is almost unchanged under similar circumstances.

11. (a) A specific active transport protein in the cytoplasmic membrane of *E. coli* is responsible for the uptake of the sugar maltose against a 100-fold concentration gradient. How much free energy must be supplied to the transporter for each mole of maltose moved across the membrane at 22°C?
    (b) What is the free-energy change for the transport of one mole of glucose at 37°C by the passive (facilitated) glucose transporter of the human erythrocyte membrane, if the glucose concentration in the blood is 5 mM and the concentration inside the cell is 0.5 mM?

12. You have isolated a new strain of bacteria and would like to know whether L-histidine and ethylene glycol enter the cell via specific transport proteins or by simple diffusion. You measure the initial rates of uptake of these molecules at increasing external concentrations and obtain the data shown on the right.
    (a) What can you conclude about the route by which L-histidine and ethylene glycol enter the cell?
    (b) Plot the data for L-histidine uptake in double reciprocal form ($1/v_0$ versus $1/[S]$) and use this to estimate K_{tr} for this amino acid.
    (c) What results might you expect if D-histidine were used in similar experiments?

Compound	Concentration (M)	Initial rate of uptake (arbitrary units)
L-Histidine	2×10^{-6}	220
	5×10^{-6}	500
	1×10^{-5}	830
	4×10^{-5}	1700
	1×10^{-4}	2100
	5×10^{-4}	2300
	1×10^{-3}	2380
Ethylene glycol	1×10^{-3}	1
	5×10^{-3}	5
	1×10^{-2}	10
	5×10^{-2}	50
	1×10^{-1}	100
	5×10^{-1}	500
	1	1000

13. Explain why agents such as 2,4-dinitrophenol, which dissipate the proton concentration gradient across the cell membrane, abolish lactose transport in *E. coli*.

14. You have produced a strain of mice that are deficient in ankyrin. Based on your knowledge of cytoskeletal structure, which of the following would you expect to see in the erythrocytes of these mice?
    (a) Exceptionally rigid cell structure
    (b) An increase in the rate of lateral diffusion of glycophorin
    (c) An increase in the rate of lateral diffusion of the anion exchange protein
    (d) Fragility of the plasma membrane
    (e) Failure of spectrin tetramers to assemble

15. Fluorescence recovery after photobleaching can be used to measure the rate of lateral diffusion (D) of a growth-factor receptor in the membrane of intact lymphocytes. In one experiment, the receptor is found to be essentially immobile ($D < 10^{-12}\,cm^2\,s^{-1}$). When the cells are treated with cytochalasin B (a compound that disrupts actin filaments), the rate of lateral diffusion of the receptor increases to $10^{-8}\,cm^2\,s^{-1}$. Explain these results.

16. Outline two means by which a G protein can be locked in the GTP-bound, active form.

17. Asthma attacks are often treated with epinephrine, which acts via the cAMP-signalling pathway to relax the muscles of the bronchioles. Aminophylline, a purine derivative structurally related to theophylline, is often coadministered with epinephrine. Explain the biochemical basis for this treatment.

Solutions

1. The plasma membrane is asymmetric with respect to lipids, proteins, and car-bohydrate. The lipids of the inner and outer leaflets of the membrane may be quite different, as shown in Figure 12·7. Membrane proteins are asymmetric, whether they are receptors (with a ligand-binding site on the external face), or transporters (with specialized domains on each face of the membrane). The carbohydrate of the plasma membrane appears to be located exclusively on the external surface, where it is covalently bound to glycoproteins and glyco-sphingolipids.

2. Ions and polar, water-soluble molecules (including many drugs) can be trapped within the aqueous inner compartment of liposomes. Liposomes loaded with drugs can be transported through the bloodstream and can then fuse with the plasma membrane of cells, releasing the drugs into the cell inte-rior.

3. The formation of lipid bilayers is driven by the hydrophobic effect, in this case, the tendency of the hydrophobic acyl chains of the phospholipids to be excluded from contact with water. The nonpolar tails of a phospholipid mole-cule in aqueous solution are surrounded by a structured, cage-like array of hydrogen-bonded water molecules. When a phospholipid bilayer is formed, the acyl chains are sequestered in the hydrophobic interior, thus releasing the ordered water molecules. This process results in a large increase in the en-tropy of these water molecules, which greatly outweighs the decrease in en-tropy that results from formation of the more ordered lipid bilayer. The in-crease in entropy, together with the van der Waals contacts among neighboring nonpolar tails in the bilayer, contributes to a favorable (negative) free-energy change, so that the overall process occurs spontaneously.

4. Dilinoleoylphosphatidylcholine, dioleoylphosphatidylcholine, dielaidoyl-phosphatidylcholine, distearoylphosphatidylcholine. A *trans* double bond does not introduce a kink into an acyl chain and thus does not lower the T_m as a *cis* double bond does. Two *cis* double bonds give rise to two kinks in the acyl chain and thus increase fluidity more than a single *cis* double bond does.

5. Either (b) or (d). At the lower growth temperature, the bacterium must syn-thesize lipid acyl chains of lower T_m (higher fluidity). Prokaryotes do not syn-thesize cholesterol.

6. The geometry of the cyclopropane ring introduces a permanent kink into the acyl chain, much like that caused by a *cis* double bond. In addition, the pres-ence of the ring prevents close packing of the acyl chains in the gel phase. The net result is that phospholipids containing cyclopropane rings have a much lower T_m than saturated acyl chains of the same length. Membranes containing these lipids are more fluid.

7. With the exception of Lys-7, the N-terminal 20 amino acids of melittin are ei-ther hydrophobic or uncharged. The six C-terminal amino acids are all charged. The peptide is thus amphipathic, with a charged region and an un-charged region. Melittin is likely to integrate into membranes in the same fashion as a detergent. Melittin interferes with lipid packing and causes per-meabilization of the membrane, which leads to leakage of the cell contents.

8. (a) Two arrangements are possible: N-terminus external and C-terminus cytoplasmic or N-terminus cytoplasmic and C-terminus external.

(b) There is very little radioactive labelling of the tyrosine residues in the receptor from intact liver cells and right-side-out sealed vesicles, indicating that the C-terminus is not available for labelling in these systems. There is a high level of labelling when liver cells are permeabilized with detergent to allow the reagents to gain access to the cytoplasmic side of the plasma membrane, and also when an inside-out plasma membrane vesicle preparation is used. These results confirm that the C-terminus of the receptor faces the cytosol, whereas the N-terminus must be located on the external surface of the plasma membrane.

9. Water, toluene, phenylalanine, galactose, chloride ion.

10. The ionophore valinomycin is a mobile carrier, ferrying potassium ions across the membrane by diffusing from one side to the other. Diffusion is relatively rapid when the membrane lipids are in the liquid-crystalline phase but becomes extremely slow when the membrane lipids are in the gel phase. Thus, the rate of potassium ion diffusion drops precipitously as the temperature is lowered below T_m.

   Gramicidin D is a channel-forming ionophore through which ions can pass directly from one side of the membrane to the other, without the need for diffusion of the ionophore itself. The movement of potassium ions through the channel is only slightly affected when the membrane changes from the liquid-crystalline phase to the gel phase.

11. (a)

$$\Delta G = RT \ln \frac{c_2}{c_1}$$

$$\Delta G = (8.315 \text{ J mol}^{-1}\text{ K}^{-1})(295 \text{ K}) \ln \frac{100}{1}$$

$$\Delta G = 11 \text{ kJ mol}^{-1}$$

(b)

$$\Delta G = (8.315 \text{ J mol}^{-1}\text{ K}^{-1})(310 \text{ K}) \ln \frac{0.5 \text{ mM}}{5 \text{ mM}}$$

$$\Delta G = -6 \text{ kJ mol}^{-1}$$

12. (a) The rate of diffusion of ethylene glycol across the plasma membrane is directly proportional to the external concentration of ethylene glycol and shows no evidence of saturation at concentrations as high as 1 M. Thus, ethylene glycol enters the cell by passive diffusion. The rate of uptake of L-histidine into the cell is nonlinear with L-histidine concentration and approaches saturation at concentrations above 1 mM. Thus, the amino acid enters the cell via a mediated pathway involving a specific membrane transporter.

(b) To plot the data, first calculate the reciprocals of L-histidine concentration ($1/[S]$) and rate of uptake ($1/v_0$).

$\dfrac{1}{[S]}$ (10^5 M^{-1})	$\dfrac{1}{v_0}$
5	4.5×10^{-3}
2	2.0×10^{-3}
1	1.2×10^{-3}
0.25	0.59×10^{-3}
0.1	0.48×10^{-3}
0.02	0.43×10^{-3}
0.01	0.42×10^{-3}

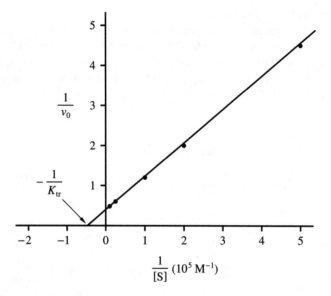

K_{tr} is obtained by taking the reciprocal of $1/K_{tr}$, whose value is obtained from the x intercept of the graph.

$$-\frac{1}{K_{tr}} = -0.5 \times 10^5 \text{ M}^{-1}$$

$$K_{tr} = 2 \times 10^{-5} \text{ M} = 20 \ \mu\text{M}$$

(c) The L-histidine transport protein would probably not be able to transport D-histidine, since D-histidine has the opposite stereochemistry.
D-Histidine would therefore enter cells very slowly by passive diffusion.

13. In *E. coli*, lactose permease carries out symport of lactose and protons in a 1:1 ratio (Figure 12·32). Lactose transport thus depends on the presence of a proton concentration gradient. Dinitrophenol acts as a proton shuttle, accepting a proton on the cell exterior, diffusing across the cell membrane, and releasing the proton on the cytosolic side. The pH gradient is thereby abolished, and the rate of lactose transport is greatly reduced.

14. (c) and (d). Ankyrin links the anion exchange protein in the plasma membrane to the spectrin-actin cytoskeletal network that underlies the membrane (Figure 12·17). If ankyrin were absent, the anion exchange protein would not be attached to the cytoskeleton, and its rate of lateral diffusion would increase. The plasma membrane would be more fragile, since it would no longer be strengthened by attachment of the cytoskeletal network.

15. The very low rate of lateral diffusion for the receptor indicates that the protein is probably linked to the cytoskeleton. When the actin-filament network of the cytoskeleton is disrupted by cytochalasin B, the receptor has greater freedom and diffuses more rapidly.

16. First, G proteins can be locked into their active forms by the binding of non-hydrolyzable GTP analogs, such as GTPγ-S. Second, ADP-ribosylation carried out by cholera toxin converts the G protein G_s to a permanently activated form lacking GTPase activity.

17. Epinephrine elevates cAMP levels in the cytosol. Aminophylline, like theophylline, inhibits the action of cAMP phosphodiesterase in the cytosol. Decreasing the rate of degradation of cAMP prolongs the time over which cellular cAMP levels are elevated by epinephrine. Aminophylline treatment thus intensifies the muscle relaxant effect of epinephrine.

13

Digestion

Summary

The major human nutrients are carbohydrates (mostly in the form of starch), proteins, and fats. Each of these biomolecules must be hydrolyzed so that its components can be assimilated. The hydrolysis, or digestion, of food polymers occurs at several locations in the gastrointestinal tract, mediated by enzymes that function under conditions that vary among the different digestive organs. The greatest amount of hydrolysis takes place in the small intestine.

Starch is partially hydrolyzed in the mouth by the action of salivary α-amylase. The action of a second α-amylase, which is synthesized in the pancreas and secreted into the small intestine, converts the oligosaccharides from starch into a mixture of maltose, maltotriose, and limit dextrins. Enzymatic hydrolysis of these intermediates and of the dietary disaccharides sucrose and lactose to monosaccharides is catalyzed by a series of glycosidases that are bound to the plasma membrane of intestinal cells. Dietary fiber and certain oligosaccharides pass through the small intestine undigested. Enzymes produced by bacteria growing in the large intestine are involved in the metabolism of some of these undigested carbohydrates.

The initial step in the digestive hydrolysis of proteins occurs in the stomach. There, the zymogen pepsinogen is activated to pepsin, an aspartic protease with an acidic pH optimum. The pancreas secretes a group of zymogens that, upon activation by limited proteolysis in the small intestine, become trypsin, chymotrypsin, elastase, and two carboxypeptidases. The product of the combined actions of these enzymes is a mixture of amino acids and peptides containing from two to six residues. Peptidases of the intestinal mucosa catalyze further hydrolysis of peptides.

Fats (triacylglycerols) are hydrolyzed to fatty acids and 2-monoacylglycerols by the action of a lipase in the mouth and pancreatic lipase. When colipase—which forms a complex with pancreatic lipase—binds to lipid, the active site of lipase becomes accessible to its substrates.

Monosaccharides, amino acids, and di- and tripeptides are absorbed by intestinal mucosa cells by cotransport with $Na^{\oplus}$. Most of the di- and tripeptides are hydrolyzed in the mucosal cells. The sugar and amino acid monomers then enter the portal blood. In the intestine, the products of fat hydrolysis are taken up by bile-salt micelles and delivered by these carriers to the mucosa, where the hydrolyzed products are absorbed. The products of fat hydrolysis are reassembled into triacylglycerols. Lipids are packaged in lipoprotein particles for transport from mucosal cells through the lymph.

Study Information

Table 13·1 Functions of the digestive organs

Organ	Major digestive roles
Salivary glands	Provide fluid and amylase
Tongue	Provides lipase
Stomach	Provides HCl and pepsin
Pancreas	Provides enzymes, zymogens, and $NaHCO_3$
Liver	Synthesizes bile salts
Gall bladder	Stores bile
Small intestine	Completes digestion; absorbs end products
Large intestine	Absorbs water, ions, and short-chain acids

Problems

1. Based on where they act in the gastrointestinal tract, indicate the approximate pH optimum for each of the following enzymes.
   (a) Salivary α-amylase
   (b) Pepsin
   (c) Pancreatic lipase

2. Would you expect all proteins to be digested completely?

3. One symptom of the disease cystic fibrosis is excessive secretion of mucus in a number of organ systems. What effect might this have on nutrient absorption?

4. Glucose absorption from the lumen of the small intestine depends on a sodium ion gradient maintained by a $Na^{\oplus}$-$K^{\oplus}$ ATPase. Glucose exits the intestinal cell via passive transport. Sketch a plot indicating the relative glucose concentrations in the lumen, the cell, and the blood plasma.

5. Most dietary folate compounds are polyglutamates (Section 8·12), which cannot easily cross plasma membranes. Suggest how dietary folate is absorbed by the intestine.

Solutions

1. (a) Neutral pH. (This may lower slightly in the absence of $Cl^\ominus$.)
   (b) pH 1–2, the pH of the gastric juice.
   (c) pH 6–8, depending on the concentration of bile salts, which are the conjugate bases of weak acids.

2. No, some proteins are likely to be less digestible than others. The digestibility of a protein depends on its susceptibility to hydrolysis by the specific proteases and peptidases in the stomach and small intestine. For example, plant proteins are not digested as well as animal proteins. Phosphoproteins and proteins with high proline contents are also less digestible. Cooking can also cause formation of covalent bonds that render proteins resistant to enzymatic digestion.

3. The organs of the digestive tract are normally coated with mucus. Excessive mucus presents a barrier to absorption of nutrients in the small intestine.

4. The glucose concentration is lowest in the lumen. An energy-requiring transport mechanism is required to bring glucose into the cell against its concentration gradient. The concentration of glucose in plasma is lower than in the cell, since glucose follows its concentration gradient as it passes out of the cell.

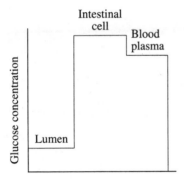

5. Just as proteins must be hydrolyzed to monomers or di- or tripeptides, folates must be hydrolyzed in order to be absorbed. The brush border has a hydrolase called folylpolyglutamate hydrolase, or conjugase, that specifically catalyzes the hydrolysis of the γ-glutamyl bonds of folates. Mono- and diglutamate products are the only forms that are absorbed.

14

Introduction to Metabolism

Summary

The chemical reactions that occur in cells are collectively called metabolism. These reactions can be classified as catabolic (degradative) or anabolic (synthetic) reactions. Metabolic activities allow cells to maintain intracellular conditions different from those of the environment, to extract energy from external sources, to grow and reproduce, and to respond to internal or external influences.

Sequences of reactions are called pathways. Although the start and end points are sometimes arbitrary, metabolic pathways may be linear, cyclic, or spiral; the pathways may also branch. Enzymes are required for cells to carry out reactions under conditions of moderate temperature, pressure, and pH. Degradative and synthetic pathways proceed in stepwise fashion, with the participation of energy carriers such as nucleoside triphosphates and nicotinamide coenzymes.

The major catabolic pathways in cells convert macromolecules to smaller energy-yielding metabolites, or fuels. The smaller compounds also serve as building blocks for the synthesis of new macromolecules. Glucose, fatty acids, and some amino acids are oxidized to form acetyl CoA, which enters the citric acid cycle, the common pathway of oxidative metabolism. The energy released in catabolic reactions is conserved in the form of nucleoside triphosphates and reduced coenzymes. The energy of the reduced compounds is used to synthesize ATP from ADP and P_i by the process of oxidative phosphorylation.

Metabolic pathways are regulated to allow the organism to use fuel sources efficiently and to respond to changing demands. Responses may involve many pathways or only a few. The flux, or flow of material through a pathway, usually depends on regulation of multiple steps, each with a particular control strength for the overall pathway. Feedback inhibition and feed-forward activation are commonly encountered. Regulation of particular enzymes may be accomplished by allosteric modulation, reversible covalent modification, and changes in the rate of enzyme synthesis or degradation.

In multicellular organisms, tissues are specialized for different metabolic tasks. Interorgan metabolism may be coordinated by hormones.

The laws of thermodynamics apply to metabolic reactions, which are in a steady state, not at equilibrium. The direction of a chemical or enzyme-catalyzed reaction depends on the change in free energy. Reactions occur spontaneously only when the free-energy change is negative. The standard free-energy change of a reaction is related to the equilibrium constant of the reaction by the formula $\Delta G^{\circ\prime} = -RT \ln K_{eq}$.

In cells, the change in free energy (ΔG) that occurs during a given reaction depends primarily on the concentrations of reactants and products and usually differs from the standard free-energy change ($\Delta G^{\circ\prime}$). Each reaction of a metabolic pathway proceeds with a negative free-energy change. The concentrations of reactants and products in many cellular reactions approach the equilibrium state; such reactions are called near-equilibrium reactions. Reactions for which the steady-state concentrations of reactants are far from equilibrium are called metabolically irreversible reactions.

ATP plays a central role in energy metabolism. The energy released by one biological process is often conserved in the form of ATP, to be used by other, energy-requiring processes. The energy of ATP is made available when a terminal phosphoryl group or a nucleotidyl group is transferred. The action of adenylate kinase helps maintain a constant concentration of ATP in cells. The concentrations of ADP and AMP are much lower.

Phosphoryl-group transfer from another energy-rich substrate to ADP forms ATP. In addition to nucleoside triphosphates, there are several other metabolites with activated phosphoryl groups, including acetyl phosphate, the phosphagens, and phosphoenolpyruvate.

Acyl-group transfer is an important metabolic reaction. The transfer of acyl groups from coenzyme A proceeds with a large negative free-energy change.

The free energy of biological oxidation reactions can be captured in the form of reduced coenzymes. This form of energy is measured as reduction potential, the quantitative measure of the ability of a molecule to donate electrons. Standard reduction potential is related to standard free-energy change by the formula $\Delta G^{\circ\prime} = -n \mathcal{F} \Delta E^{\circ\prime}$. Under nonstandard conditions, reduction potential is given by the Nernst equation.

Study Information

Table 14·3 Standard free energies of hydrolysis for common metabolites

Metabolite	$\Delta G^{\circ\prime}_{hydrolysis}$ (kJ mol^{-1})
Phosphoenolpyruvate	−62
1,3-*Bis*phosphoglycerate	−49
Acetyl phosphate	−43
Phosphocreatine	−43
Pyrophosphate	−33
Phosphoarginine	−32
ATP to AMP + PP$_i$	−32
ATP to ADP + P$_i$	−30
Glucose 1-phosphate	−21
Glucose 6-phosphate	−14
Glycerol 3-phosphate	− 9

Table 14·4 Standard reduction potentials for some important biological half-reactions

Reduction half-reaction	$E^{\circ\prime}$ (V)
Acetyl CoA + CO_2 + $H^{\oplus}$ + $2\,e^{\ominus}$ $\longrightarrow$ Pyruvate + CoA	−0.48
Ferredoxin (spinach), Fe^{3+} + $e^{\ominus}$ $\longrightarrow$ Fe^{2+}	−0.43
$2\,H^{\oplus}$ + $2\,e^{\ominus}$ $\longrightarrow$ H_2	−0.42
α-Ketoglutarate + CO_2 + $2\,H^{\oplus}$ + $2\,e^{\ominus}$ $\longrightarrow$ Isocitrate	−0.38
Lipoyl dehydrogenase (FAD) + $2\,H^{\oplus}$ + $2\,e^{\ominus}$ $\longrightarrow$ Lipoyl dehydrogenase ($FADH_2$)	−0.34
$NADP^{\oplus}$ + $2\,H^{\oplus}$ + $2\,e^{\ominus}$ $\longrightarrow$ NADPH + $H^{\oplus}$	−0.32
$NAD^{\oplus}$ + $2\,H^{\oplus}$ + $2\,e^{\ominus}$ $\longrightarrow$ NADH + $H^{\oplus}$	−0.32
Lipoic acid + $2\,H^{\oplus}$ + $2\,e^{\ominus}$ $\longrightarrow$ Dihydrolipoic acid	−0.29
Glutathione (oxidized) + $2\,H^{\oplus}$ + $2\,e^{\ominus}$ $\longrightarrow$ 2 Glutathione (reduced)	−0.23
FAD + $2\,H^{\oplus}$ + $2\,e^{\ominus}$ $\longrightarrow$ $FADH_2$	−0.22
FMN + $2\,H^{\oplus}$ + $2\,e^{\ominus}$ $\longrightarrow$ $FMNH_2$	−0.22
Acetaldehyde + $2\,H^{\oplus}$ + $2\,e^{\ominus}$ $\longrightarrow$ Ethanol	−0.20
Pyruvate + $2\,H^{\oplus}$ + $2\,e^{\ominus}$ $\longrightarrow$ Lactate	−0.18
Oxaloacetate + $2\,H^{\oplus}$ + $2\,e^{\ominus}$ $\longrightarrow$ Malate	−0.17
Cytochrome b_5 (microsomal), Fe^{3+} + $e^{\ominus}$ $\longrightarrow$ Fe^{2+}	0.02
Fumarate + $2\,H^{\oplus}$ + $2\,e^{\ominus}$ $\longrightarrow$ Succinate	0.03
Ubiquinone (Q) + $2\,H^{\oplus}$ + $2\,e^{\ominus}$ $\longrightarrow$ QH_2	0.04
Cytochrome b (mitochondrial), Fe^{3+} + $e^{\ominus}$ $\longrightarrow$ Fe^{2+}	0.08
Cytochrome c_1, Fe^{3+} + $e^{\ominus}$ $\longrightarrow$ Fe^{2+}	0.22
Cytochrome c, Fe^{3+} + $e^{\ominus}$ $\longrightarrow$ Fe^{2+}	0.23
Cytochrome a, Fe^{3+} + $e^{\ominus}$ $\longrightarrow$ Fe^{2+}	0.29
Cytochrome f, Fe^{3+} + $e^{\ominus}$ $\longrightarrow$ Fe^{2+}	0.36
$NO_3^{\ominus}$ + $e^{\ominus}$ $\longrightarrow$ $NO_2^{\ominus}$	0.42
Photosystem P700	0.43
Fe^{3+} + $e^{\ominus}$ $\longrightarrow$ Fe^{2+}	0.77
$\frac{1}{2}O_2$ + $2\,H^{\oplus}$ + $2\,e^{\ominus}$ $\longrightarrow$ H_2O	0.82

[Most values are from Loach, P. A. (1968). Oxidation-reduction potentials, absorbance bands and molar absorbance of compounds used in biochemical studies. In *Handbook of Biochemistry: Selected Data for Molecular Biology*, H. A. Sober, ed. (Cleveland, Ohio: CRC Press).]

Problems

1. In phosphorolysis, a bond is attacked and cleaved by inorganic phosphate, rather than by water, as in hydrolysis. The bacterium *Pseudomonas saccharophila* contains sucrose phosphorylase, an enzyme that catalyzes the phosphorolytic cleavage of sucrose:

    Sucrose + P_i $\longrightarrow$ Glucose 1-phosphate + Fructose

    (a) From the following data, calculate the standard free-energy change for the phosphorolysis of sucrose.

    H_2O + Sucrose $\longrightarrow$ Glucose + Fructose
    $$\Delta G^{\circ\prime} = -29 \text{ kJ mol}^{-1}$$

    H_2O + Glucose 1-phosphate $\longrightarrow$ Glucose + P_i
    $$\Delta G^{\circ\prime} = -21 \text{ kJ mol}^{-1}$$

    (b) Calculate the equilibrium constant for the phosphorolysis of sucrose.

2. The standard free-energy change of hydrolysis for phosphotyrosine (tyrosine phosphate) is $-10\ kJ\ mol^{-1}$.

(a) Draw the structures of phosphotyrosine and phosphoarginine that would predominate at pH 7.

(b) Could phosphotyrosine, like phosphoarginine, be used as an energy-rich metabolite for replenishment of cellular ATP from ADP?

3. Indicate which of the following phosphate compounds are energy-rich and explain why (that is, identify any high-energy bonds or linkages).

(a)

$$^{\ominus}O-\overset{\overset{\displaystyle O}{\|}}{\underset{\underset{\displaystyle O^{\ominus}}{|}}{P}}-O-\overset{\overset{\displaystyle O}{\|}}{C}-CH_2-O-\overset{\overset{\displaystyle O}{\|}}{\underset{\underset{\displaystyle O^{\ominus}}{|}}{P}}-O^{\ominus}$$

(b)

$$H_3C-\overset{\overset{\displaystyle O}{\|}}{C}-CH_2-O-\overset{\overset{\displaystyle O}{\|}}{\underset{\underset{\displaystyle O^{\ominus}}{|}}{P}}-O^{\ominus}$$

(c)

$$\text{(phenyl)}-\overset{}{\underset{\underset{\displaystyle H}{|}}{N}}-\overset{\overset{\displaystyle O}{\|}}{C}-O-\overset{\overset{\displaystyle O}{\|}}{\underset{\underset{\displaystyle O^{\ominus}}{|}}{P}}-O^{\ominus}$$

(d)

ring structure with $COO^{\ominus}$, H_3C, H_3C-C, H_3C, $P=O$, $O^{\ominus}$, O linkages

(e)

ring structure: (phenyl)$-CH_2-N$, H_2C-O, $P=O$ with $O^{\ominus}$, $C-N$ with H, $H_2N^{\oplus}$

4. The standard reduction potential for ubiquinone (Q) is +0.04 V, and the standard reduction potential for flavin adenine dinucleotide (FAD) is -0.22 V. Show that the oxidation of $FADH_2$ by Q theoretically liberates enough energy to drive the synthesis of ATP from $ADP + P_i$ under standard conditions.

5. In a rat hepatocyte, the concentrations of ATP, ADP, and P_i are 3.4 mM, 1.3 mM, and 4.8 mM, respectively. Calculate the free-energy change for hydrolysis of ATP in this cell. How does this compare to the standard free-energy change?

Solutions

1. (a) To calculate the standard free-energy change for the phosphorolysis of sucrose, add the standard free-energy changes of the two reactions whose combination yields the net reaction.

	$\Delta G^{\circ\prime}(kJ\ mol^{-1})$
H_2O + Sucrose $\longrightarrow$ Glucose + Fructose	-29
Glucose + P_i $\longrightarrow$ Glucose 1-phosphate + H_2O	$+21$
Net: Sucrose + P_i $\longrightarrow$ Glucose 1-phosphate + Fructose	-8

Thus, the standard free-energy change for phosphorolysis of sucrose is $-8\ kJ\ mol^{-1}$.

(b) To calculate the equilibrium constant, use Equation 14·12.

$$\Delta G^{\circ\prime} = -RT \ln K_{eq}$$

$$\ln K_{eq} = -\frac{\Delta G^{\circ\prime}}{RT}$$

$$\ln K_{eq} = \frac{8000 \text{ J mol}^{-1}}{(8.315 \text{ J K}^{-1}\text{mol}^{-1})(298\text{K})} = 3.2$$

$$K_{eq} = 25$$

2. (a)

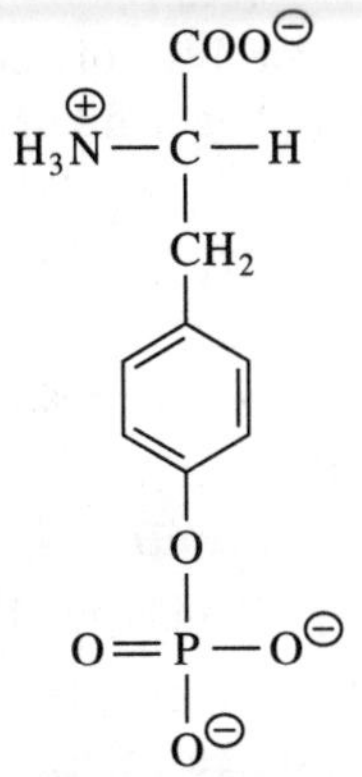

Phosphotyrosine

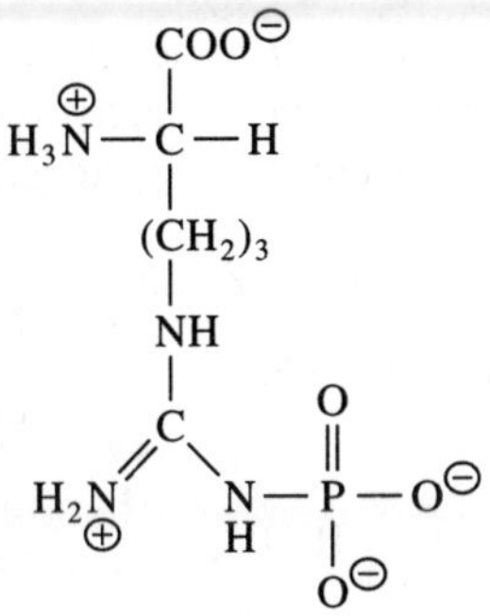

Phosphoarginine

(b) The phosphate linkage in phosphotyrosine does not contain sufficient energy to drive ATP replenishment from ADP. Since the standard free-energy change for hydrolysis of ATP to ADP + P_i is -30 kJ mol^{-1} (Table 14·1), the standard free-energy change for ATP synthesis is $+30$ kJ mol^{-1}. The standard free-energy change for hydrolysis of phosphotyrosine is -10 kJ mol^{-1}, much less than the standard free-energy required for the transfer of the phosphoryl group of phosphotyrosine to ATP. Phospho-arginine can serve as a source of energy for ATP replenishment because its standard free-energy change for hydrolysis is -32 kJ mol^{-1}.

3. The term *high energy* is commonly applied to compounds that release considerable energy upon hydrolysis. The bond that is cleaved during breakdown of such a molecule is called a "high-energy bond." However, the high reactivities of these compounds in group-transfer reactions are not due to just a single bond but to the overall structures.

(a) This is an energy-rich molecule, but only one phosphoryl group, the carboxylate-phosphate anhydride has high energy; the phosphate ester does not.

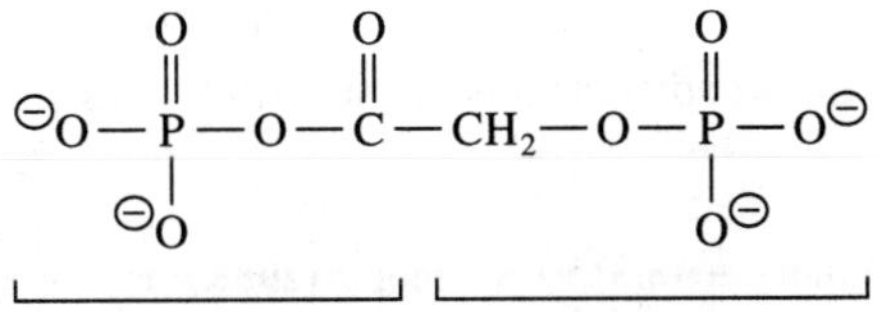

Carboxylate-phosphate anhydride

Phosphate ester

(b) This is not an energy-rich molecule because it is a phosphate ester.

(c) This is an energy-rich molecule that resembles carbamoyl phosphate. The carbamate-phosphate anhydride has high energy.

(d) This is an energy-rich molecule. Like phosphoenolpyruvate, it contains a high-energy enol phosphate linkage.

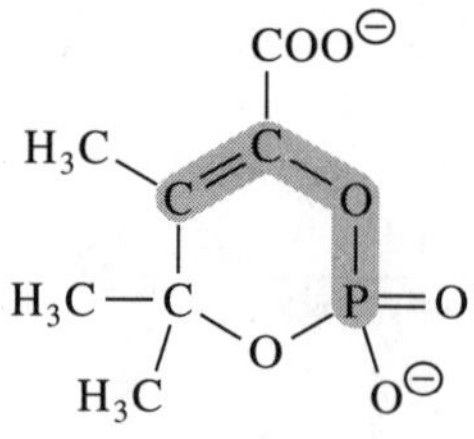

Enol phosphate 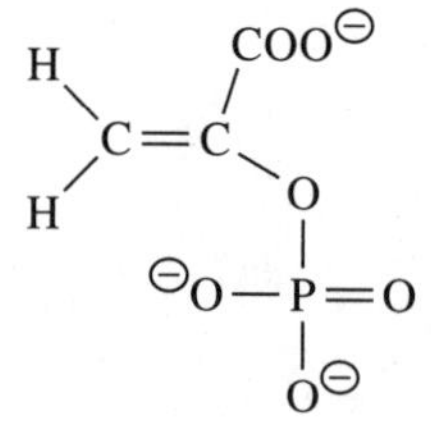   Phosphoenolpyruvate (PEP)

(e) This is an energy-rich molecule. Like the hydrolysis of phosphocreatine and phosphoarginine, hydrolysis of the phosphoguanidine N—P bond of this molecule releases considerable energy.

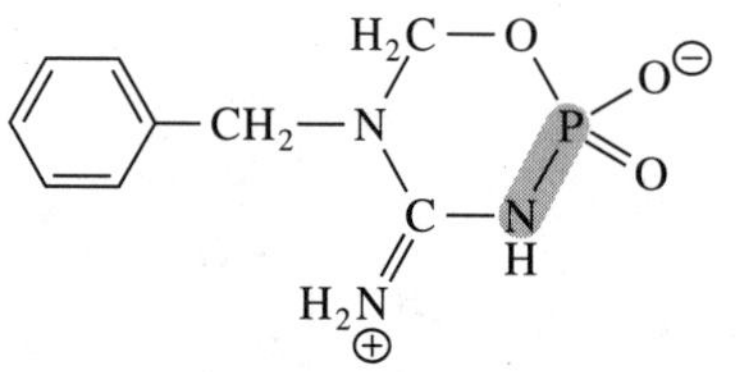

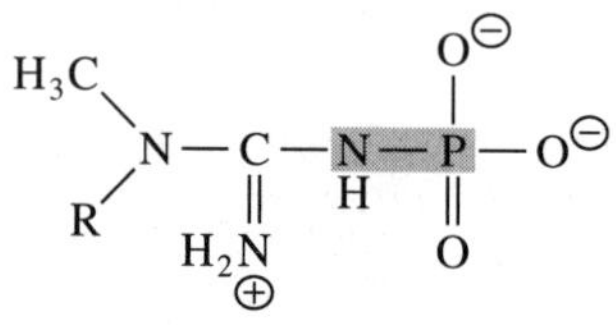

Phosphoguanidine

4. First calculate the standard reduction potential for the coupled oxidation-reduction reaction and then calculate the standard free-energy change using Equation 14·22.

$$
\begin{array}{ll}
 & E^{\circ\prime}\,(V) \\
Q + 2\,H^{\oplus} + 2\,e^{\ominus} \longrightarrow QH_2 & +0.04 \\
FADH_2 \longrightarrow FAD + 2\,H^{\oplus} + 2\,e^{\ominus} & +0.22 \\
\hline
Q + FADH_2 \longrightarrow QH_2 + FAD \qquad \Delta E^{\circ\prime}\,(V) = & +0.26
\end{array}
$$

$$\Delta G^{\circ\prime} = -n\mathcal{F}\Delta E^{\circ\prime}$$

$$\Delta G^{\circ\prime} = -(2)(96.48\ \text{kJ V}^{-1}\text{mol}^{-1})(0.26\ \text{V})$$

$$\Delta G^{\circ\prime} = -50\ \text{kJ mol}^{-1}$$

Since the standard free-energy change for the hydrolysis of ATP to ADP + P_i is -30 kJ mol^{-1} (Table 14·1), the standard free-energy change for ATP synthesis is $+30$ kJ mol^{-1}. Thus, oxidation of FADH$_2$ by ubiquinone theoretically liberates more than enough free energy to drive ATP synthesis from ADP + P_i.

5. The actual free-energy change is calculated according to Equation 14·14.

$$\Delta G = \Delta G^{\circ\prime} + RT \ln \frac{[\text{ADP}][P_i]}{[\text{ATP}]}$$

When known values and constants are substituted, assuming pH 7.0 and 25°C,

$$\Delta G = -30\,000\ \text{J mol}^{-1} + (8.315\ \text{J K}^{-1}\text{mol}^{-1})(298\text{K})\ \ln \frac{(1.3 \times 10^{-3})\,(4.8 \times 10^{-3})}{(3.4 \times 10^{-3})}$$

$$\Delta G = -30\,000\ \text{J mol}^{-1} + 2480\ \text{J mol}^{-1}\ln\,(1.8 \times 10^{-3})$$

$$\Delta G = -30\,000\ \text{J mol}^{-1} - 16\,000\ \text{J mol}^{-1}$$

$$\Delta G = -46\,000\ \text{J mol}^{-1} = -46\ \text{kJ mol}^{-1}$$

The actual free-energy change is about one and a half times the standard free-energy change. This exercise illustrates how cellular conditions influence free energy.

15

Glycolysis

Summary

Glycolysis is a ubiquitous pathway for the catabolism of monosaccharides. The 10 enzymes that catalyze the individual steps of glycolysis are located in the cytosol. For each molecule of hexose that is converted to pyruvate, there is a net production of two molecules of ATP from $ADP + P_i$, and two molecules of $NAD^\oplus$ are reduced to NADH. Glycolysis can be divided into two stages: a hexose stage, in which ATP is consumed, and a triose stage, in which a net gain of ATP is realized.

Anaerobic reoxidation of glycolytically produced NADH to $NAD^\oplus$ can occur through reductive metabolism of pyruvate. During alcoholic fermentation in yeast, pyruvate is cleaved to acetaldehyde in a reaction catalyzed by pyruvate decarboxylase, and acetaldehyde is reduced to ethanol, with concomitant oxidation of NADH to $NAD^\oplus$, in a reaction catalyzed by alcohol dehydrogenase. In anaerobic glycolysis to lactate, NADH is oxidized to $NAD^\oplus$ during the reduction of pyruvate to lactate, which is catalyzed by lactate dehydrogenase.

Three of the glycolytic reactions are metabolically irreversible in cells. These are the steps catalyzed by hexokinase, phosphofructokinase-1, and pyruvate kinase. Regulation of these enzymes, and thus of glycolysis, involves both allosteric interactions and covalent modifications. Regulation by covalent modification is mediated by enzymes that are not part of the glycolytic pathway but that act on enzymes of the pathway.

Study Information

Please see the next page.

Figure 15·3
Conversion of glucose to pyruvate by glycolysis. Following interconversion of glyceraldehyde 3-phosphate and dihydroxyacetone phosphate, the remaining reactions of glycolysis are traversed by two triose molecules for each hexose molecule metabolized. ATP is consumed in the hexose stage and generated in the triose stage.

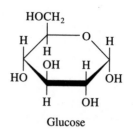

Glucose

Transfer of a phosphoryl group from ATP to glucose

① Hexokinase, glucokinase

ATP ⟶
ADP ⟵

Glucose 6-phosphate

Isomerization

② Glucose 6-phosphate isomerase

Fructose 6-phosphate

Transfer of a second phosphoryl group from ATP to fructose 6-phosphate

③ Phosphofructokinase-1

ATP ⟶
ADP ⟵

Fructose 1,6-*bis*phosphate

C-3—C-4 bond cleavage, yielding two triose phosphates

④ Aldolase

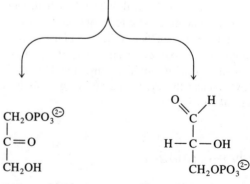

Dihydroxyacetone phosphate   Glyceraldehyde 3-phosphate

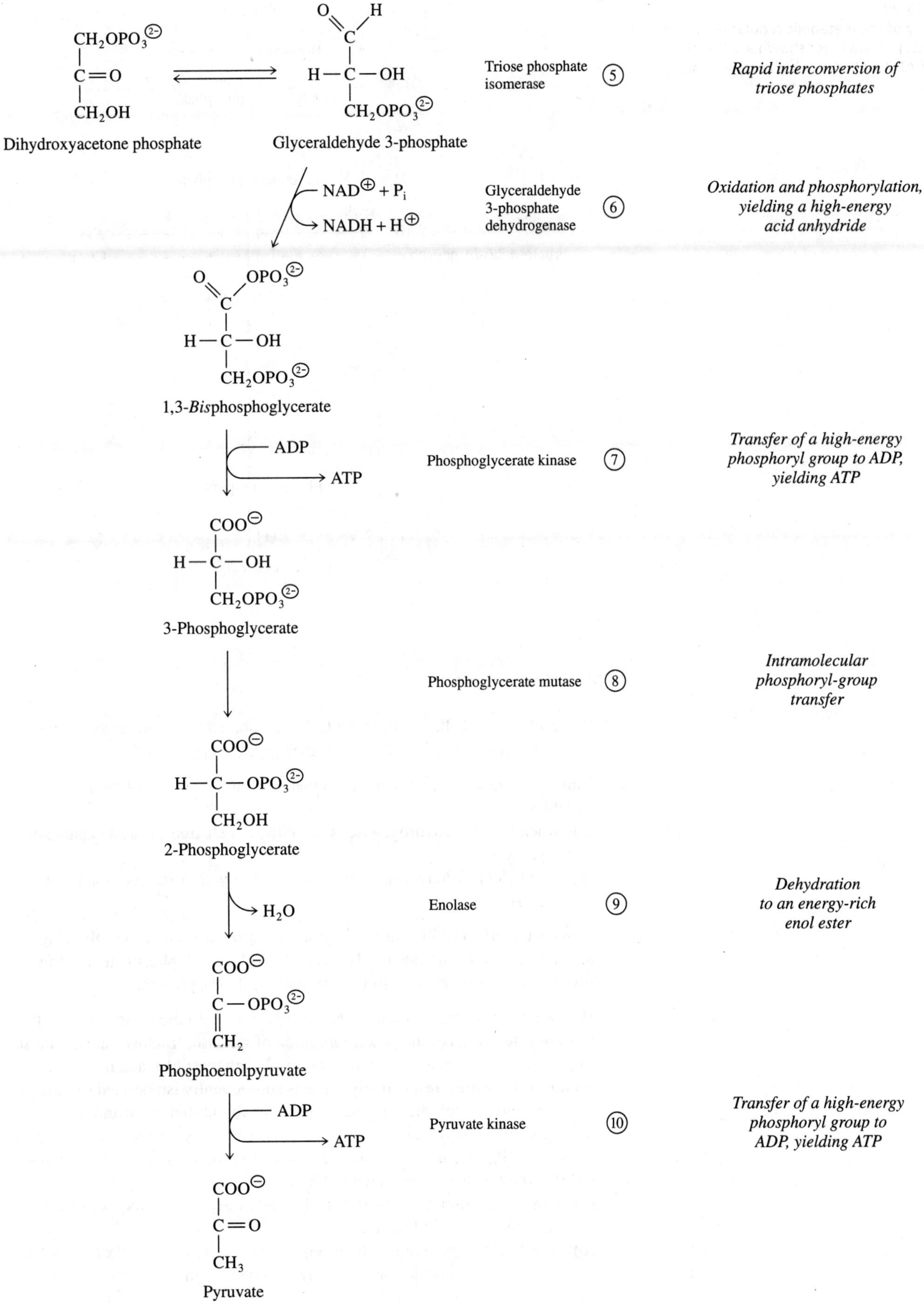

Dihydroxyacetone phosphate

$CH_2OPO_3^{2-}$
|
$C=O$
|
CH_2OH

⇌

Glyceraldehyde 3-phosphate

$O\diagdown\ \diagup H$
C
|
$H-C-OH$
|
$CH_2OPO_3^{2-}$

Triose phosphate isomerase ⑤ *Rapid interconversion of triose phosphates*

$NAD^{\oplus} + P_i$ → $NADH + H^{\oplus}$

Glyceraldehyde 3-phosphate dehydrogenase ⑥ *Oxidation and phosphorylation, yielding a high-energy acid anhydride*

1,3-*Bis*phosphoglycerate

$O=C-OPO_3^{2-}$
|
$H-C-OH$
|
$CH_2OPO_3^{2-}$

ADP → ATP

Phosphoglycerate kinase ⑦ *Transfer of a high-energy phosphoryl group to ADP, yielding ATP*

3-Phosphoglycerate

$COO^{\ominus}$
|
$H-C-OH$
|
$CH_2OPO_3^{2-}$

Phosphoglycerate mutase ⑧ *Intramolecular phosphoryl-group transfer*

2-Phosphoglycerate

$COO^{\ominus}$
|
$H-C-OPO_3^{2-}$
|
CH_2OH

H_2O

Enolase ⑨ *Dehydration to an energy-rich enol ester*

Phosphoenolpyruvate

$COO^{\ominus}$
|
$C-OPO_3^{2-}$
‖
CH_2

ADP → ATP

Pyruvate kinase ⑩ *Transfer of a high-energy phosphoryl group to ADP, yielding ATP*

Pyruvate

$COO^{\ominus}$
|
$C=O$
|
CH_3

101

Figure 15·29
Summary of the metabolic regulation of the glycolytic pathway. Not shown are the effects of ADP on PFK-1, which vary among species.

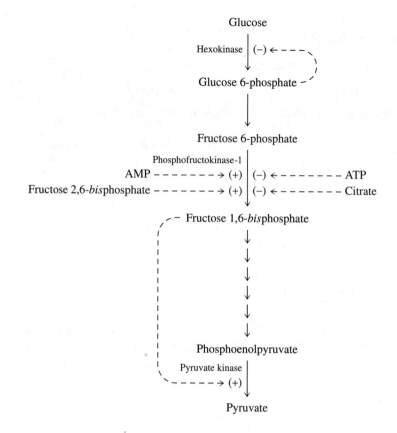

Problems

1. When glucose labelled with ^{14}C at C-1 is incubated with cell-free extracts capable of glycolysis, where does the label appear in pyruvate?

2. Name the metabolic intermediate(s) that accumulate in the following situations:
   (a) when lactate dehydrogenase is inhibited in cell-free extracts capable of glycolysis.
   (b) when alcohol dehydrogenase is inhibited in yeast extracts capable of fermentation.

3. What is the effect of increasing the concentration of each of the following metabolites on the net rate of glycolysis? (a) glucose 6-phosphate, (b) fructose 1,6-*bis*phosphate, (c) citrate, (d) fructose 2,6-*bis*phosphate.

4. Hexokinase has a rather broad substrate specificity. In the presence of ATP, this enzyme catalyzes the phosphorylation of glucose, fructose, and mannose to glucose 6-phosphate (G6P), fructose 6-phosphate (F6P), and mannose 6-phosphate (M6P), respectively. M6P is subsequently isomerized to F6P.
   Glycogen (a polymer of glucose residues found in muscle and liver) undergoes phosphorolysis—a phosphorylase catalyzes the attack of inorganic phosphate (P_i) on a terminal glucose residue to produce glucose 1-phosphate (G1P), which is then isomerized to G6P.
   (a) Draw a diagram showing the catabolism of glycogen, glucose, fructose, and mannose to F6P and lactate.
   (b) Are there any differences between these carbohydrates in the total number of ATP molecules produced by the conversion of sugar to lactate?

5. The Entner-Doudoroff pathway of glucose degradation, a variation on the glycolytic pathway, is found in an anaerobic bacterium from the genus *Pseudomonas.* This pathway has several reactions not found in glycolysis, listed below (Steps 1–3) along with the unfamiliar structures.

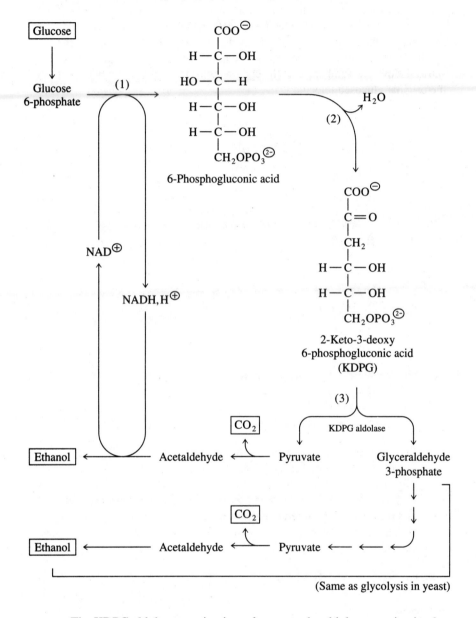

The KDPG aldolase reaction is analogous to the aldolase reaction in glycolysis. The net products of the Entner-Doudoroff pathway are two ethanol molecules and two CO_2 molecules per molecule of glucose. (The NADH produced is recycled to $NAD^\oplus$, as it must be in any anaerobic process.)

(a) Show the labelling pattern for the carbon atoms in the products starting with glucose. Is this the same labelling pattern that you would expect for alcoholic fermentation by yeast?

(b) How many ATP molecules per glucose molecule are produced via this pathway?

Solutions

1. Labelled glucose enters the glycolytic pathway via glucose 6-phosphate. The label remains at the C-1 position until fructose 1,6-*bis*phosphate is cleaved by the action of aldolase to produce glyceraldehyde 3-phosphate and dihydroxyacetone phosphate. Since dihydroxyacetone phosphate contains C-1 through C-3 of the original glucose molecule, it bears the label. Dihydroxyacetone phosphate is then converted to glyceraldehyde 3-phosphate. Ultimately, the labelled carbon appears in the methyl group of pyruvate. (In the pathway shown below, the numbering of the carbon atoms reflects the numbering of the original glucose molecule.)

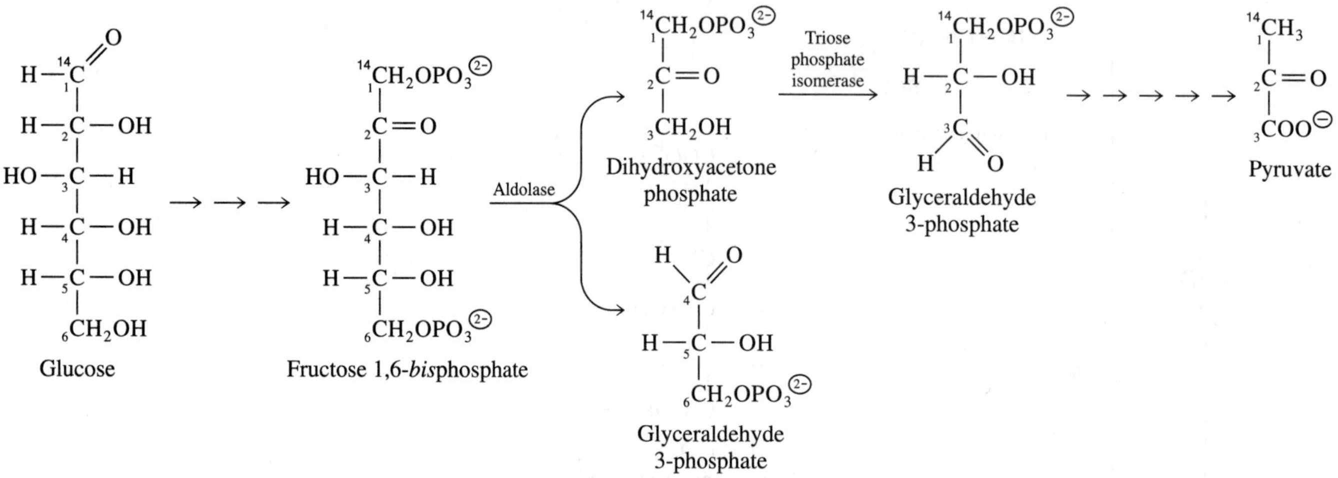

2. See Figure 15·3 for the steps of the glycolytic pathway.
   (a) Lactate dehydrogenase catalyzes the following reaction.

   $$NADH + H^{\oplus} + Pyruvate \rightleftharpoons Lactate + NAD^{\oplus}$$

   When lactate dehydrogenase is inhibited, $NAD^{\oplus}$ is not regenerated and glycolysis is inhibited at the glyceraldehyde 3-phosphate dehydrogenase step since this step requires $NAD^{\oplus}$. The accumulation of glyceraldehyde 3-phosphate, dihydroxyacetone phosphate, fructose 1,6-*bis*phosphate, and prior glycolytic intermediates is likely.

   (b) Alcohol dehydrogenase catalyzes the following reaction under anaerobic conditions.

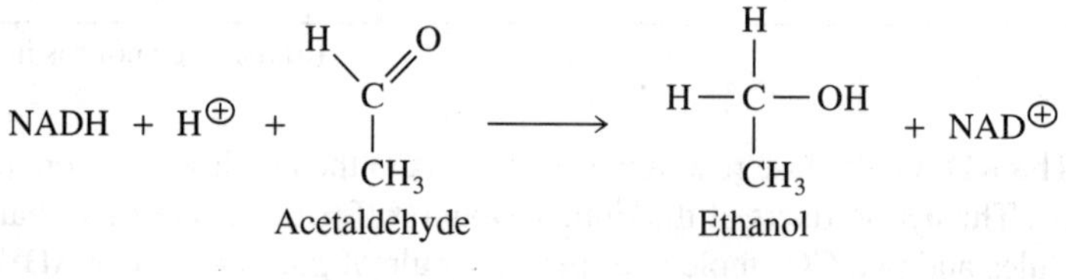

   When alcohol dehydrogenase is inhibited, $NAD^{\oplus}$ is not produced, and glycolysis is inhibited as described in part (a).

3. (a) Initial increases in the concentration of glucose 6-phosphate increase the rate of glycolysis by raising the level of substrate for glucose 6-phosphate isomerase and for all subsequent reactions of the glycolytic pathway. However, glucose 6-phosphate is also an allosteric inhibitor of hexokinase; thus, high concentrations of glucose 6-phosphate could inhibit glycolysis by decreasing the entry of glucose into the pathway (left).

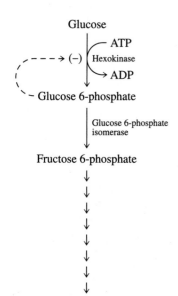

The activity of key enzymes in multistep pathways is commonly affected by negative feedback. For example, high concentrations of glucose 6-phosphate signal hexokinase to decrease the flux of metabolites through glycolysis. A decrease in the net rate of glycolysis at this early stage prevents the synthesis of glucose metabolites that may not be needed and permits storage of excess glucose as glycogen.

(b) Fructose 1,6-*bis*phosphate is a product of the reaction catalyzed by phosphofructokinase-1, which appears to be the primary regulatory point of glycolysis. Increasing the concentration of fructose 1,6-*bis*phosphate increases the rate of glycolysis by raising the level of substrates for all subsequent reactions of the glycolytic pathway.

(c) Citrate, an intermediate of the citric acid cycle, is a feedback inhibitor of phosphofructokinase-1. Thus, increasing the concentration of citrate decreases the rate of glycolysis.

(d) Fructose 2,6-*bis*phosphate, which is formed from fructose 6-phosphate in a reaction catalyzed by phosphofructokinase-2 (PFK-2), increases the rate of glycolysis, since it is an activator of phosphofructokinase-1 (PFK-1).

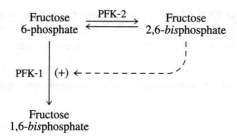

4. (a)

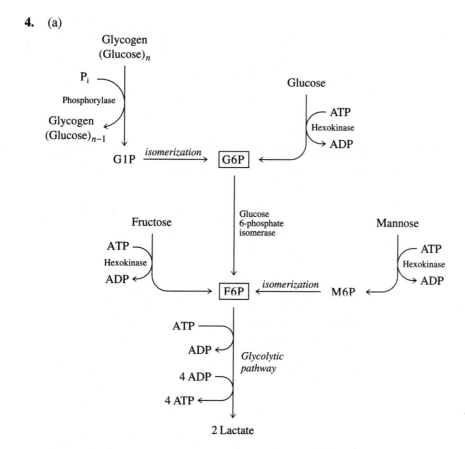

(b) Glucose, fructose, and mannose all consume one molecule of ATP in their conversion to F6P; the subsequent conversion of F6P to lactate requires one molecule of ATP and produces four molecules of ATP. Thus, the catabolism of each molecule of glucose, fructose, or mannose to lactate has a net yield of two molecules of ATP.

$$\begin{matrix} \text{Glucose} \\ \text{Fructose} \\ \text{Mannose} \end{matrix} + 2\,\text{ADP} \longrightarrow 2\,\text{Lactate} + 2\,\text{ATP}$$

Since phosphorolysis of glycogen and concomitant production of G1P do not require ATP, there is a net yield of three molecules of ATP for each glucose residue catabolized to lactate.

5. (a) The Entner-Doudoroff pathway does not give the same labelling pattern as fermentation by yeast (in which ethanol carbon atoms arise from carbons 1, 2, 5, and 6, and CO_2 carbon atoms arise from carbons 3 and 4 of glucose). In the Entner-Doudoroff pathway, C-1 and C-4 of glucose end up in CO_2 and carbons 2, 3, 5, and 6 end up in ethanol (see opposite page).

(b) There is a net production of one ATP per glucose molecule (one consumed and two produced) in this pathway. In contrast, glycolysis yields two molecules of ATP per glucose molecule consumed.

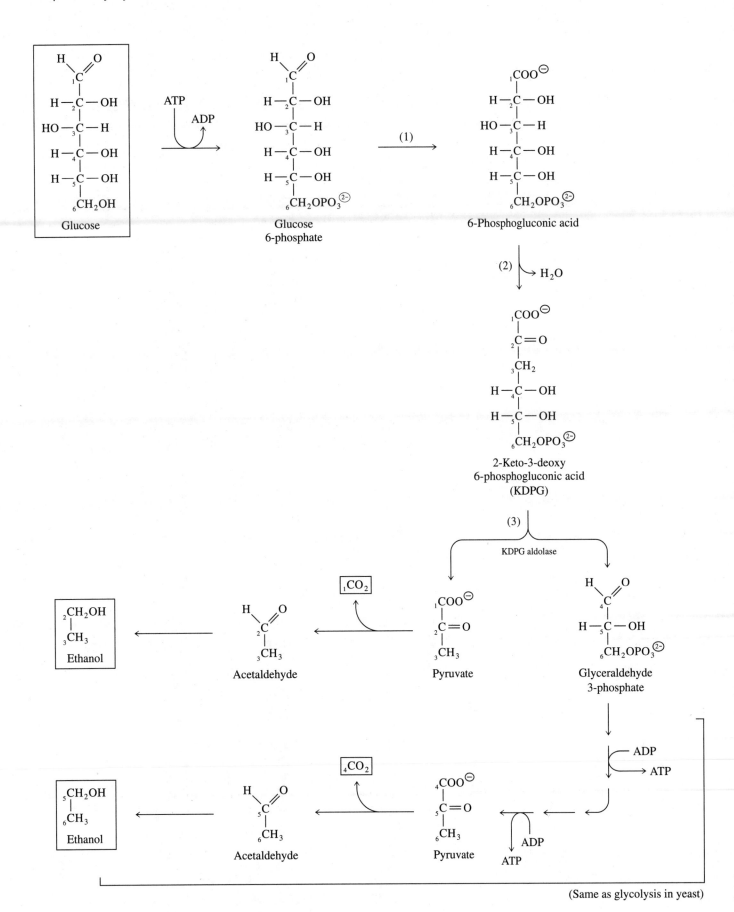

(Same as glycolysis in yeast)

16

The Citric Acid Cycle

Summary

The citric acid cycle is the final stage in the aerobic catabolism of carbohydrates, amino acids, and fatty acids. In eukaryotic cells, pyruvate generated by glycolysis in the cytosol is transported into the mitochondrial matrix where a multienzyme complex, the pyruvate dehydrogenase complex, catalyzes its oxidation to acetyl CoA and CO_2. The pyruvate dehydrogenase complex consists of enzyme components E_1 (pyruvate dehydrogenase), E_2 (dihydrolipoamide acetyltransferase), and E_3 (dihydrolipoamide dehydrogenase) and requires as cofactors thiamine pyrophosphate, lipoamide, CoASH, FAD, and $NAD^{\oplus}$.

The citric acid cycle consists of eight enzyme-catalyzed reactions. Citrate synthase catalyzes the condensation of the acetyl group of acetyl CoA with oxaloacetate, generating the tricarboxylic acid citrate. Citrate is a tertiary alcohol that must be converted to a secondary alcohol before further oxidation can occur. Aconitase catalyzes the conversion of prochiral citrate to the chiral molecule $2R,3S$-isocitrate. Successive oxidative decarboxylation reactions catalyzed by isocitrate dehydrogenase and the α-ketoglutarate dehydrogenase complex produce two molecules of NADH and two molecules of CO_2, leading to the formation of succinyl CoA. Succinyl-CoA synthetase catalyzes substrate-level phosphorylation of GDP to GTP (or ADP to ATP, depending on the organism) as the thioester of succinyl CoA is cleaved to succinate and CoASH. The eukaryotic flavoprotein succinate dehydrogenase, embedded in the inner mitochondrial membrane, is a component of a complex that catalyzes oxidation of succinate to fumarate, with electrons transferred from $FADH_2$ to ubiquinone (Q), forming ubiquinol (QH_2), a lipid-soluble mobile carrier of reducing equivalents. The double bond of fumarate is hydrated by the action of fumarase to produce malate, which can be oxidized to oxaloacetate by $NAD^{\oplus}$-dependent malate dehydrogenase. The regeneration of oxaloacetate completes one turn of the citric acid cycle.

For each molecule of acetyl CoA oxidized via the citric acid cycle, three molecules of $NAD^{\oplus}$ are reduced to NADH (in the reactions catalyzed by isocitrate dehydrogenase, the α-ketoglutarate dehydrogenase complex, and malate dehydrogenase), one molecule of Q is reduced to QH_2 (succinate dehydrogenase), and one molecule of GTP is generated from $GDP + P_i$ or one molecule of ATP is generated from $ADP + P_i$ (succinyl-CoA synthetase). Oxidation via the respiratory electron-transport chain of the reduced coenzymes NADH and QH_2 produced during one turn of the citric acid cycle leads to the formation of about 10 ATP molecules per molecule of acetyl CoA that enters the cycle. Complete oxidation of one molecule of glucose by glycolysis, the pyruvate dehydrogenase complex, the citric acid cycle, and the electron-transport chain leads to the formation of approximately 32 molecules of ATP by a combination of substrate-level and oxidative phosphorylation.

The citric acid cycle has several control points. The pyruvate dehydrogenase complex is regulated by the levels of its end products, acetyl CoA and NADH. Isocitrate dehydrogenase and the α-ketoglutarate dehydrogenase complex are allosterically regulated. In mammals, the pyruvate dehydrogenase complex is further controlled by covalent modification. In *E. coli,* isocitrate dehydrogenase is subject to covalent modification.

In addition to its role in oxidative catabolism, the citric acid cycle provides precursors for biosynthetic pathways. Citrate, α-ketoglutarate, succinyl CoA, and oxaloacetate are the principal branch-point metabolites. The pathway is replenished by formation of oxaloacetate from pyruvate, by reversible reactions forming oxaloacetate and α-ketoglutarate, and by pathways for the degradation of some amino acids and fatty acids, which can contribute succinyl CoA.

The glyoxylate cycle is a pathway closely related to the citric acid cycle that allows plants and some microorganisms to use acetyl CoA to generate four-carbon intermediates for gluconeogenesis and other biosynthetic pathways. Two enzymes unique to the glyoxylate cycle, isocitrate lyase and malate synthase, provide a bypass around the CO_2-producing reactions of the citric acid cycle. These enzymes are not present in animal cells; thus, animals cannot synthesize anabolic metabolites from acetyl CoA or two-carbon molecules such as acetate. Isocitrate lyase catalyzes the cleavage of isocitrate to succinate and glyoxylate. Succinate enters the citric acid cycle, and glyoxylate condenses with CoASH to form malate, catalyzed by malate synthase. Malate can serve as a precursor for the formation of glucose.

Study Information

Please see the next page.

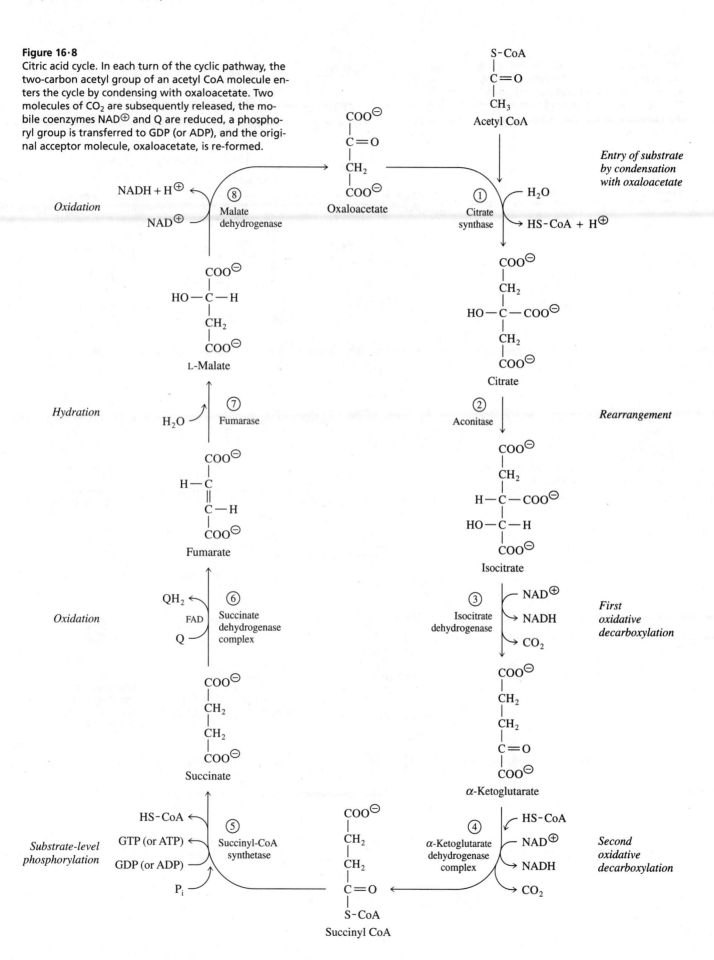

Figure 16·8
Citric acid cycle. In each turn of the cyclic pathway, the two-carbon acetyl group of an acetyl CoA molecule enters the cycle by condensing with oxaloacetate. Two molecules of CO_2 are subsequently released, the mobile coenzymes $NAD^\oplus$ and Q are reduced, a phosphoryl group is transferred to GDP (or ADP), and the original acceptor molecule, oxaloacetate, is re-formed.

Figure 16·24
ATP production from the catabolism of one molecule of glucose by glycolysis, the citric acid cycle, and oxidative phosphorylation. The complete oxidation of glucose leads to the formation of approximately 32 molecules of ATP.

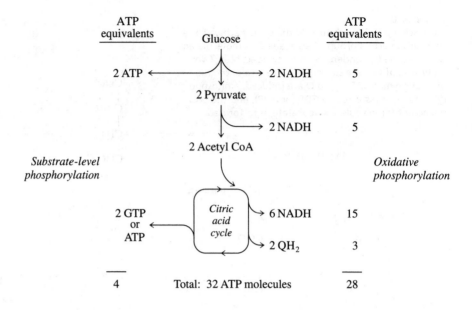

Figure 16·28
Regulation of the pyruvate dehydrogenase complex and the citric acid cycle.

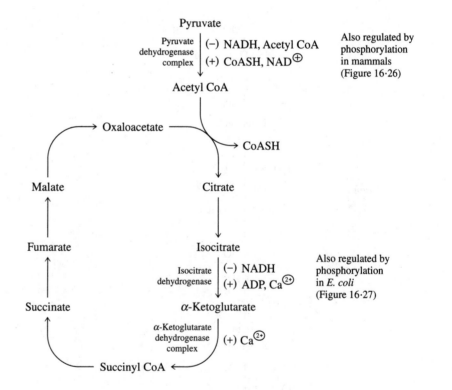

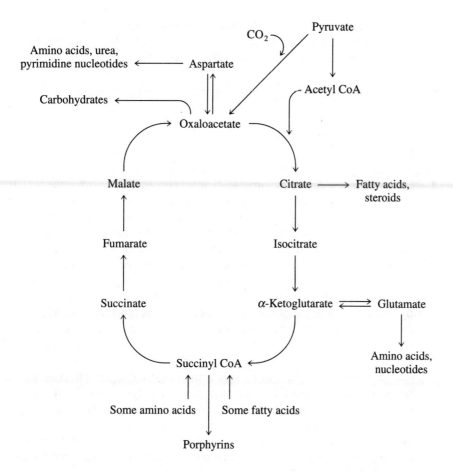

Figure 16·29
Routes leading to and from the citric acid cycle. Intermediates of the citric acid cycle are precursors for carbohydrates, lipids, and amino acids, as well as nucleotides and porphyrins. Reactions feeding into the cycle replenish the pool of cycle intermediates.

Problems

1. Early experiments with minced tissue showed that the citric acid cycle is an aerobic pathway (i.e., requires O_2 to function) and that the compounds metabolized by this cycle are eventually oxidized to CO_2. However, addition of cycle intermediates caused more O_2 to be consumed than expected. For example, addition of 1 μmol of fumarate to a muscle homogenate caused about 25 μmol of O_2 to be consumed, but the oxidation reaction below indicates that only 3 μmol of O_2 are needed to completely oxidize 1 μmol of fumarate. Similar observations were made when succinate, malate, and oxaloacetate were added to the muscle homogenate. Explain why addition of these pathway intermediates caused such an unexpectedly large oxygen consumption.

$$HOOC-CH=CH-COOH \ + \ 3\,O_2 \ \longrightarrow \ 4\,CO_2 \ + \ 2\,H_2O$$
$$\text{Fumarate}$$

2. What is the effect of increasing the concentration of each of the following on the rate of the citric acid cycle? (a) Coenzyme A, (b) acetyl CoA, (c) $NAD^{\oplus}$.

3. Trace the isotopic label of 2-[^{18}O]-pyruvate through the pyruvate dehydrogenase complex and one round of the citric acid cycle.

4. Ingested fatty acids are metabolically degraded to acetyl CoA. Acetyl CoA activates pyruvate carboxylase, which catalyzes the conversion of pyruvate to oxaloacetate. How does this activation of pyruvate carboxylase contribute to the recovery of energy from fatty acids?

113

5. Yeast can grow both aerobically and anaerobically on glucose. Explain why the rate of glucose consumption decreases when yeast cells that have been maintained under anaerobic conditions are exposed to oxygen.

6. (a) What is the major function of the glyoxylate cycle?
   (b) Can the glyoxylate shunt be considered an anaplerotic pathway?

Solutions

1. O_2 consumption during the citric acid cycle is necessary to oxidize the NADH and QH_2 produced during conversion of pyruvate to CO_2. When the rate of the citric acid cycle increases, the rate of O_2 consumption increases. Because the citric acid cycle is cyclic, the intermediates of the citric acid cycle *catalytically* stimulate O_2 utilization. Fumarate is not oxidized in a linear series of reactions to four CO_2 (where three O_2 would be required); instead it enters the citric acid cycle and is metabolized to a molecule of oxaloacetate. The oxaloacetate combines with a molecule of acetyl CoA in a reaction catalyzed by citrate synthase, yielding a molecule of citrate. A fumarate molecule is then regenerated from citrate, and two CO_2 are produced (the two carbons entered the cycle via acetyl CoA). There is no *net* consumption of fumarate since the four carbons originally provided continue in the cycle, and the amount of O_2 consumed is much greater than expected for the direct oxidation of fumarate to four CO_2. Of course, to observe these catalytic effects there must be a supply of pyruvate or acetyl CoA present in the tissue.

   Other intermediates such as succinate, malate, and oxaloacetate also enter the citric acid cycle and *catalytically* stimulate O_2 consumption by increasing cycle intermediate concentrations and thus the overall rate of the citric acid cycle.

2. (a) Coenzyme A is an allosteric activator of the dihydrolipoamide acetyltransferase (E_2) component of the pyruvate dehydrogenase complex in mammalian cells and in *E. coli*. Increasing the concentration of coenzyme A thus increases production of acetyl CoA from pyruvate and the amount of acetyl CoA entering the citric acid cycle, thereby initially increasing the rate of the cycle. For more information on the effects of increasing the concentration of acetyl CoA, see part (b).
   (b) Increasing the concentration of acetyl CoA could have two effects. Acetyl CoA is a substrate for citrate synthase, which catalyzes the first reaction of the citric acid cycle. If the initial concentration of acetyl CoA is low, increasing it increases the rate of the reaction catalyzed by citrate synthase. Above a certain concentration, however, acetyl CoA directly inhibits the dihydrolipoamide acetyltransferase (E_2) component of the pyruvate dehydrogenase complex, resulting in a decrease in the concentration of acetyl CoA formed from pyruvate. Such a decrease could decrease the rate of the reactions of the citric acid cycle.

      In mammalian cells, high concentrations of acetyl CoA also allosterically activate pyruvate dehydrogenase kinase, an enzyme that catalyzes phosphorylation of the pyruvate dehydrogenase component of the pyruvate dehydrogenase complex. This covalent modification inhibits the decarboxylation of pyruvate, decreasing the rate of citric acid cycle reactions.
   (c) $NAD^{\oplus}$ activates the dihydrolipoamide dehydrogenase (E_3) component of the pyruvate dehydrogenase complex and a cosubstrate of isocitrate dehydrogenase, malate dehydrogenase, and E_3 of both the α-ketoglutarate dehydrogenase complex and the pyruvate dehydrogenase complex. Increasing the concentration of $NAD^{\oplus}$ thus increases the rate of the citric acid cycle.

3. The figure below indicates which oxygen atoms may be labelled. Note, though, that only one oxygen atom per molecule is actually labelled. The ^{18}O may appear in either of the carboxylate moieties of oxaloacetate (that is, oxygen bound to carbon atoms 1 and 4 of this metabolite) after one round of the citric acid cycle.

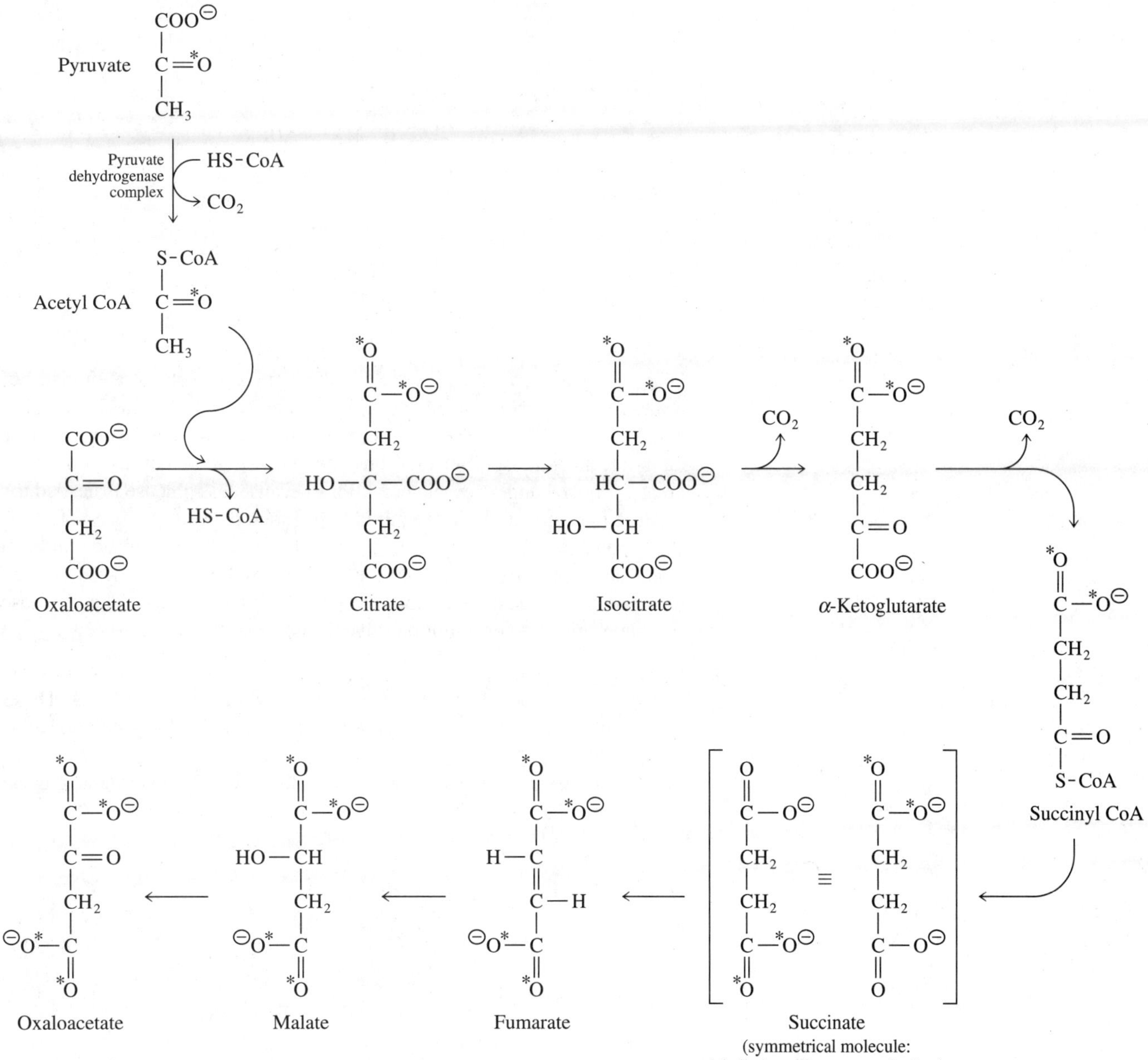

Pyruvate
$$COO^{\ominus}$$
$$|$$
$$C = \overset{*}{O}$$
$$|$$
$$CH_3$$

Pyruvate dehydrogenase complex — HS-CoA → CO_2

Acetyl CoA
$$S\text{-}CoA$$
$$|$$
$$C = \overset{*}{O}$$
$$|$$
$$CH_3$$

Oxaloacetate
$$COO^{\ominus}$$
$$|$$
$$C = O$$
$$|$$
$$CH_2$$
$$|$$
$$COO^{\ominus}$$

HS-CoA

Citrate
$$\overset{*}{O}$$
$$\|$$
$$C - \overset{*}{O}^{\ominus}$$
$$|$$
$$CH_2$$
$$|$$
$$HO - C - COO^{\ominus}$$
$$|$$
$$CH_2$$
$$|$$
$$COO^{\ominus}$$

Isocitrate
$$\overset{*}{O}$$
$$\|$$
$$C - \overset{*}{O}^{\ominus}$$
$$|$$
$$CH_2$$
$$|$$
$$HC - COO^{\ominus}$$
$$|$$
$$HO - CH$$
$$|$$
$$COO^{\ominus}$$

CO_2

α-Ketoglutarate
$$\overset{*}{O}$$
$$\|$$
$$C - \overset{*}{O}^{\ominus}$$
$$|$$
$$CH_2$$
$$|$$
$$CH_2$$
$$|$$
$$C = O$$
$$|$$
$$COO^{\ominus}$$

CO_2

Succinyl CoA
$$\overset{*}{O}$$
$$\|$$
$$C - \overset{*}{O}^{\ominus}$$
$$|$$
$$CH_2$$
$$|$$
$$CH_2$$
$$|$$
$$C = O$$
$$|$$
$$S\text{-}CoA$$

Oxaloacetate
$$\overset{*}{O}$$
$$\|$$
$$C - \overset{*}{O}^{\ominus}$$
$$|$$
$$C = O$$
$$|$$
$$CH_2$$
$$|$$
$$^{\ominus}\overset{}{O} - \overset{*}{C}$$
$$\|$$
$$\overset{*}{O}$$

Malate
$$\overset{*}{O}$$
$$\|$$
$$C - \overset{*}{O}^{\ominus}$$
$$|$$
$$HO - CH$$
$$|$$
$$CH_2$$
$$|$$
$$^{\ominus}\overset{}{O} - \overset{*}{C}$$
$$\|$$
$$\overset{*}{O}$$

Fumarate
$$\overset{*}{O}$$
$$\|$$
$$C - \overset{*}{O}^{\ominus}$$
$$|$$
$$H - C$$
$$\|$$
$$C - H$$
$$|$$
$$^{\ominus}\overset{}{O} - \overset{*}{C}$$
$$\|$$
$$\overset{*}{O}$$

Succinate
(symmetrical molecule: labelled positions are equivalent)
$$O$$
$$\|$$
$$C - O^{\ominus}$$
$$|$$
$$CH_2$$
$$|$$
$$CH_2$$
$$|$$
$$C - \overset{*}{O}^{\ominus}$$
$$\|$$
$$\overset{*}{O}$$
$$\equiv$$
$$\overset{*}{O}$$
$$\|$$
$$C - \overset{*}{O}^{\ominus}$$
$$|$$
$$CH_2$$
$$|$$
$$CH_2$$
$$|$$
$$C - O^{\ominus}$$
$$\|$$
$$O$$

4. By activating pyruvate carboxylase, acetyl CoA increases the amount of oxaloacetate produced directly from pyruvate. The increased amounts of oxaloacetate can react with the large amounts of acetyl CoA produced by the degradation of fatty acids. As a result, the citric acid cycle proceeds at a faster rate to recover the energy stored in the fatty acids.

5. The decrease in glucose consumption when oxygen is introduced is known as the Pasteur effect (Section 15·7E). Under anaerobic conditions, glucose is not completely oxidized to CO_2 and H_2O but is converted to ethanol and CO_2. Under these conditions, there is a net production of only two molecules of ATP for each glucose molecule converted to ethanol and CO_2 by the glycolytic pathway.

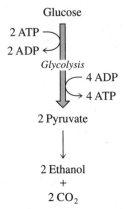

Glucose

2 ATP

2 ADP

Glycolysis

4 ADP

4 ATP

2 Pyruvate

2 Ethanol
+
2 CO_2

   When oxygen is available, yeast can utilize glucose much more efficiently, producing many more molecules of ATP for each glucose molecule oxidized completely to CO_2 and H_2O by glycolysis, the citric acid cycle, and the respiratory electron-transport chain. Thus, the rate of glucose consumption is greatly reduced under aerobic conditions because less glucose is needed to produce the amount of ATP required to maintain the cell.

   Upon exposure of yeast cells to oxygen, most of the pyruvate will be metabolized aerobically to CO_2 and acetyl CoA, resulting in increased synthesis of citrate, a key allosteric inhibitor of phosphofructokinase-1. This decreases the rate of glucose consumption (glycolysis).

6. (a) The glyoxylate cycle enables many plants and microorganisms to synthesize four-carbon dicarboxylic acids from two-carbon precursors. These organisms can thereby convert ethanol or fat-derived acetate into carbohydrates.

   (b) Yes. It results in the *net* synthesis of malate, which is a citric acid cycle intermediate.

17

Additional Pathways in Carbohydrate Metabolism

Summary

Dietary saccharides can be converted to metabolites that can enter the glycolytic pathway. Fructose is converted to glyceraldehyde 3-phosphate, galactose is converted to glucose 1-phosphate, and mannose is converted to fructose 6-phosphate.

Glycogen is the glucose-storage polysaccharide of animals. Glycogen phosphorylase catalyzes degradation of intracellular glycogen to form glucose 1-phosphate, which can be converted to glucose 6-phosphate in a near-equilibrium reaction catalyzed by phosphoglucomutase. In mammalian liver, glucose 6-phosphate is hydrolyzed to glucose and P_i.

In liver and muscle, glycogen degradation and glycogen synthesis are reciprocally regulated pathways that are controlled by the hormones epinephrine, glucagon, and insulin. Both epinephrine and glucagon stimulate the production of cAMP, which activates protein kinase A. This leads to phosphorylation and inactivation of glycogen synthase. In addition, protein kinase A catalyzes phosphorylation of glycogen phosphorylase kinase, which in turn catalyzes phosphorylation of glycogen phosphorylase, converting glycogen phosphorylase to its active form. Epinephrine can also increase cytosolic calcium, which activates phosphorylase kinase, leading to activation of glycogen phosphorylase. The hormonally induced phosphorylations in both liver and muscle are reversed by a protein phosphatase that is itself regulated.

Gluconeogenesis is the pathway for glucose synthesis from noncarbohydrate precursors, such as lactate and amino acids. Many of the gluconeogenic reactions are simply the reverse of the near-equilibrium reactions of glycolysis. Enzymes specific to gluconeogenesis catalyze the metabolically irreversible conversion of pyruvate to phosphoenolpyruvate (pyruvate carboxylase and phosphoenolpyruvate carboxykinase), fructose 1,6-*bis*phosphate to fructose 6-phosphate (fructose 1,6-*bis*phosphatase), and glucose 6-phosphate to glucose (glucose 6-phosphatase). The principal regulatory site of gluconeogenesis is the conversion of pyruvate to phosphoenolpyruvate; a secondary site is the reaction catalyzed by fructose 1,6-*bis*phosphatase. Because pyruvate carboxylase is located in mitochondria, pyruvate must be transported into the mitochondria, and oxaloacetate must be transported to the cytosol via shuttle systems.

Sometimes, glucose can be converted to sorbitol or fructose via the sorbitol pathway. In the mammary gland, glucose reacts with UDP-galactose to form lactose. Lactose synthase consists of galactosyltransferase and α-lactalbumin.

The pentose phosphate pathway provides an alternate pathway for glucose 6-phosphate metabolism, initiated by the action of glucose 6-phosphate dehydrogenase. Glucose 6-phosphate dehydrogenase is also the major regulatory site of the pathway; the enzyme is allosterically inhibited by NADPH. The oxidative stage of the pentose phosphate pathway generates two molecules of NADPH per molecule of glucose 6-phosphate converted to ribulose 5-phosphate and CO_2. The nonoxidative stage of pentose phosphate metabolism includes isomerization of ribulose 5-phosphate to ribose 5-phosphate, a metabolite required for the biosynthesis of nucleotides and nucleic acids. Further metabolism of pentose phosphate molecules via the nonoxidative stage of the pentose phosphate pathway provides a mechanism for their conversion to triose-phosphate and hexose-phosphate intermediates of glycolysis and gluconeogenesis.

Study Information

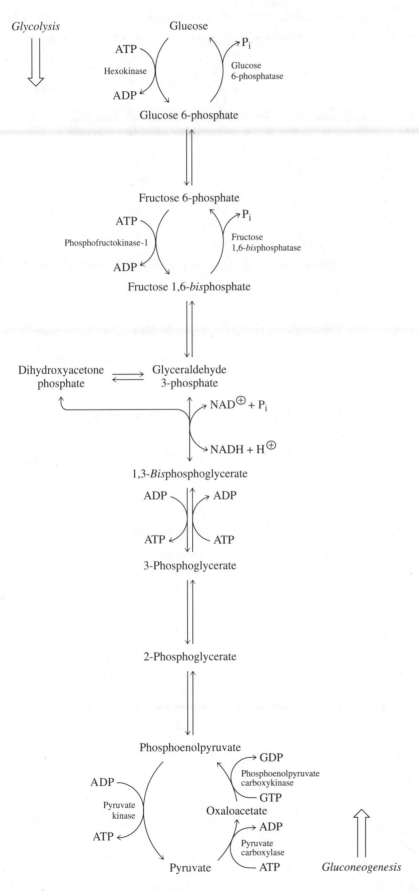

Figure 17·18
Comparison of gluconeogenesis and glycolysis. In both pathways, the triose phase is traversed by two molecules for each glucose molecule.

Figure 17·30
Pentose phosphate pathway. The pathway can be divided into an oxidative and a nonoxidative stage. The oxidative stage produces the five-carbon sugar phosphate ribulose 5-phosphate, with concomitant production of NADPH. The nonoxidative stage produces the glycolytic intermediates glyceraldehyde 3-phosphate and fructose 6-phosphate.

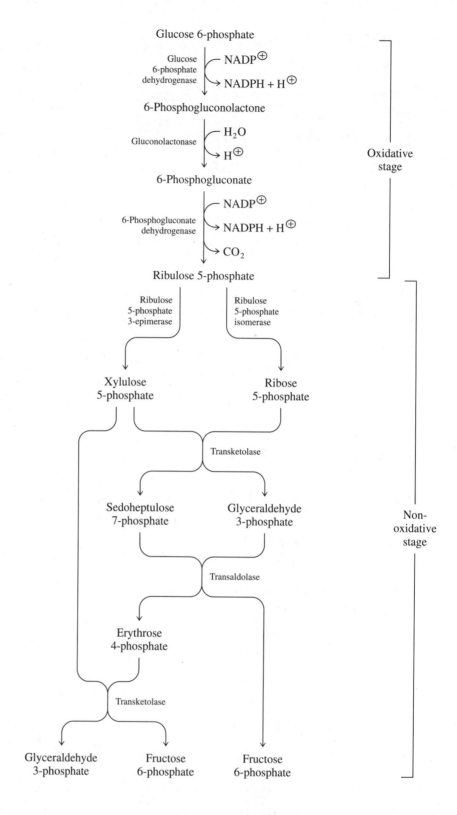

Problems

1. What is the effect of increasing the concentration of each of the following on the rate of glycogen degradation in the liver? (a) cytosolic Ca^{2+}, (b) plasma glucagon, (c) blood glucose.

2. Is the energy required to synthesize glycogen from glucose 6-phosphate the same as the energy required to degrade glycogen to glucose 6-phosphate?

3. Explain the effects of each of the following on gluconeogenesis in liver cells: (a) decreasing the concentration of acetyl CoA, (b) increasing the concentration of fructose 2,6-*bis*phosphate, (c) increasing the concentration of fructose 6-phosphate.

4. Predict the metabolic fate of glucose 6-phosphate under each of the following metabolic conditions: (a) more NADPH than ribose 5-phosphate is required, (b) both ribose 5-phosphate and NADPH are required.

5. In many tissues, one of the earliest responses to cellular injury is a rapid increase in the levels of the enzymes involved in the pentose phosphate pathway. Ten days after an injury, heart tissue has levels of glucose 6-phosphate dehydrogenase and 6-phosphogluconate dehydrogenase that are 20 to 30 times higher than normal, whereas the levels of glycolytic enzymes are only 10 to 20% of normal. Suggest an explanation for this phenomenon.

Solutions

1. (a) Increasing the concentration of cytosolic Ca^{2+} increases glycogen degradation. Ca^{2+} partially activates phosphorylase kinase by binding to the calmodulin subunit of the enzyme. Phosphorylase kinase in turn catalyzes the phosphorylation and thus the conversion of inactive glycogen phosphorylase *b* to active glycogen phosphorylase *a*, the enzyme responsible for breaking down glycogen (Figure 17·14).

   (b) Increasing the concentration of plasma glucagon stimulates glycogen degradation in the liver. When glucagon binds to specific receptors in the plasma membrane of liver cells, these receptors stimulate the activity of adenylate cyclase, which catalyzes the production of cAMP. cAMP in turn activates protein kinase A, which activates phosphorylase kinase and inactivates glycogen synthase (Figure 17·14).

   (c) Increasing the concentration of glucose in the blood decreases glycogen degradation and increases glycogen synthesis. By entering the liver cells and binding to the active form of glycogen phosphorylase, glucose deactivates this enzyme by making it a better substrate for protein phosphatase-1 (Figure 17·17). Since activated glycogen phosphorylase inhibits protein phosphatase-1, inactivation of the phosphorylase lifts this inhibition. As a result, protein phosphatase-1 activates glycogen synthase via dephosphorylation and glycogen synthesis increases.

2. More energy is required to synthesize glycogen from glucose 6-phosphate. During glycogen synthesis, one high-energy phosphoanhydride linkage is hydrolyzed since PP_i formed by the action of UDP-glucose pyrophosphorylase is rapidly converted to $2\,P_i$. In contrast, no cleavage of a high-energy phosphoanhydride linkage is needed to degrade glycogen to glucose 6-phosphate because the glucose residue is removed by a phosphorolysis reaction.

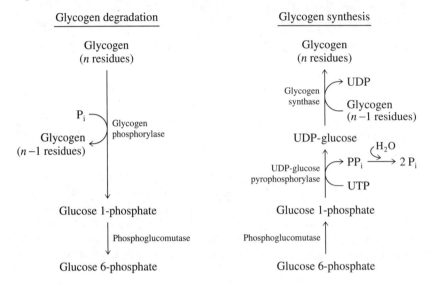

3. (a) Acetyl CoA allosterically activates pyruvate carboxylase, the enzyme that converts pyruvate to oxaloacetate in the first step of gluconeogenesis from pyruvate. Thus, decreasing the concentration of acetyl CoA decreases the rate of gluconeogenesis.

    (b) Increasing the concentration of fructose 2,6-*bis*phosphate (F2,6BP) decreases the rate of gluconeogenesis and increases the rate of glycolysis. F2,6BP inhibits fructose 1,6-*bis*phosphatase in the gluconeogenesis pathway and activates the glycolytic enzyme phosphofructokinase-1.

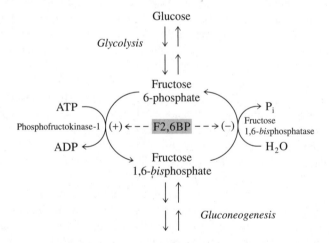

    (c) Increasing the concentration of fructose 6-phosphate decreases the rate of gluconeogenesis. Fructose 6-phosphate is a substrate not only for glucose 6-phosphate isomerase but also for both phosphofructokinase-1 and phosphofructokinase-2, which convert fructose 6-phosphate to fructose 1,6-*bis*phosphate and fructose 2,6-*bis*phosphate, respectively. Fructose 2,6-*bis*phosphate is an allosteric activator of phosphofructokinase-1 and an allosteric inhibitor of fructose 1,6-*bis*phosphatase. Therefore, an increase in the concentration of fructose 6-phosphate, and thus fructose 2,6-*bis*phosphate, favors glycolysis over gluconeogenesis.

4. The metabolism of glucose 6-phosphate by the glycolytic pathway or pentose phosphate pathway depends on the cellular demands for NADPH, ribose 5-phosphate, and ATP.

    (a) When more NADPH than ribose 5-phosphate is required, the pentose phosphate pathway operates as a cycle. Each 6-carbon sugar that enters the pathway is degraded to a 5-carbon sugar and CO_2. The actions of transketolase and transaldolase then convert the 5-carbon intermediates to 6-carbon sugars that can reenter the pathway. The net result is that 12 NADPH are produced by the oxidation of 6 glucose 6-phosphate molecules.

    $$\text{6 Glucose 6-phosphate } + \text{ 12 NADP}^{\oplus} \longrightarrow$$
    $$\text{5 Glucose 6-phosphate } + \text{ 12 NADPH } + \text{ 6 CO}_2 + \text{ P}_i$$

    (b) When considerable amounts of ribose 5-phosphate as well as NADPH are required, only the oxidative portion of the pentose phosphate pathway is used. Each molecule of glucose 6-phosphate yields two molecules of NADPH, one molecule of CO_2, and one molecule of ribulose 5-phosphate (Figure 17·30), which is converted to ribose 5-phosphate by the action of ribulose 5-phosphate isomerase.

5. Repair of tissue injury requires cell proliferation and synthesis of scar tissue. NADPH is needed for synthesis of cholesterol and fatty acids (components of cellular membranes), and ribose 5-phosphate is needed for synthesis of DNA and RNA. Since the pentose phosphate pathway is the primary source of NADPH and ribose 5-phosphate, the tissue responds to the increased demand for these products following injury by increasing the level of synthesis of the enzymes in the pentose phosphate pathway.

18

Electron Transport and Oxidative Phosphorylation

Summary

The energy in reduced coenzymes is recovered in the mitochondria by the process of oxidative phosphorylation. This process consists of two tightly coupled phenomena: (1) oxidation of substrates by the respiratory electron-transport chain, accompanied by the translocation of protons across the inner mitochondrial membrane to generate a proton concentration gradient; and (2) formation of ATP driven by the energy of protons flowing into the matrix through a channel in ATP synthase. The chemiosmotic theory explains how ADP phosphorylation is coupled to electron transport. It also explains the effects of uncoupling agents that allow substrate oxidation without ADP phosphorylation. The postulates of the theory include an intact mitochondrial membrane, a proton concentration gradient, and a membrane-bound ATPase that operates in reverse.

As electrons move through the electron-transport complexes in the mitochondrial membrane, proton translocation results in a potential difference across the membrane. The protonmotive force (Δp) depends on the difference in concentration of the protons (ΔpH) and the difference in charge ($\Delta \psi$) on either side of the membrane.

The five assemblies of proteins and cofactors involved in oxidative phosphorylation include the electron-transferring Complexes I–IV and ATP synthase. Electron flow through the complexes generally proceeds according to the relative reduction potentials of the various components. Electrons flow from NADH through Complexes I, III, and IV. Electrons from succinate are introduced via Complex II. Several cofactors participate in electron transfer, including FMN, FAD, iron-sulfur clusters, copper atoms, cytochromes, and ubiquinone. The cytochromes are differentiated by their absorption spectra and reduction potentials. The mobile carrier ubiquinone links Complexes I and II with Complex III, and cytochrome c links Complex III with Complex IV.

For every two electrons transferred from NADH to Q by Complex I, four protons are translocated to the intermembrane space. Electron transfer by Complex II does not contribute to the proton concentration gradient. The transfer of electrons from QH$_2$ to cytochrome c by Complex III results in the net translocation of two protons from the matrix. The sequence of electron transfers in Complex III is described by the two-step Q cycle. Complex IV transfers electrons from cytochrome c to O$_2$, the ultimate oxidizing agent of the electron-transport chain. Electron transfer and reduction of O$_2$ to H$_2$O contributes four protons to the gradient.

Protons reenter the mitochondrial matrix by flowing through the multimeric F$_O$ portion of the F$_O$F$_1$ ATP synthase. The mechanism by which proton flow drives the synthesis of ATP from ADP + P$_i$ catalyzed by F$_1$ ATPase is not understood. The binding-change mechanism has been proposed to explain how the flow of three protons is linked to conformational changes in the synthase.

Transport of ADP and P$_i$ into and ATP out of the mitochondrial matrix consumes the equivalent of one proton. The ATP yield per pair of electrons transferred by Complexes I–IV can be calculated from the number of protons translocated. Oxidation of mitochondrial NADH generates 2.5 ATP; oxidation of succinate generates 1.5 ATP. These values are lower than those traditionally used. Cytosolic NADH can contribute to oxidative phosphorylation when the reducing power is transferred to the mitochondria by the action of the glycerol phosphate shuttle or the malate-aspartate shuttle.

In certain cells, oxidation of reduced substrates is uncoupled from phosphorylation to generate heat. Phagocytes use the reducing power of NADPH to generate superoxide for fighting bacterial infections. Cells use reduced molecules such as glutathione, vitamin E, and ascorbic acid to protect against oxidative damage by reactive oxygen species.

Study Information

Figure 18·8
Mitochondrial electron transport. Each of the four complexes of the electron-transport chain, composed of several protein subunits and cofactors, undergoes cyclic reduction and oxidation. The complexes are linked by the mobile carrier ubiquinone (Q) and by cytochrome c. The height of each complex indicates the $\Delta E^{o\prime}$ between its reducing agent (substrate) and its oxidizing agent (product). Transfer of electrons from NADH and succinate to O$_2$ is accompanied by translocation of protons from the matrix to the intermembrane space, resulting in a proton concentration gradient across the membrane.

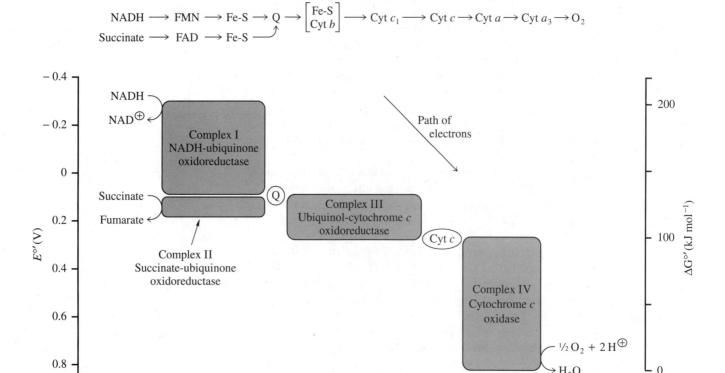

Problems

1. Using data from Table 14·4 , calculate the standard reduction potential and standard free-energy change for each of the following oxidation-reduction reactions.
    (a) Acetaldehyde + NADH + $H^{\oplus}$ ⟶ Ethanol + $NAD^{\oplus}$
    (b) Ubiquinol (QH_2) + 2 Cytochrome $c(Fe^{③})$ ⟶ Ubiquinone (Q) + 2 Cytochrome $c(Fe^{②})$ + 2 $H^{\oplus}$
    (c) Succinate + ½ O_2 ⟶ Fumarate + H_2O

2. The six cytochromes found in the electron-transport chain all catalyze one-electron transfers via the reversible oxidation-reduction reaction $Fe^{③}$ ⇌ $Fe^{②}$. Although iron is the electron carrier in each case, the $E^{\circ\prime}$ values for the reduction half-reaction vary from about 0.05 V for cytochrome b to 0.39 V for cytochrome a_3. Explain.

3. Electron transfer between cytochromes involves transfer of one electron from one iron atom to another.

    $$\text{Cytochrome } b \ (Fe^{②}) \qquad \text{Cytochrome } c_1 \ (Fe^{③})$$
    $$\text{Cytochrome } b \ (Fe^{③}) \qquad \text{Cytochrome } c_1 \ (Fe^{②})$$

    A second method of electron transfer involves hydrogen atoms as well as electrons. The *net* transfer of a hydrogen molecule (H_2 or H:H) is common in biological oxidation-reduction reactions. What are the two mechanisms by which H_2 is transferred? Give an example of each.

4. The complexes of the respiratory electron-transport chain are associated with the inner mitochondrial membrane. What conclusions can be drawn about the location in the membrane of the electron carriers based on their electron sources and role in pumping protons?

5. Functional electron-transport systems can be reconstituted from purified respiratory electron-transport chain components and membrane vesicles. For each of the following sets of components, determine the final electron acceptor. Assume O_2 is present.
    (a) NADH, Q, Complexes I, III, and IV
    (b) NADH, Q, cytochrome c, Complexes II and III
    (c) Succinate, Q, cytochrome c, Complexes II, III, and IV
    (d) Succinate, Q, cytochrome c, Complexes II and III
    (e) Succinate, Q, Complexes I and III

6. The c subunits of the F_O component of F_OF_1 ATP synthase form an ion channel across the inner mitochondrial membrane. When essential glutamate or aspartate residues of a c subunit react with dicyclohexylcarbodiimide (DCCD), the subunit is unable to participate in proton transport.
    (a) What effect will DCCD have on electron transport and respiration in suspensions of intact mitochondria?
    (b) What will happen when dinitrophenol is subsequently added to DCCD-treated mitochondria?

7. Submitochondrial particles are pieces of the inner mitochondrial membrane, generated by sonication, that reclose *inside out* to form closed spherical membrane particles. These particles can synthesize ATP in the presence of an electron source such as NADH or QH_2. Draw a diagram that illustrates how electron transport from NADH and subsequent proton transport occur in these particles.

8. The barbiturate amytal and the plant-derived toxin rotenone (which has been used as an insecticide) both block Complex I. Do these compounds affect the P:O ratio when succinate is the substrate added to a suspension of mitochondria?

9. Can ATP be produced by oxidative phosphorylation in the presence of antimycin? If so, what is P:O ratio for the reduction of two-electron donors?

Solutions

1. The standard reduction potentials listed in Table 14·4 refer to half-reactions that are written as

$$\text{Oxidized species} + n\,e^{\ominus} \longrightarrow \text{Reduced species}$$

Two half-reactions can be added to obtain the coupled oxidation-reduction reaction by reversing the direction of the half-reaction involving the reduced species in the full reaction and reversing the sign of its reduction potential. The standard free-energy change for the net reaction can be calculated using Equation 18·12.

(a)

	$E^{\circ\prime}$ (V)
Acetaldehyde + $2\,H^{\oplus} + 2\,e^{\ominus} \longrightarrow$ Ethanol	-0.20
NADH + $H^{\oplus} \longrightarrow NAD^{\oplus} + 2\,H^{\oplus} + 2\,e^{\ominus}$	$+0.32$
Acetaldehyde + NADH + $H^{\oplus} \longrightarrow$ Ethanol + $NAD^{\oplus}$ $\quad \Delta E^{\circ\prime}$(V) = $+0.12$	

$$\Delta G^{\circ\prime} = -n\mathcal{F}\Delta E^{\circ\prime}$$
$$\Delta G^{\circ\prime} = -(2)(96.48 \text{ kJ V}^{-1}\text{mol}^{-1})(0.12 \text{ V})$$
$$\Delta G^{\circ\prime} = -23 \text{ kJ mol}^{-1}$$

(b)

	$E^{\circ\prime}$ (V)
$2\,\text{Cyt } c(\text{Fe}^{3+}) + 2\,e^{\ominus} \longrightarrow 2\,\text{Cyt } c(\text{Fe}^{2+})$	$+0.23$
$QH_2 \longrightarrow Q + 2\,H^{\oplus} + 2\,e^{\ominus}$	-0.04
$2\,\text{Cyt } c(\text{Fe}^{3+}) + QH_2 \longrightarrow 2\,\text{Cyt } c(\text{Fe}^{2+}) + Q + 2\,H^{\oplus}$ $\quad \Delta E^{\circ\prime}$(V) = $+0.27$	

$$\Delta G^{\circ\prime} = -(2)(96.48 \text{ kJ V}^{-1}\text{mol}^{-1})(0.15 \text{ V})$$
$$\Delta G^{\circ\prime} = -52 \text{ kJ mol}^{-1}$$

(c)

	$E^{\circ\prime}$ (V)
$\frac{1}{2}O_2 + 2\,H^{\oplus} + 2\,e^{\ominus} \longrightarrow H_2O$	$+0.82$
Succinate $\longrightarrow$ Fumarate + $2\,H^{\oplus} + 2\,e^{\ominus}$	-0.03
$\frac{1}{2}O_2$ + Succinate $\longrightarrow H_2O$ + Fumarate $\quad \Delta E^{\circ\prime}$(V) = $+0.79$	

$$\Delta G^{\circ\prime} = -(2)(96.48 \text{ kJ V}^{-1}\text{mol}^{-1})(0.79 \text{ V})$$
$$\Delta G^{\circ\prime} = -150 \text{ kJ mol}^{-1}$$

2. Cytochromes are electron-transport proteins that contain heme groups. The reduction potential of each iron atom in the porphyrin ring depends upon the

surrounding protein environment. Since the protein component of each cyto-chrome is different, the iron atom of each cytochrome has a different reduction potential. This difference in potential allows a series of cytochromes to pass electrons down a potential gradient.

3. A hydrogen molecule contains 2 protons and 2 electrons and is biologically transferred from oxidizable substrates by one of two mechanisms:

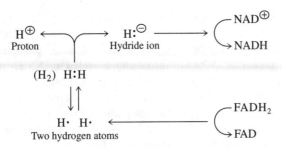

In the reduction of $NAD^{\oplus}$, a hydride ion ($H:^{\ominus}$) is transferred to the nicotin-amide ring and $H^{\oplus}$ is released into solution. In the oxidation of $FADH_2$, transfer is carried out by the equivalent of two hydrogen radicals ($H\cdot$). The $H^{\oplus}$ and electrons that make up the radical are transferred in a stepwise fash-ion ($H^{\oplus} + e^{\ominus} = H\cdot$).

4. Both NADH and QH_2 provide electrons for the electron-transport chain. The citric acid cycle is the main source of NADH and a source of QH_2. Since the citric acid cycle is in the mitochondrial matrix, both Complex I (NADH-ubiquinone oxidoreductase) and Complex II (succinate-ubiquinone oxido-reductase) must have access to the matrix. Since Complex I also pumps pro-tons from the matrix to the intermembrane space, this complex must also have access to the intermembrane space side of the inner mitochondrial mem-brane. Thus, Complex I must span the inner mitochondrial membrane.

Ubiquinone (Q) accepts electrons from both Complex I and Complex II and transports them to Complex III (ubiquinol-cytochrome *c* reductase). Therefore, ubiquinone, which is hydrophobic, must move freely within the membrane bilayer.

Complex III and Complex IV (cytochrome *c* oxidase) both pump protons from the matrix to the intermembrane space and must therefore span the inner mitochondrial membrane. Cytochrome *c,* which is hydrophilic, transports electrons between these two complexes; therefore, it could lie on either side of the membrane. It is, in fact, found in the intermembrane space.

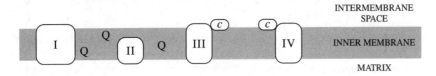

5. Refer to Figure 18·8 for the order of electron transfer. If an electron carrier or complex is missing, electrons will not flow past that point.
   (a) Complex III. The absence of cytochrome *c* prevents further electron flow.
   (b) No reaction will occur since Complex I, which accepts electrons from NADH, is missing.
   (c) O_2
   (d) Cytochrome *c*. The absence of Complex IV prevents further electron flow.
   (e) No reaction will occur since Complex II, which accepts electrons from succinate, is missing.

6. (a) In intact mitochondria, ATP synthesis via $F_O F_1$ ATP synthase is tightly coupled to electron transport and oxygen utilization. When treatment with DCCD inhibits the inward flow of protons through the F_O channel, the proton gradient generated by the proton-translocating complexes of the respiratory electron-transport chain becomes so large that proton translocation will cease. Electron flow is inhibited, and respiration will stop.

  (b) Dinitrophenol uncouples electron transport from ATP synthesis by transporting protons from the intermembrane space into the mitochondrial matrix, thus dissipating the proton gradient across the inner mitochondrial membrane. When dinitrophenol is added to DCCD-treated mitochondria, electron transport will resume since protons continue to be translocated into the intermembrane space but return to the matrix bound to dinitrophenol. Oxygen consumption will increase due to increased electron flow, but ATP will not be synthesized because the inward flow of protons through F_O is prevented by DCCD.

7. The proteins of the electron-transport chain are embedded in the inner mitochondrial membrane. When the membrane is turned inside out, the proton-pumping oxidoreductases transport $H^\oplus$ *into* the submitochondrial particle (the pH inside decreases) as electrons are transferred to oxygen. ATP synthase, which is now oriented with the F_1 component outside the particle, responds to the proton gradient by transporting $H^\oplus$ *out* of the particle via the F_O channel. ATP is then synthesized on the outside of the particle.

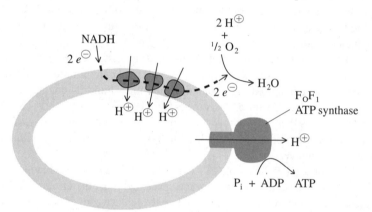

8. No. The oxidation of succinate produces QH_2, whose electrons enter the electron-transport chain via Complex II and thus bypass a block at Complex I. The P:O ratio is not affected.

9. Antimycin blocks the electron-transport chain at Complex III. ATP can only be produced by oxidative phosphorylation in the presence of antimycin if the electron source is a substrate that interacts with cytochrome *c*. For example, ascorbate can donate two electrons to cytochrome *c,* and these electrons are transported via cytochrome *c* oxidase to oxygen to yield P:O = 1.0.

19

Photosynthesis

Summary

The photosynthetic processes encompass the light-dependent electron-transport re-actions that form NADPH and ATP and the subsequent utilization of NADPH and ATP in the conversion of atmospheric CO_2 to carbohydrate (CH_2O).

The chloroplast is the organelle in which photosynthesis occurs in algae and plants. In the chloroplast, the thylakoid membrane contains electron-transport components, including photosystem I (PSI) and photosystem II (PSII), and ATP synthase. The thylakoid membrane is suspended in the aqueous stroma. The stroma contains all the enzymes and metabolites that participate in the conversion of CO_2 to (CH_2O) by the reductive pentose phosphate (RPP) cycle.

Chlorophyll and other pigments capture light energy for photosynthesis. These pigments are components of photosystems, the functional units of the light-dependent reactions of photosynthesis. Each photosystem is composed of light-absorbing antenna pigments and a reaction center. The reaction center contains two chlorophyll molecules called the special pair. Plants have two types of photosystems, designated PSI and PSII, whereas each photosynthetic bacterium has only one type of photosystem, which is different from the plant photosystems.

The absorption of light energy by chlorophyll and accessory pigments culminates in the excitation of the special pair of reaction-center chlorophyll molecules, which lowers their reduction potential. The excitation of the reaction-center primary-donor chlorophylls occurs in each of the two photosystems of plants and drives the noncyclic transfer of electrons from water through the photosystems to $NADP^\oplus$. The lowering of the reduction potential of a reaction-center chlorophyll by photoexcitation makes electron transfer to carriers of lower reduction potential possible. Cyclic electron transport, in which electrons are not transferred to $NADP^\oplus$ but rather are cycled back to PSI, also occurs. In both noncyclic and cyclic electron transport, protons are translocated across the photosynthetic membrane to generate a proton concentration gradient that drives the chemiosmotic conversion of ADP and P_i to ATP.

An elaborate short-term regulatory mechanism controls the energy distribution between PSI and PSII through phosphorylation and subsequent migration of the mobile LHC. Long-term changes in photosynthetic capacity (e.g., changes in the number and proportions of the various pigment complexes and modifications in the stoichiometries of PSI, PSII, and electron carriers) can occur in response to long-term changes in light intensity and spectral quality.

Each photosynthetic bacterium (except cyanobacteria) has only one type of photosystem, which is located in the plasma membrane. Protons are translocated across the plasma membrane into the periplasmic space. Different species of bacteria carry out either cyclic or noncyclic electron transfer using electrons from various reduced compounds.

The products of the light-dependent reactions, NADPH and ATP, are consumed during the reductive pentose phosphate (RPP) cycle. There are three stages in the RPP cycle: 1) the fixation of CO_2 by RuBisCO, 2) the reduction of CO_2 to (CH_2O), and 3) the regeneration of the CO_2 acceptor molecule, ribulose 1,5-*bis*-phosphate. The ATP and NADPH formed during the electron-transport reactions are consumed during the reduction and regeneration phases of the RPP cycle.

Coordination of the RPP cycle with the rate of ATP and NADPH synthesis depends on the regulation of several RPP-cycle enzymes by light, the stromal concentration of Mg^{2+}, and stromal pH. This coordination is essential to maintain balanced pools of RPP-cycle intermediates during changing photosynthetic conditions. The rate of photosynthesis can be limited by one or more of the following factors: light, the supply of CO_2, the rate of photosynthetic electron transport, RuBisCO activity, and the availability of inorganic phosphate for photophosphorylation.

There are three additional metabolic pathways associated with the RPP cycle. The first pathway is directly related to the oxygenase activity of RuBisCO. The reaction products that arise from the oxygenation of ribulose 1,5-*bis*phosphate by RuBisCO are catabolized by the photorespiratory pathway. This reaction sequence, which takes place in three different cellular organelles, recovers half of the carbon that is consumed by the oxygenation of RuBisCO. The other two pathways, the C_4 and CAM pathways, essentially eliminate the oxygenation of ribulose 1,5-*bis*phosphate by concentrating CO_2 at the site of RuBisCO. The C_4 pathway is found in such plants as maize (corn), sorghum, sugarcane, and many weeds, whereas the CAM pathway is found in such plants as cacti, bromeliads, and orchids. Both the C_4 and CAM pathways occur in addition to the RPP cycle. In both pathways, there is a preliminary PEP-carboxylation step, which yields a four-carbon acid. The subsequent decarboxylation of the four-carbon acid releases CO_2, which is refixed by RuBisCO. In both pathways, the preliminary carboxylation reaction has the effect of increasing the ratio of CO_2 to oxygen at the site of RuBisCO. In the C_4 pathway, the processes of CO_2 fixation by PEP carboxylase and CO_2 fixation by RuBisCO occur in two different cell types. In the CAM pathway, these two processes occur in the same cell type but at different times during the day/night cycle.

The end products of carbon assimilation are starch and sucrose. Starch is synthesized in the chloroplast. Sucrose is formed in the cytosol from triose phosphates that are exported from the chloroplast to the cytosol via the phosphate translocator. The major regulatory control for the rate of synthesis of starch and sucrose is the ratio of phosphate esters to inorganic phosphate. This ratio is a general indicator of the rate of photophosphorylation and carbon assimilation.

Study Information

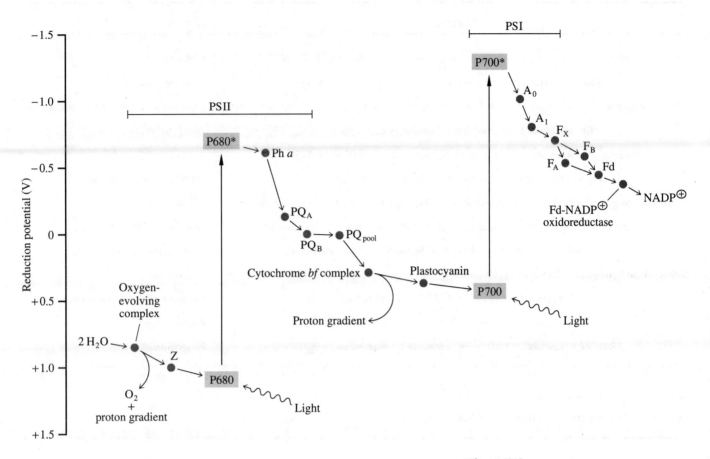

Figure 19·9
Z-scheme. The Z-scheme, so called because of its shape, is a widely accepted model illustrating the reduction potentials associated with electron flow through photosynthetic electron carriers. Because the reduction potentials of the carriers vary with experimental conditions, the values shown are approximate. Abbreviations: Z, electron donor to P680; Ph a, pheophytin a, electron acceptor of PSII; PQ_A, plastoquinone tightly bound to PSII; PQ_B, reversibly bound PQ undergoing reduction by PSII; PQ_{pool}, plastoquinone pool made up of PQ and PQH_2; A_0, chlorophyll a, the primary electron acceptor of PSI; A_1, phylloquinone, or vitamin K_1; F_X, F_B, and F_A, iron-sulfur clusters; and Fd, ferredoxin.

Table 19·1 Reactions of the RPP cycle, involving the fixation of three molecules of carbon dioxide for each molecule of triose phosphate to be used in carbohydrate synthesis

Reaction	Enzyme
3 Ribulose 1,5-*bis*phosphate + 3 CO_2 + 3 H_2O $\longrightarrow$ 6 3-Phosphoglycerate + 6 $H^{\oplus}$	RuBisCO
6 3-Phosphoglycerate + 6 ATP $\longrightarrow$ 6 1,3-*Bis*phosphoglycerate + 6 ADP	Phosphoglycerate kinase
6 1,3-*Bis*phosphoglycerate + 6 NADPH + 6 $H^{\oplus}$ $\longrightarrow$ 6 Glyceraldehyde 3-phosphate + 6 $NADP^{\oplus}$ + 6 P_i	Glyceraldehyde 3-phosphate dehydrogenase
2 Glyceraldehyde 3-phosphate $\rightleftarrows$ 2 Dihydroxyacetone phosphate	Triose phosphate isomerase
Dihydroxyacetone phosphate + Glyceraldehyde 3-phosphate $\rightleftarrows$ Fructose 1,6-*bis*phosphate	Aldolase
Fructose 1,6-*bis*phosphate + H_2O $\longrightarrow$ Fructose 6-phosphate + P_i	Fructose 1,6-*bis*phosphatase
Fructose 6-phosphate + Glyceraldehyde 3-phosphate $\rightleftarrows$ Erythrose 4-phosphate + Xylulose 5-phosphate	Transketolase
Erythrose 4-phosphate + Dihydroxyacetone phosphate $\rightleftarrows$ Sedoheptulose 1,7-*bis*phosphate	Aldolase
Sedoheptulose 1,7-*bis*phosphate + H_2O $\longrightarrow$ Sedoheptulose 7-phosphate + P_i	Sedoheptulose 1,7-*bis*phosphatase
Sedoheptulose 7-phosphate + Glyceraldehyde 3-phosphate $\rightleftarrows$ Xylulose 5-phosphate + Ribose 5-phosphate	Transketolase
2 Xylulose 5-phosphate $\rightleftarrows$ 2 Ribulose 5-phosphate	Ribulose 5-phosphate 3-epimerase
Ribose 5-phosphate $\rightleftarrows$ Ribulose 5-phosphate	Ribose 5-phosphate isomerase
3 Ribulose 5-phosphate + 3 ATP $\longrightarrow$ 3 Ribulose 1,5-*bis*phosphate + 3 ADP	Phosphoribulokinase

Net reaction

$3 CO_2$ + 9 ATP + 6 NADPH + 5 H_2O $\longrightarrow$ 9 ADP + 8 P_i + 6 $NADP^{\oplus}$ + Triose phosphate (G3P or DHAP)

Problems

1. Can a suspension of chloroplasts in the dark synthesize glucose from CO_2 and H_2O? If not, what must be added for glucose synthesis to occur? Assume that all the intermediates of the reductive pentose phosphate (RPP) cycle are present, including activated RPP cycle enzymes.

2. Cyclic electron transport occurs at certain times in green plants along with noncyclic electron transport. Is any ATP, O_2, or NADPH produced by cyclic electron transport?

3. Explain how the following changes in metabolic conditions alter the RPP cycle: (a) an increase in stromal pH, (b) a decrease in the stromal concentration of Mg^{2+}, and (c) an increase in O_2 pressure.

4. The synthesis of glucose from CO_2 via the C_4 pathway requires more molecules of ATP (or ATP equivalents) per molecule of CO_2 reduced than does the synthesis of glucose via the RPP (C_3) cycle. Explain why the extra energy is needed.

5. Describe the immediate consequences of a limitation in the following factors on the rate of photosynthesis: (a) light, (b) CO_2, (c) the rate of electron transport, (d) the activity of RuBisCO, and (e) the availability of inorganic phosphate in the chloroplast.

Solutions

1. Because synthesis of glucose from CO_2 and H_2O is driven by the products of the light reactions, isolated chloroplasts cannot synthesize glucose in the dark. If the products of the light reactions, NADPH and ATP, are supplied, chloroplasts can synthesize glucose in the absence of light. The net chemical reaction is

$$6\,CO_2 \ + \ 18\,ATP \ + \ 12\,NADPH \ + \ 12\,H_2O \ \longrightarrow$$
$$Glucose\ (C_6H_{12}O_6) \ + \ 18\,ADP \ + \ 18\,P_i \ + \ 12\,NADP^{\oplus} \ + \ 6\,H^{\oplus}$$

2. During cyclic electron transport, reduced ferredoxin donates its electrons back to the plastoquinone (PQ) pool. As electrons are recycled through photosystem I, a proton gradient is generated, leading to ATP synthesis. However, no NADPH is produced because there is no *net* flow of electrons from H_2O. No O_2 is produced because photosystem II, the site of O_2 production, is not involved in cyclic electron transport.

   (a) An increase in stromal pH increases the rate of the RPP cycle in two ways.

       (1) An increase in stromal pH increases the activity of ribulose 1,5-*bis*-phosphate carboxylase-oxygenase (RuBisCO), the central regulatory enzyme of the RPP cycle, and the activities of fructose 1,6-*bis*phosphatase and sedoheptulose 1,7-*bis*phosphatase. It also increases the activity of phosphoribulokinase. Phosphoribulokinase is inhibited by 3-phosphoglycerate (3PG) in the $3PG^{\textcircled{2\ominus}}$ ionization state but not in the $3PG^{\textcircled{3\ominus}}$ ionization state, which predominates at higher pH.

       (2) An increase in stromal pH also increases the proton gradient that drives the synthesis of ATP in chloroplasts. Since the reactions of the RPP cycle are driven by ATP, an increase in ATP production increases the rate of the RPP cycle.

   (b) A decrease in the stromal concentration of $Mg^{\textcircled{2+}}$ decreases the rate of the RPP cycle by decreasing the activity of RuBisCO, fructose 1,6-*bis*-phosphatase, and sedoheptulose 1,7-*bis*phosphatase.

   (c) An increase in O_2 pressure decreases the rate of the RPP cycle by increasing the oxygenase activity of RuBisCO, which is in direct competition with the carboxylase activity of the enzyme. The oxygenase activity of RuBisCO lowers the efficiency of CO_2 fixation by converting ribulose 1,5-*bis*phosphate to phosphoglycolate and 3-phosphoglycerate (Figure 19·19).

   In C_4 plants, assimilation of CO_2 requires not only the RPP cycle but also transport of CO_2 from the mesophyll cells to the bundle sheath cells, where the enzymes of the RPP cycle are located. C_4 plants are classified into three subgroups based on the decarboxylase present in their bundle sheath cells.

   In two subgroups of C_4 plants (the subgroup containing $NADP^{\oplus}$-malic enzyme and the subgroup containing $NAD^{\oplus}$-malic enzyme), one molecule of ATP is cleaved by pyruvate-phosphate dikinase to yield AMP and PP_i (Figure 19·22). Since the PP_i is subsequently cleaved to $2\,P_i$ by the action of a pyrophosphatase, a total of two additional ATP equivalents are required per molecule of CO_2 reduced. In the third subgroup of C_4 plants (the subgroup containing PEP carboxykinase), only one additional molecule of ATP is required per molecule of CO_2 reduced.

   Thus, plants using the C_4 pathway require at least six more molecules of ATP per molecule of glucose synthesized than do plants using the C_3 pathway alone.

5. (a) A limitation in light slows the rate of electron transfer and thus the rates of photophosphorylation and NADPH formation, thereby lowering the capacity for CO_2 reduction.

(b) A limitation in CO_2 concentration slows carbohydrate formation and, in the light, can lead to a build up of oxygen and reactive oxygen species that can damage photosynthetic membranes.

(c) A limitation in the rate of electron transport slows the rates of photophosphorylation and NADPH formation, thereby lowering the capacity for CO_2 reduction.

(d) When RuBisCO activity is limited, ATP and NADPH are consumed at a lower rate, and ADP and $NADP^{\oplus}$ become less available as substrates for the light-dependent reactions. Electron carriers can become over-reduced, and photosynthetic pigments may eventually be damaged.

(e) When the concentration of inorganic phosphate is limited, the rate of photophosphorylation is decreased. The transmembrane proton gradient increases, eventually slowing the rate of electron transport.

20

Lipid Metabolism

Summary

Fatty acids are released from adipocytes by epinephrine-stimulated activation of hormone-sensitive lipase, which catalyzes the conversion of triacylglycerols stored in fat cells to free fatty acids and monoacylglycerols. Another more specific and active monoacylglycerol lipase catalyzes the hydrolysis of monoacylglycerol to free fatty acids and glycerol. The free fatty acids and glycerol then enter the bloodstream. Fatty acids bind to albumin and are delivered to tissues where they are oxidized for energy.

Fatty acids are degraded to acetyl CoA by the sequential removal of two-carbon fragments, a process called β oxidation. Before oxidation can take place, fatty acids are activated by esterification to coenzyme A and then transported into mitochondria. In eukaryotic cells, β oxidation occurs in the mitochondrial matrix. Oxidation of fatty acids produces large amounts of ATP.

The β-oxidation pathway for saturated fatty acids consists of four enzyme-catalyzed steps: oxidation, hydration, further oxidation, and thiolysis. Oxidation of unsaturated fatty acids follows the same pathway until a double bond is reached. Then additional enzymes are required, an isomerase and a reductase.

β oxidation of odd-chain fatty acids produces propionyl CoA rather than acetyl CoA in the final reaction of the degradative pathway. Three enzyme-catalyzed reactions convert the propionyl CoA to succinyl CoA, an intermediate of the citric acid cycle.

The ketone bodies β-hydroxybutyrate and acetoacetate are fuel molecules produced in the liver by the condensation of acetyl CoA molecules. They are water soluble and serve as a readily available fuel. A third ketone body, acetone, is produced in trace amounts by nonenzymatic decarboxylation of acetoacetate.

Fatty acid synthesis, which in animal cells takes place in the cytosol, occurs by a different pathway than fatty acid oxidation. Acetyl CoA needed for fatty acid synthesis is shuttled from the mitochondrion, where it is produced, to the cytosol via the citrate transport system. In addition, the citrate transport system, along with the pentose phosphate pathway, provides NADPH needed for the reactions of fatty acid biosynthesis.

The formation of long-chain fatty acids from malonyl CoA and acetyl CoA occurs in five stages: loading, condensation, reduction, dehydration, and further reduction. This sequence is repeated until a long-chain fatty acid is released. In *E. coli,* the reactions of fatty acid synthesis are carried out by separate enzymes; in mammals, they are carried out by a multienzyme complex. The most common product of fatty acid synthesis is palmitate. Longer-chain and unsaturated fatty acids are produced by additional reactions.

Fatty acid oxidation is regulated at the release of fatty acids from adipocytes and the transport of fatty acids into mitochondria. Epinephrine stimulates fatty acid release from adipocytes. The cAMP-linked hormones glucagon and epinephrine stimulate transport of fatty acids into mitochondria for oxidation. The conversion of acetyl CoA to malonyl CoA is the first committed step of fatty acid synthesis and the key regulatory step. The enzyme that catalyzes this reaction, acetyl-CoA carboxylase, is controlled by reversible phosphorylation, responding both to hormone signals and the presence of fatty acyl CoA.

Eicosanoids are metabolic regulators derived from arachidonate by two branched pathways. The cyclooxygenase pathway leads to prostacyclin, prostaglandins, and thromboxane. Among the products of the lipoxygenase pathway are the leukotrienes.

Triacylglycerols and neutral phospholipids are synthesized by a common pathway, both being derived from 1,2-diacylglycerol. Phosphatidate is the precursor of the acidic phospholipids phosphatidylserine and phosphatidylinositol. Members of a separate class of phospholipids called plasmalogens contain an ether bond at C-1 of their glycerol backbone. Sphingolipids have a backbone derived from palmitoyl CoA and serine. Acylation of sphingosine produces a ceramide, which can be modified by addition of phosphatidylcholine to form a sphingomyelin or by addition of a sugar moiety to form a cerebroside. More complex sugar-lipid conjugates constitute the gangliosides.

Cholesterol is an essential component of animal cell membranes and a precursor of a large number of cell constituents. All the carbon atoms of cholesterol arise from acetyl CoA. The major regulatory step in cholesterol biosynthesis is the conversion of 3-hydroxy-3-methylglutaryl CoA to mevalonate.

Study Information

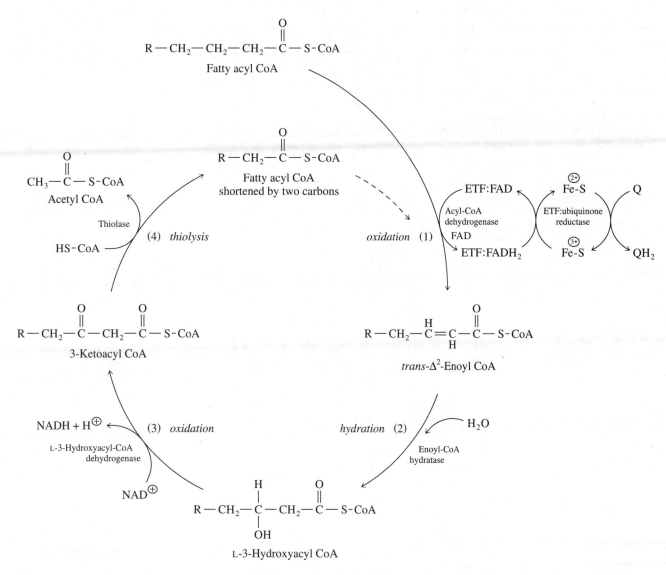

Figure 20·5
β Oxidation of saturated fatty acids. One round of β oxidation consists of four enzyme-catalyzed reactions. Each round generates one molecule each of QH_2, NADH, and acetyl CoA and produces a fatty acyl CoA molecule two carbon atoms shorter than the molecule that entered the round.

Figure 20·14
Stages in the biosynthesis of fatty acids from acetyl CoA and malonyl CoA in *E. coli*. In the loading stage, acetyl CoA and malonyl CoA are esterified to ACP. In the condensation stage, ketoacyl-ACP synthase (also called the condensing enzyme) accepts an acetyl group from acetyl-ACP, releasing ACP-SH. Ketoacyl-ACP synthase then catalyzes transfer of the acetyl group to malonyl-ACP to form aceto-acetyl-ACP and CO_2. In the first reduction, acetoacetyl-ACP is converted to D-β-hydroxy-butyryl-ACP in a reaction catalyzed by NADPH-dependent ketoacyl-ACP reductase. Dehydration of D-β-hydroxybutyryl-ACP results in the formation of a double bond, producing *trans*-butenoyl-ACP. Finally, reduction of *trans*-butenoyl-ACP produces butyryl-ACP. Synthesis continues by repeating the last four stages, with butyryl-ACP substituting for acetyl-ACP in the next condensation stage.

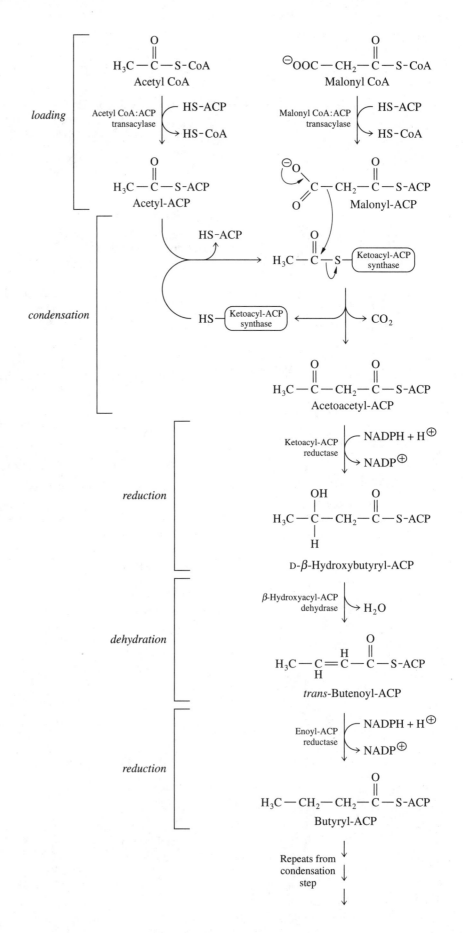

Problems

1. Why is more energy available from oxidation of stored fatty acids than from oxidation of an equivalent weight of glycogen?

2. How does inhibition of the following enzymes affect lipid metabolism?
   (a) HMG-CoA reductase
   (b) cyclooxygenase
   (c) phosphatidate phosphatase

3. Where is the labelled carbon found when the following compounds are added to a liver homogenate carrying out palmitate synthesis?

$$\text{(a) } H^{14}CO_3^{\ominus} \qquad \text{(b) } H_3{}^{14}C-\overset{\displaystyle O}{\overset{\displaystyle \|}{C}}-S\text{-}CoA$$

4. How much ATP is generated by the oxidation of the following fatty acids? Assume that the citric acid cycle is functioning. (Section 16·4 quantifies the conversion of high-energy coenzymes and ATP equivalents.)
   (a) stearate (octadecanoate; Figure 11·3)
   (b) oleate (cis-Δ^9-octadecenoate; Figure 11·3)
   (c) heptadecanoate, a C_{17} fatty acid

Solutions

1. There are two reasons why more energy is available from the oxidation of fatty acids than from the oxidation of carbohydrates.

   First, fatty acids are more highly reduced than carbohydrates. Carbohydrates contain oxygen and are therefore partly oxidized. Since the oxidation of fuel molecules yields energy based on the electrons and protons it provides to the respiratory electron-transport chain to support oxidative phosphorylation, the more hydrogens that are attached to each carbon in a substrate, the more energy the substrate can provide. Saturated fatty acids have two hydrogens attached to each carbon (excluding the $-COO^{\ominus}$ and $-CH_3$ ends), whereas carbohydrates generally have only one hydrogen attached to each carbon. The hydroxyl hydrogens in carbohydrates yield $H^{\oplus}$ but no electrons.

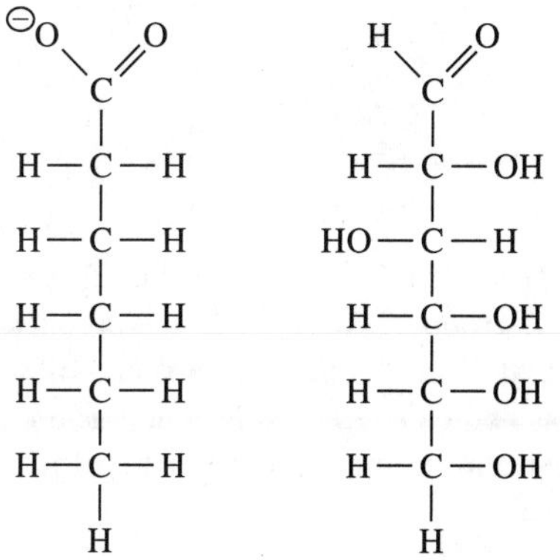

Fatty acid (Hexanoate) 11 H / 6 C   Glucose (Hexose) 7 H / 6 C

   Second, fatty acids are stored in a nearly anhydrous form as fat globules in adipose and other tissues; in contrast, carbohydrates and proteins are more polar and are stored in a hydrated form. Each gram of glycogen associates with about two grams of water.

The net result of the higher reduction of fatty acids and their anhydrous storage is that a gram of fat yields several times as much energy as a gram of glycogen. Thus, it is much more efficient to store excess energy as fats than as carbohydrates.

2. (a) Since HMG-CoA reductase catalyzes the rate-limiting step of cholesterol synthesis, inhibition of this enzyme decreases cholesterol synthesis.

   (b) Cyclooxygenase catalyzes the cyclization of arachidonate, leading to the production of prostaglandins, prostacyclin, and thromboxane (Figure 20·21); therefore, inhibition of cyclooxygenase prevents synthesis of these compounds.

   (c) Phosphatidate phosphatase catalyzes dephosphorylation of phosphatidate; thus, inhibition of this enzyme suppresses synthesis of 1,2-diacylglycerol, the precursor of triacylglycerol, phosphatidylcholine, and phosphatidylethanolamine (Figure 20·23).

3. (a) The labelled carbon remains in $H^{14}CO_3^{\ominus}$; none is incorporated into palmitate. Although $H^{14}CO_3^{\ominus}$ is incorporated into malonyl CoA, the same carbon atom is lost as CO_2 during the ketoacyl-ACP synthase reaction in *each* turn of the cycle.

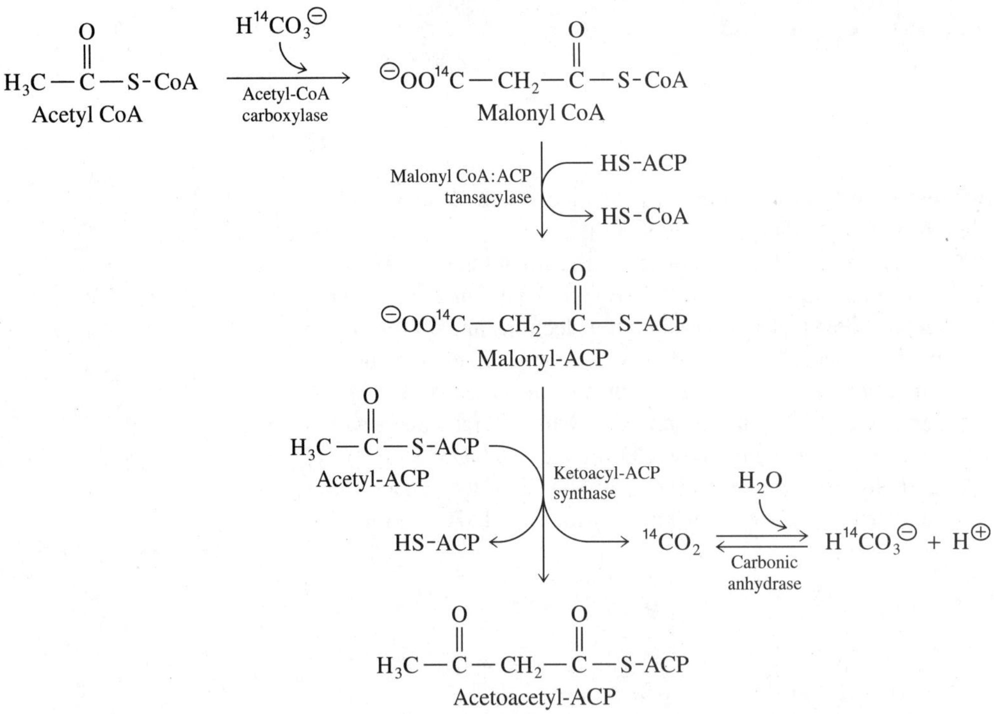

   (b) As shown in the figure on the facing page, the labelled acetyl CoA is carboxylated to produce labelled malonyl CoA. During the condensation step, the acetyl group of acetyl-ACP undergoes a condensation reaction with malonyl-ACP, producing a 4-carbon molecule containing two labelled carbons and CO_2. Each subsequent round of synthesis adds a labelled acetyl group. In the final product (palmitate), all of the even-numbered carbons are labelled.

4. (a) Approximately 120 ATP are generated. Two ATP equivalents are consumed when the fatty acid is activated (Figure 20·3). Eight rounds of β oxidation generate $8\,QH_2$ and 8 NADH, which lead to the production of 12 ATP and 20 ATP, respectively, through oxidative phosphorylation. The C_{18} fatty acid is converted to nine acetyl CoA, and each acetyl CoA generates 10 ATP by the reactions of the citric acid cycle and oxidative phosphorylation (Section 16·4). Thus, the net yield of ATP is $-2 + 12 + 20 + 90 = 120$.

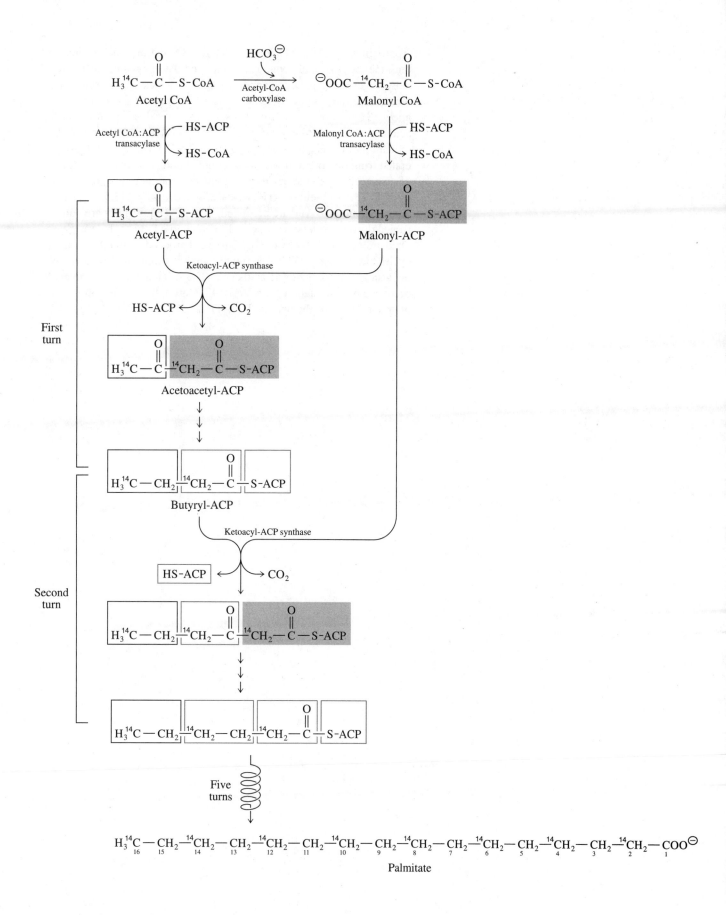

(b) Approximately 118.5 ATP are generated. β Oxidation proceeds normally until the *cis*-β,γ double bond is encountered. At this point, the acyl-CoA dehydrogenase step, which generates QH_2, is bypassed, and the double bond of the fatty acyl CoA is isomerized to a *trans*-α,β structure, which allows β oxidation to proceed. Thus, 2 ATP equivalents are used for fatty acid activation; 28 ATP are generated from seven normal rounds of β oxidation of the C_{18} fatty acid, and about 2.5 ATP (instead of 4) are generated from the round in which the double bond is encountered. A total of nine acetyl CoA molecules are produced, leading to the production of 90 ATP. The net yield of ATP is $-2 + 28 + 2.5 + 90 = 118.5$.

(c) Approximately 96 ATP are produced. Two ATP equivalents are consumed during conversion of heptadecanoate to a fatty acyl CoA. β Oxidation of the odd-chain (C_{17}) fatty acid proceeds through seven cycles, producing 28 ATP. The final product is propionyl CoA, which is not a substrate for the citric acid cycle. The seven acetyl CoA molecules produced by β oxidation generate 70 ATP through oxidative phosphorylation. Thus, the net yield of ATP is $-2 + 28 + 70 = 96$.

21

Amino Acid Metabolism

Summary

Amino acid metabolism can be approached from two points of view: the origins and fates of the amino acid nitrogen atoms, and the origins and fates of their carbon skeletons. Nitrogen is introduced into biological systems by reduction of chemically unreactive N_2 from the atmosphere to ammonia. This reaction, catalyzed by nitrogenase, is carried out by a few species of bacteria and algae. An alternate route for the assimilation of nitrogen, carried out by plants and microorganisms, is the reduction of nitrate to nitrite, catalyzed by nitrate reductase. Nitrite is reduced by the action of nitrite reductase to form ammonia, which can be used by all organisms.

Ammonia is assimilated into metabolites by several routes. Glutamate dehydrogenase catalyzes the reductive amination of α-ketoglutarate in a reversible reaction forming glutamate, which can then enter the central pathways of metabolism. Another important carrier of nitrogen is glutamine, which can be formed from glutamate and ammonia by the action of glutamine synthetase. Glutamate and glutamine are nitrogen donors in many reactions.

Because glutamine is a precursor of many metabolites, glutamine synthetase plays a critical role in nitrogen metabolism. *E. coli* glutamine synthetase is regulated by a sophisticated mechanism that includes feedback inhibition, covalent modification, and regulation at the level of protein synthesis.

The amino group of glutamate can be transferred to many α-keto acids in reversible transamination reactions that form α-ketoglutarate and the corresponding α-amino acids. In the reverse direction, transamination reactions convert a number of amino acids to glutamate or aspartate, which contribute amino groups to pathways for nitrogen disposal.

Nonessential amino acids are those that an organism can produce in sufficient quantity for growth. Essential amino acids are those that must be supplied in the diet. The nonessential amino acids are generally those formed by short, energetically inexpensive pathways. Of the nonessential amino acids in mammals, glutamate, alanine, and aspartate are formed by simple transamination, and glutamine and asparagine are formed by transfer of amide groups to the side chains of glutamate and aspartate. Serine and glycine are derived from 3-phosphoglycerate. In bacteria and plants, cysteine is formed from serine; in mammals, it is formed from the breakdown of methionine. Proline is formed from glutamate. Tyrosine is formed in a single reaction from the essential amino acid phenylalanine. In mammals, arginine is synthesized from glutamate γ-semialdehyde in a pathway that involves reactions in two tissues—the intestine and the kidney.

Of the amino acids that are essential in the diets of mammals, lysine, threonine, and methionine are synthesized in bacteria from aspartate. Pathways sharing certain enzymatic steps lead to the branched-chain amino acids isoleucine, valine, and leucine. The aromatic amino acids arise from a pathway in which chorismate is formed in seven steps from phosphoenolpyruvate and erythrose 4-phosphate, followed by conversion of chorismate to phenylalanine, tyrosine, or tryptophan. In plants, simpler eukaryotes, and bacteria, histidine is formed from phosphoribosyl pyrophosphate, ATP, and glutamine.

Protein molecules in all living cells are continuously being synthesized and degraded. The rate of turnover differs for each protein. In eukaryotes, intracellular proteins are degraded by nonselective hydrolysis in lysosomes and by selective ATP-dependent and ATP-independent hydrolysis in the cytosol. A major pathway for ATP-dependent hydrolysis requires that the target proteins be covalently attached to the protein ubiquitin. Proteins are selected for ubiquitination and subsequent degradation by a complex regulatory system that is not fully understood.

Catabolism of amino acids often begins with deamination, followed by modification of the remaining carbon chains for entry into the central pathways of carbon metabolism. Glutamate or aspartate is the amino-group acceptor at the end of one or a series of transamination reactions. In mammals, nitrogen is disposed of via urea formed by the urea cycle in the liver. Urea is then transported to the kidneys and excreted. The carbon atom of urea is derived from bicarbonate. One amino group is derived from ammonia, and the other, from the α-amino group of aspartate. Fumarate is released in the cytosol as a product of the urea cycle and enters the pathway of gluconeogenesis. Aspartate needed for continued operation of the urea cycle arises from the catabolism of other amino acids.

The pathways for degradation of amino acids lead to pyruvate, acetyl CoA, or intermediates of the citric acid cycle. Amino acids that are degraded to citric acid cycle intermediates can supply the pathway of gluconeogenesis and are called glucogenic. Those that form acetyl CoA can contribute to the formation of fatty acids or ketone bodies and are called ketogenic. Amino acids degraded to pyruvate can be metabolized to either oxaloacetate or acetyl CoA and so can be either glucogenic or ketogenic.

Alanine, aspartate, and glutamate are degraded by reversal of the transamination reactions by which they were formed. Degradation of glutamine and asparagine begins with their hydrolysis to glutamate and aspartate, respectively. Proline, arginine, and histidine are degraded to glutamate. Serine is converted to glycine, which is catabolized by the multienzyme glycine-cleavage system. Threonine can be converted by alternate routes to succinyl CoA or glycine and acetyl CoA. The branched-chain amino acids are degraded by pathways that share common steps leading to different fates; leucine is degraded to acetyl CoA, valine to succinyl CoA, and isoleucine to succinyl CoA and acetyl CoA. Methionine is degraded to succinyl CoA, with formation of cysteine as a by-product. Cysteine is degraded to pyruvate. Of the aromatic amino acids, phenylalanine is converted to tyrosine, which is degraded to fumarate and acetoacetate. Tryptophan is degraded to acetyl CoA, with formation of alanine as a by-product; alanine is converted to pyruvate and then to either oxaloacetate or acetyl CoA. Lysine catabolism merges with the pathway of tryptophan catabolism, leading to acetyl CoA.

The metabolism of glutamine in the kidney provides bicarbonate to buffer acids formed in the catabolism of some ketone bodies and amino acids.

Amino acids are converted to a number of important biomolecules including heme prosthetic groups, nitric oxide, and some hormones and neurotransmitters.

Table 21·1 Essential and nonessential amino acids for mammals, with energetic requirements for their biosynthesis

Amino acid	Moles of ATP required per mole of amino acid produced[a]	
	Nonessential	*Essential*
Glycine	12	
Serine	18	
Cysteine	19[b]	
Alanine	20	
Aspartate	21	
Asparagine	22	
Glutamate	30	
Glutamine	31	
Threonine		31
Proline	39	
Valine		39
Histidine		42
Arginine	44[c]	
Methionine		44
Leucine		47
Lysine		50 or 51
Isoleucine		55
Tyrosine	62[d]	
Phenylalanine		65
Tryptophan		78

[a] Moles of ATP required includes ATP used for synthesis of precursors and conversion of precursors to products.
[b] Formed from serine and homocysteine, a metabolite of the essential amino acid methionine.
[c] Essential for some mammals.
[d] Formed from the essential amino acid phenylalanine.
[Adapted from Atkinson, D. E. (1977). *Cellular Energy Metabolism and Its Regulation* (New York: Academic Press).]

Study Information

Please see also Table 21·1, Page 144.

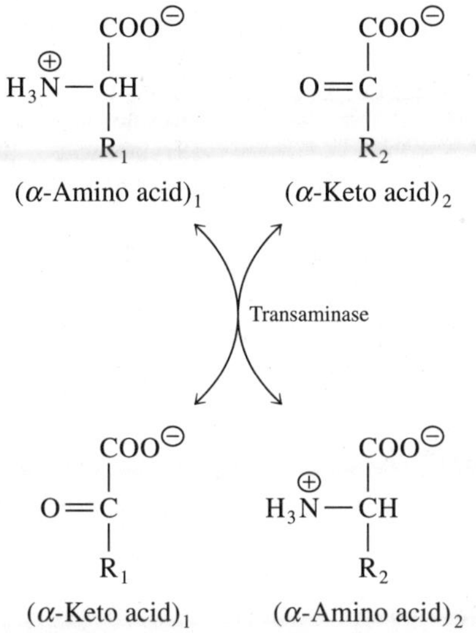

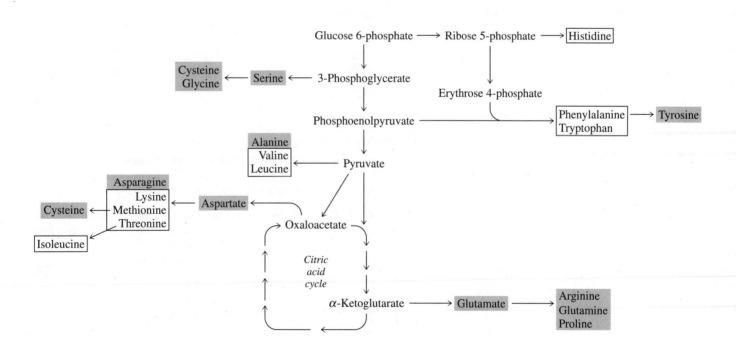

Figure 21·7
Transfer of an amino group from an α-amino acid to an α-keto acid, catalyzed by a transaminase. In biosynthetic reactions, (α-amino acid)₁ is often glutamate, with its carbon skeleton producing α-ketoglutarate, (α-keto acid)₁. (α-Keto acid)₂ represents the precursor of a variety of newly formed amino acids, (α-amino acid)₂.

Figure 21·10
Biosynthesis of amino acids. Shown are the connections to glycolysis, the pentose phosphate pathway, and the citric acid cycle. Amino acids essential for humans are boxed; nonessential amino acids are shaded.

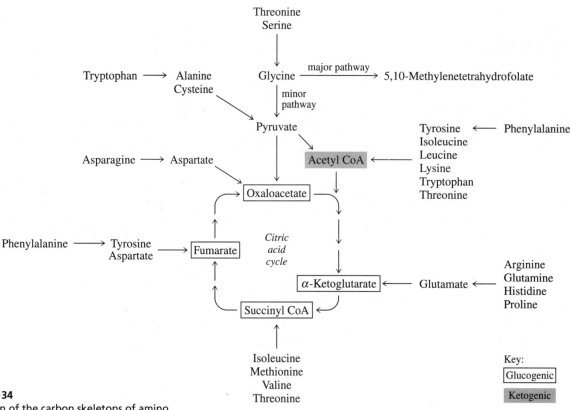

Figure 21·34
Conversion of the carbon skeletons of amino acids to pyruvate, acetyl CoA, or citric acid cycle intermediates for further catabolism.

Problems

1. When ^{15}N-labelled aspartate is fed to animals, many amino acids labelled with ^{15}N quickly appear. Explain this observation.

2. Most amino acids are products of multistep pathways; however, 3 of the 20 common amino acids are synthesized by simple transamination of carbohydrate metabolites of central pathways.
   (a) Write the equations for the three transamination reactions.
   (b) One of these amino acids can also be synthesized directly by reductive amination. Write the equation for this reaction.

3. In the winter, the African clawed toad *Xenopus laevis* lives in an aqueous environment and excretes much of its excess nitrogen as ammonia. In the summer, when ponds dry up, the animal is trapped in the mud and becomes inactive, or estivates. Suggest how *Xenopus* alters its nitrogen metabolism during estivation to avoid accumulation of toxic ammonia.

4. (a) Identify the cellular location of each of the five urea cycle enzymes listed below as either the cytosol (C) or mitochondria (M).
   (1) Carbamoyl phosphate synthetase I
   (2) Ornithine transcarbamoylase
   (3) Argininosuccinate synthetase
   (4) Argininosuccinate lyase
   (5) Arginase
   (b) Since the reactions of the urea cycle occur in two different cellular compartments, which urea cycle intermediates must be transported across the inner mitochondrial membrane?

5. The left-hand column below lists molecules whose carbon skeletons can be at least partially derived from degradation of amino acids listed in the column on the right. Match the molecules on the left with the amino acids from which they can be derived. (Ignore any citric acid cycle interconversions of molecules in the left-hand column.)

(a) Fumarate — (1) Isoleucine, methionine, threonine, valine
(b) α-Ketoglutarate — (2) Arginine, glutamine, histidine
(c) Succinyl CoA — (3) Phenylalanine, tyrosine

6. Serine dehydratase catalyzes the degradation of serine to pyruvate and ammonia (Reaction 21·16). This reaction, which requires pyridoxal phosphate (PLP) as a coenzyme, illustrates how PLP catalyzes elimination reactions, in this case β-elimination of water and α-elimination of ammonia from serine. Propose a mechanism for this reaction.

Solutions

1. The labelled amino group is transferred from aspartate to glutamate in a reaction catalyzed by aspartate transaminase. Since transaminase reactions are reversible and many transaminases use glutamate as the α-amino group donor, the ^{15}N atoms of ^{15}N-glutamate are quickly distributed among the other amino acids that are substrates of glutamate-dependent transaminases, that is, those other than lysine and threonine (Section 21·5).

2. (a) Glutamate is produced from α-ketoglutarate, alanine from pyruvate, and aspartate from oxaloacetate.

α-Ketoglutarate — Glutamate

Pyruvate — Alanine

Oxaloacetate — Aspartate

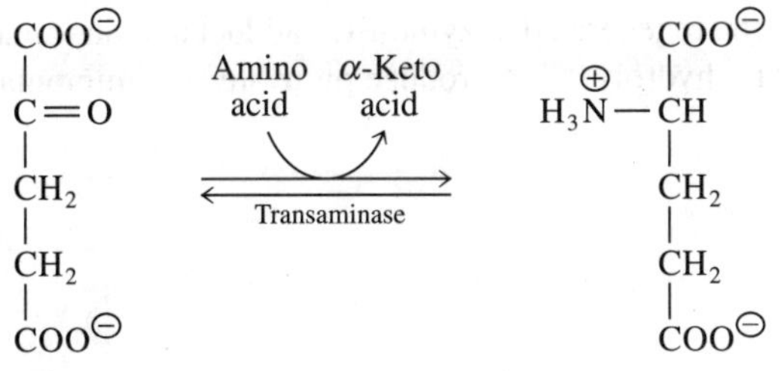

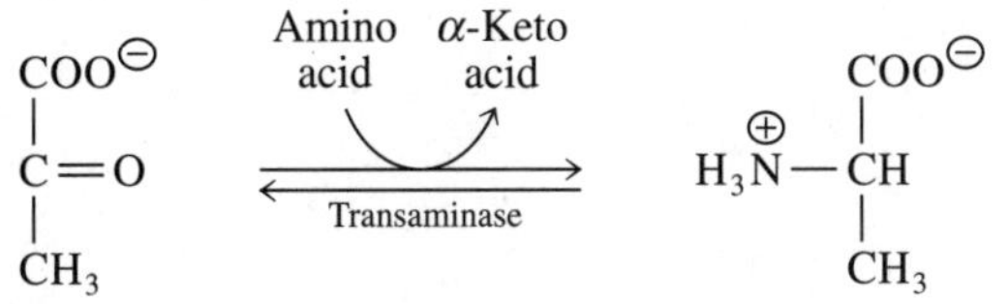

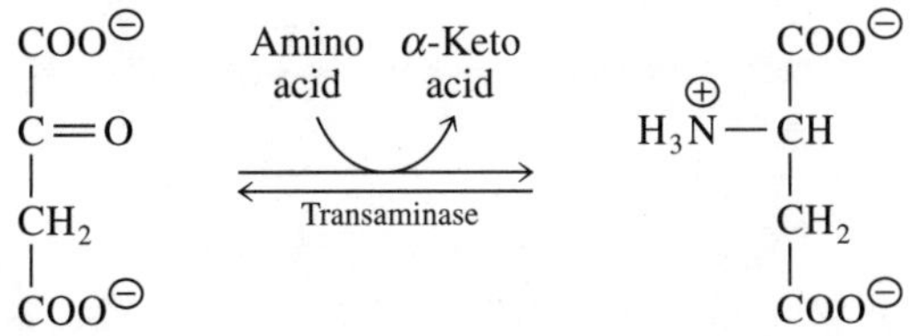

(b) Glutamate is also produced from α-ketoglutarate by the action of glutamate dehydrogenase.

$$NH_4^{\oplus} \; + \; \begin{array}{c} COO^{\ominus} \\ | \\ C=O \\ | \\ CH_2 \\ | \\ CH_2 \\ | \\ COO^{\ominus} \end{array} \quad \underset{\text{Glutamate dehydrogenase}}{\overset{NAD(P)H,H^{\oplus} \quad NAD(P)^{\oplus}}{\rightleftharpoons}} \quad \begin{array}{c} COO^{\ominus} \\ | \\ H_3\overset{\oplus}{N}-CH \\ | \\ CH_2 \\ | \\ CH_2 \\ | \\ COO^{\ominus} \end{array} \; + \; H_2O$$

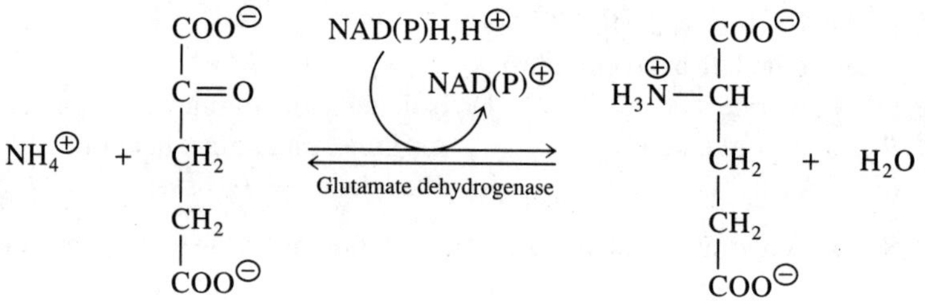

α-Ketoglutarate $\qquad\qquad\qquad\qquad\qquad$ Glutamate

3. During estivation, *Xenopus* becomes ureotelic; that is, it uses the reactions of the urea cycle for elimination of waste nitrogen. The activity of the enzymes related to the urea cycle drastically increases during estivation.

4. See Figure 21·27.
   (a) (1) M, (2) M, (3) C, (4) C, (5) C.
   (b) Citrulline and ornithine.

5. See Figure 21·34. (a) 3, (b) 2, (c) 1.

6. In the first step of the conversion of serine to pyruvate and ammonia (see opposite page), serine reacts with the enzyme-PLP adduct in a transimination reaction to form a serine Schiff base. Next, the α-hydrogen of the serine Schiff base is removed as a proton to produce a resonance-stabilized carbanion (a quinonoid intermediate). This intermediate undergoes acid-catalyzed elimination of the hydroxyl group from the skeleton of serine, and an aminoacrylate Schiff base is formed. Hydrolysis of the aminoacrylate Schiff base yields the regenerated enzyme-PLP adduct and an imine that is non-enzymatically hydrolyzed to produce pyruvate and ammonia.

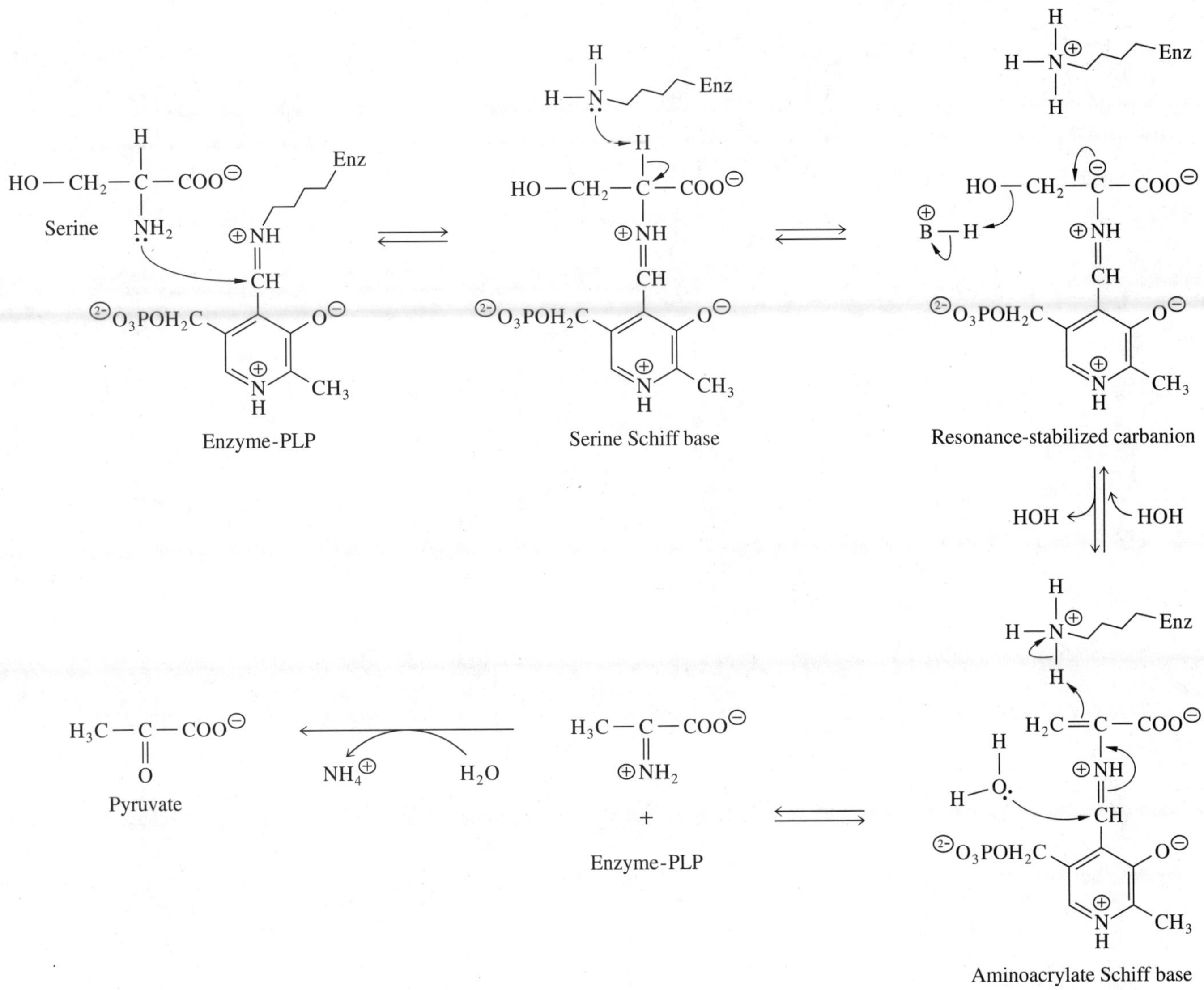

Serine

Enzyme-PLP

Serine Schiff base

Resonance-stabilized carbanion

HOH ⇌ HOH

Aminoacrylate Schiff base

Pyruvate

$NH_4^{\oplus}$ H_2O

Enzyme-PLP

22

Nucleotide Metabolism

Summary

Most organisms can synthesize purines and pyrimidines de novo. This biosynthesis involves nucleotides as key intermediates.

The de novo synthesis of purine nucleotides is achieved by a 10-step pathway that leads to the formation of IMP (inosinate). Isotopic labelling experiments showed that the carbon and nitrogen atoms of the purine ring are derived from glycine, 10-formyltetrahydrofolate, glutamine, aspartate, and carbon dioxide. The purine molecule is assembled by the stepwise addition of precursor units to a series of intermediates built upon a foundation of ribose 5′-phosphate donated by 5-phosphoribosyl 1-pyrophosphate (PRPP). The first purine-containing product of the de novo pathway, IMP, can be converted to either AMP or GMP. These compounds are precursors of nucleic acids and a variety of nucleotide coenzymes.

PRPP can also react directly with adenine, guanine, or hypoxanthine in salvage reactions to yield AMP, GMP, or IMP, respectively. A deficiency of hypoxanthine-guanine phosphoribosyltransferase, the enzyme that catalyzes the salvage of hypoxanthine and guanine, is the cause of Lesch-Nyhan syndrome, a disease that leads to spasticity and severe mental disability.

In the synthesis of the pyrimidine nucleotide UMP, PRPP enters the pathway *after* completion of the ring structure. The metabolic precursors of the pyrimidine ring are bicarbonate, glutamine, and aspartate. In eukaryotes, the regulation of this pathway occurs at the first step, namely, the formation of carbamoyl phosphate. In prokaryotes, the enzyme that catalyzes the second step—aspartate transcarbamoylase (ATCase)—is inhibited allosterically by the ultimate products CTP and UTP. In eukaryotes, two multifunctional proteins catalyze key steps in the biosynthesis of UMP. CTP is formed by the ATP- and glutamine-dependent amidation of UTP.

Deoxyribonucleotides are synthesized by reduction of ribonucleotides at C-2′ of the ribose moiety. The reaction, catalyzed by ribonucleotide reductase, requires the reducing power of NADPH. Reducing equivalents from NADPH are transferred via a protein coenzyme, either thioredoxin or glutaredoxin. Deoxyribose formation occurs at the nucleoside diphosphate level in most organisms, but in certain microorganisms reduction takes place at the triphosphate level. In both *E. coli* and mammalian cells, ribonucleotide reductase is closely regulated. Allosteric effectors control not only the overall catalytic activity of the enzyme but also its substrate specificity. Through the action of ribonucleotide reductase and nucleoside diphosphate kinases, dATP, dGTP, and dCTP are made available for the synthesis of DNA.

Thymidylate (dTMP) is formed from deoxyuridylate (dUMP) by a methylation reaction in which 5,10-methylenetetrahydrofolate donates both a one-carbon group and a hydride ion. 7,8-Dihydrofolate, the other product of this methylation, is recycled by reduction to the active coenzyme tetrahydrofolate. Inhibition of this reduction by methotrexate can prevent the synthesis of dTMP and thus of DNA.

In birds and reptiles, waste nitrogen from both amino acid and purine nucleotide catabolism is incorporated into IMP and eventually excreted as uric acid. Primates (for whom urea is the main excretion product of amino acid catabolism) also degrade surplus purines to uric acid. Most organisms catabolize uric acid further to allantoin, allantoate, urea, or even ammonia. In humans, overproduction of uric acid leads to gout.

Pyrimidines can be catabolized to ammonia, bicarbonate, and either β-alanine (from cytosine or uracil) or β-aminoisobutyrate (from thymine). Further breakdown produces acetyl CoA and succinyl CoA, respectively.

Study Information

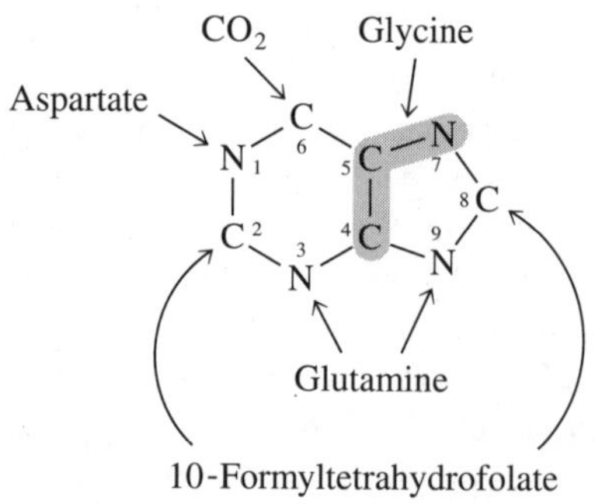

Figure 22·3
Sources of the ring atoms in purines synthesized de novo. Note that the nitrogen atom and both carbon atoms of a glycine molecule are incorporated into positions 4, 5, and 7 of the purine; the amide nitrogen atoms of two glutamine molecules provide N-3 and N-9; C-6 arises from CO_2; aspartate contributes its amino nitrogen to position 1; and C-2 and C-8 originate from formate via 10-formyltetrahydrofolate.

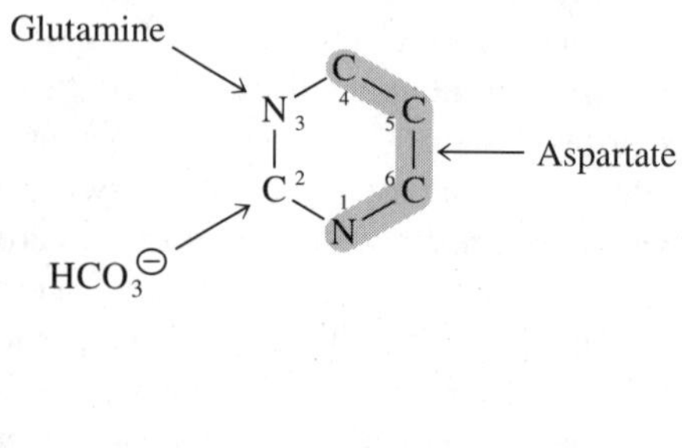

Figure 22·12
Sources of the ring atoms of pyrimidines synthesized de novo. Note that C-4, C-5, C-6, and N-1 are all contributed by aspartate.

Problems

1. Where do the isotopic labels appear in IMP when each of the following precursors is used? (a) Glutamine labelled with ^{15}N in its amide group, (b) 2-$[^{14}C]$-glycine.

2. How much energy (expressed as ATP or ATP equivalents) is needed to synthesize one molecule of AMP, starting from ribose 5-phosphate? Assume that all other precursors are available.

3. Where do the isotopic labels appear in UMP when each of the following precursors is used? (a) 2-$[^{14}C]$-Aspartate, (b) $H^{14}CO_3^{\ominus}$.

4. Describe the sequence of reactions by which carbons 1 and 6 of glucose can be incorporated into the 5-methyl substituent of thymidylate.

5. Are purines and pyrimidines major sources of energy in eukaryotic cells?

Solutions

1. See Figure 22·6 for the reactions in the pathway of IMP synthesis.
   (a) The labelled amide nitrogen from glutamine appears at N-3 and N-9 in purines synthesized de novo.
   (b) Purines synthesized de novo from 2-$[^{14}C]$-glycine contain the label at C-5.

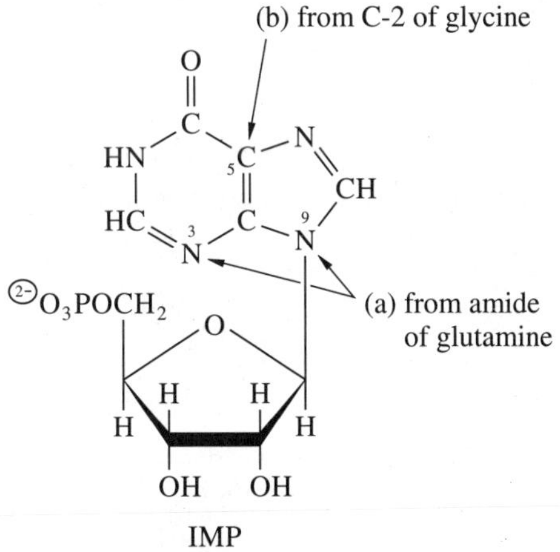

IMP

2. Seven ATP molecules or ATP equivalents are required. Synthesis of phosphoribosyl pyrophosphate (PRPP) involves transfer of a pyrophosphoryl group from ATP to ribose 5-phosphate (Figure 22·5). This pyrophosphoryl group is released as PP_i and hydrolyzed to $2\,P_i$ in Step 1 of the pathway for synthesis of IMP (Figure 22·6), thus accounting for consumption of two ATP equivalents. Four ATP molecules are consumed in Steps 2, 4, 5, and 7, each of which involves the conversion of ATP to ADP and P_i. When IMP is converted to AMP (Figure 22·9), one additional ATP equivalent (GTP) is consumed in the reaction catalyzed by adenylosuccinate synthetase.

3. See Figure 22·13 for the reactions in the pathway of UMP synthesis.
   (a) Labelled C-2 from aspartate, which is incorporated into carbamoyl aspartate, appears at C-6 of the uracil of UMP.
   (b) The labelled carbon from $HCO_3^{\ominus}$, which is incorporated into carbamoyl phosphate, appears at C-2 of the pyrimidine ring of UMP.

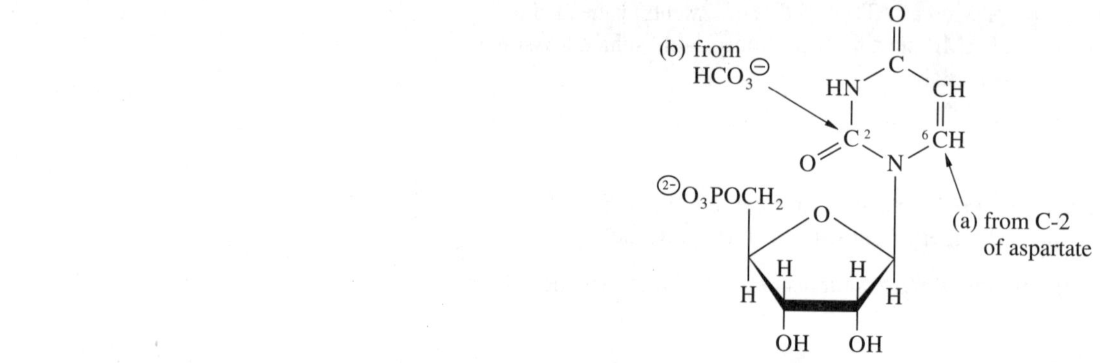

UMP

4. When glucose is converted to 3-phosphoglycerate by the glycolytic pathway, C-1 and C-6 of glucose become C-3 of 3-phosphoglycerate, which is subsequently converted to C-3 of serine (Figure 21·11). This carbon is then transferred to tetrahydrofolate in a reaction catalyzed by serine hydroxymethyltransferase that produces 5,10-methylenetetrahydrofolate (Figure 21·12), which provides the carbon group of thymidylate (Figure 22·17).

5. In eukaryotes, purines and pyrimidines are not significant sources of energy. Whereas the carbon atoms in fatty acids and carbohydrates can be oxidized to yield ATP, there are no comparable energy-yielding pathways for nitrogen-containing purines and pyrimidines. No ATP is produced by substrate-level phosphorylation during purine degradation to uric acid (Figure 22·23) or ammonia (Figure 22·25), nor during pyrimidine degradation (Figure 22·28). However, NADH produced during the conversion of hypoxanthine to uric acid may indirectly generate ATP via oxidative phosphorylation.

23

Integration of Fuel Metabolism in Mammals

Summary

In mammals, metabolic tasks are distributed among different tissues. Some organs, such as the liver, are suppliers of fuels, whereas others, such as the brain, are simply consumers. The principal fuels in mammals are glucose, lactate, fatty acids, ketone bodies, triacylglycerols, and amino acids. Glycerol and ethanol are fuels of minor importance. Not all cells can use each of the major fuels, and not all fuels are available at all times. Glucose is the most important fuel, and control of glucose metabolism is central to overall metabolic control.

The maintenance of constant levels of glucose in the circulation (glucose homeostasis) is achieved by balancing glucose synthesis or absorption against utilization. When large amounts of glucose are present, tissues absorb glucose. As the supply of glucose diminishes, tissues switch to other fuels. The use of glucose is governed by hormones, principally insulin and glucagon. The liver plays a central role in glucose homeostasis.

During starvation, the body responds to the interruption in the supply of exogenous fuels with unified changes in metabolic activity. The coordinated nature of the changes reveals the degree of integration of metabolic pathways. Five phases of glucose homeostasis have been identified in the course of starvation: the absorptive and postabsorptive phases, followed by early, intermediate, and prolonged starvation.

In the absorptive phase, extending from 2 to 4 hours after a meal, the level of glucose in the blood rises. Glucose is rapidly taken up by the liver, skeletal muscle, and brain. If the level of glucose in the blood rises above the ability of the kidney to reabsorb filtered glucose, the excess is lost to the urine.

High blood glucose triggers the release of insulin, which has many physiological effects. In liver, the effects of insulin include stimulation of glycolysis, fatty acid synthesis and esterification, and protein synthesis, as well as inhibition of glycogenolysis, gluconeogenesis, fatty acid oxidation, ketogenesis, and proteolysis. Glucagon, which acts only on liver, has effects that are, in general, opposite those of insulin. Insulin levels are high in the fed state; glucagon levels are high in the fasted state.

157

Lipids entering the body during the absorptive phase are packaged in chylomicrons. Triacylglycerols in chylomicrons are delivered to peripheral tissues, where triacylglycerols are hydrolyzed outside the cell and fatty acids are taken up, esterified, and stored.

Large amounts of dietary amino acids arrive at the liver during the absorptive phase. These amino acids are either catabolized, used for protein synthesis, or permitted to pass unaltered to peripheral tissues. Oxidation of the amino acids in liver generates a large amount of ATP, much of which is immediately consumed by the pathways of gluconeogenesis and urea synthesis, which remove the carbon and ammonia, respectively, generated by amino acid catabolism.

In the transition to the postabsorptive and early starvation phases, the absorption of dietary glucose slows, the level of insulin drops, and the level of glucagon rises. The liver responds to the hormonal changes by breaking down glycogen and releasing glucose. Low levels of insulin are also accompanied by an increase in the rate of lipolysis in adipose tissue and an increase in the release of free fatty acids into the circulation. The glucose–fatty acid cycle is a mechanism that spares the use of glucose when fatty acids are available: when glucose is low, insulin is low, release and catabolism of fatty acids is high, and products of fatty acid catabolism (NADH, acetyl CoA, and citrate) inhibit glucose degradation and spare the use of glucose.

The Cori cycle is an important aspect of fuel metabolism in which glucose is metabolized to lactate in muscle and other tissues, and lactate is returned to liver for conversion back to glucose. The energy for gluconeogenesis is derived from the catabolism of other fuels, such as fatty acids. The glucose-alanine cycle is a variation of the Cori cycle that includes conversion of pyruvate to alanine in muscle and transport of alanine to liver. This cycle has the advantages of transport of nitrogen to liver and greater energy yield in muscle, since pyruvate is not converted to lactate at the expense of NADH.

Amino acids are major gluconeogenic precursors, although the use of endogenous amino acids requires degradation of proteins and accompanying loss of protein function.

As starvation progresses, fatty acids are converted to ketone bodies, which, unlike fatty acids, can cross the blood-brain barrier and serve as a fuel for the brain, further sparing the use of glucose. Less use of glucose at this stage means less proteolysis of muscle protein for gluconeogenesis.

Gluconeogenesis in kidney becomes important in the late stage of starvation. High levels of ketone bodies result in metabolic acidosis, and gluconeogenesis in kidney is linked to the production of ammonia and bicarbonate. These products of renal gluconeogenesis are used to help maintain acid-base homeostasis.

Refeeding reverses the adaptations that develop during starvation. The concentration of glucose in the blood rises, the level of insulin rises, gluconeogenesis in liver and lipolysis in adipose tissue decreases, glucose is taken up by tissues, and fuel stores are replenished.

Diabetes mellitus is a metabolic disease characterized by extreme hyperglycemia. Insulin-dependent diabetes mellitus (IDDM) arises from a lack of insulin due to damage of the pancreatic β cells. In non–insulin-dependent diabetes mellitus (NIDDM), insulin levels may be normal or even elevated, but tissues are insulin-resistant due to decreased sensitivity, poor responsiveness, or both.

Study Information

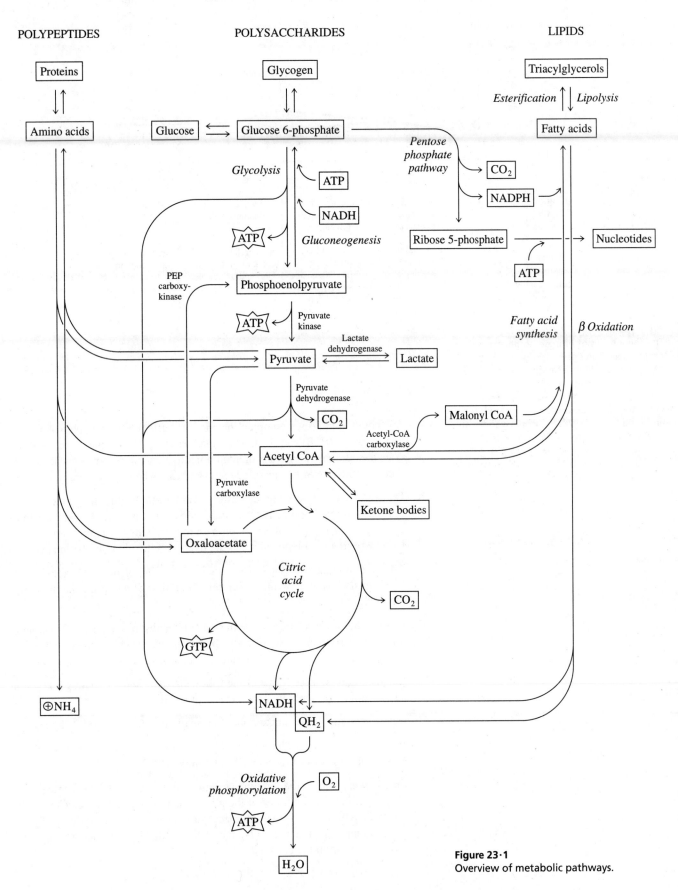

Figure 23·1
Overview of metabolic pathways.

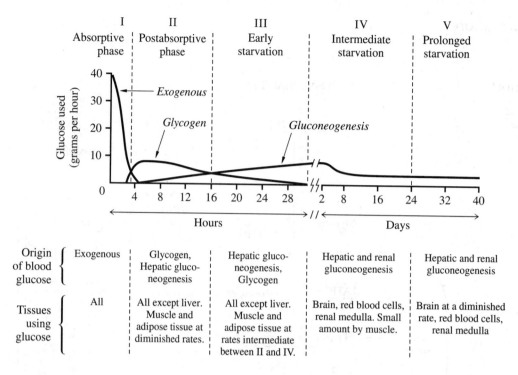

Figure 23·5
The five phases of glucose homeostasis. The graph, based on observations of a number of individuals, illustrates glucose utilization in a 70 kg man consuming a 100 g oral glucose load and then fasting for 40 days. [Adapted from Ruderman, N. B., Aoki, T. T., and Cahill, G. F. (1978). Gluconeogenesis and its disorders in man. In *Gluconeogenesis: Its Regulation in Mammalian Species*, R. W. Hanson and M. A. Mehlman, eds. (New York: John Wiley & Sons), p. 515.]

Problems

1. The compound 3-mercaptopicolinate is a specific inhibitor of phospho-enolpyruvate (PEP) carboxykinase, a key enzyme of gluconeogenesis. Why does the administration of 3-mercaptopicolinate to starved rats cause a decrease in the rate of hepatic glycogen synthesis when the rats are refed?

2. During starvation, when many tissues switch to fuels other than glucose, mammalian red blood cells continue to derive energy from the catabolism of glucose to lactate (glycolysis). Compare the efficiency of interorgan metabolism involving red blood cells and the liver with the glucose-alanine cycle.

3. The predominant hexokinase isozyme in liver differs from the hexokinases found in skeletal muscle and other peripheral tissues. Describe the main difference.

4. List the major source of glucose during (a) the absorptive phase (2–4 hours after a meal), (b) the postabsorptive phase and early starvation phase (4–24 hours after a meal), and (c) late starvation (two weeks since the last meal).

5. List the major storage forms of the following fuels and the main sites of storage (a) lipids, (b) carbohydrates, (c) amino acids.

6. Muscle glycogen does not supply glucose directly to the blood. Why not?

7. What is the function of lipoprotein lipase in adipose tissue? When would you expect the activity of this enzyme to be highest?

8. Enterocytes, cells of the immune system, and fetal cells show high rates of glucose utilization, but most of the glucose is released as lactate. These cells utilize an amino acid as their major respiratory fuel. Which amino acid do they use, and where in the body is it synthesized? Give another function of this amino acid in the body.

Solutions

1. During starvation, the major source of glucose is hepatic gluconeogenesis. On refeeding, liver glycogen stores are replenished by two pathways. In the direct pathway, glucose is phosphorylated by the action of glucokinase, and the glucose 6-phosphate produced is incorporated into glycogen (Section 23·5B). In the indirect pathway, glucose 6-phosphate synthesized via gluconeogenesis is not released as free glucose but is diverted into glycogen synthesis. Since this latter pathway involves gluconeogenesis, it is decreased when PEP carboxykinase is inhibited by 3-mercaptopicolinate.

2. The interorgan pathway linking red blood cells and the liver is the Cori cycle. Lactate produced by red blood cells (and other glycolytic tissues) is transported to the liver, where it is used as a substrate for glucose synthesis. Glucose from gluconeogenesis can then be delivered to the red blood cells for use as fuel. The energy for the synthesis of glucose in the liver is derived from the oxidation of other fuels, such as fatty acids.

    The glucose-alanine cycle is a modification of the Cori cycle in which pyruvate produced from glucose in peripheral tissues is transaminated to alanine. Alanine is transported to the liver and used for gluconeogenesis. The nitrogen from alanine enters the pathway for urea synthesis. The glucose-alanine cycle has two advantages over the Cori cycle. Like the Cori cycle, it transports carbon between tissues, but the glucose-alanine cycle also transports nitrogen in a nontoxic form. In addition, the glucose-alanine cycle yields more energy in peripheral tissues. In the Cori cycle, glucose is converted to lactate in peripheral tissues, with a yield of two ATP molecules per molecule of glucose metabolized. When the glucose-alanine cycle operates, the energy yield per molecule of glucose metabolized in peripheral tissues is two ATP and two NADH, since the conversion of pyruvate to lactate, which occurs in the Cori cycle but not the glucose-alanine cycle, consumes one molecule of NADH per molecule of pyruvate reduced. The reducing equivalents not consumed when the glucose-alanine cycle operates can be used for the synthesis of ATP via oxidative phosphorylation. Obviously, this energy cannot be recovered in tissues that lack mitochondria, such as red blood cells. The glucose-alanine cycle is limited to those tissues capable of oxidative metabolism.

3. The isozyme of hexokinase in liver (and the β cells of the pancreas) is known as glucokinase or hexokinase IV. It differs from other hexokinases in two important ways. First, it is not subject to inhibition by physiological levels of the product glucose 6-phosphate and second, it has a high apparent K_m for glucose (5–10 mM). The hexokinases of other tissues are very sensitive to inhibition by glucose 6-phosphate and show a K_m for glucose of about 0.1 mM. In the body, liver glucokinase is only active when glucose is abundant, and glucokinase catalyzes the phosphorylation of glucose even when the level of glucose 6-phosphate is high, allowing the liver to form glycogen when glucose is present in high concentrations.

4. (a) During the absorptive phase, the major source of glucose is dietary glucose absorbed directly from the intestine.
   (b) During the postabsorptive and early starvation phases of glucose homeostasis, the major source of glucose is hepatic glycogenolysis.
   (c) During late starvation, the major source of glucose is hepatic and renal gluconeogenesis.

5. (a) Triacylglycerols are the major storage forms of lipids. Large amounts of triacylglycerols are found in adipose tissue, where they are stored in anhydrous form in fat droplets.

   (b) Glycogen is the major storage form of carbohydrates in animals. The main depots of glycogen are liver and skeletal muscle. In plants, carbohydrates are stored as starch.

   (c) There is no purely storage form of amino acids. During starvation, mobilization of amino acids occurs when degradation of proteins releases amino acids that are used as fuel and as substrates for gluconeogenesis. However, the proteins degraded are not storage molecules but functional proteins. All protein degradation during starvation means a loss of functional capacity, which is not the case when fuels are mobilized from storage in fat droplets or glycogen granules.

6. Muscle does not release glucose derived from the breakdown of muscle glycogen because muscle lacks the enzyme glucose 6-phosphatase, which catalyzes the reaction glucose 6-phosphate $\longrightarrow$ glucose + P_i. Since glucose 6-phosphate cannot cross the plasma membrane of muscle cells, it is used within the cells for energy production. Muscle glycogen is therefore a source of fuel for local requirements only.

7. Lipoprotein lipase is an extracellular enzyme that catalyzes the hydrolysis of triacylglycerols. Its activity generates free fatty acids that can be taken up into the cell and reesterified for storage. The lipoprotein lipase activity of adipose tissue is high in the absorptive phase and low during starvation. Not surprisingly, there is good evidence that the expression of lipoprotein lipase in adipose tissue increases when the level of insulin is high.

8. Glutamine is utilized as the major respiratory fuel in a number of cells. The reason for this special role of glutamine is not known. Most glutamine in the body is synthesized in skeletal muscle by the enzyme glutamine synthetase. In addition to serving as a respiratory fuel, glutamine also functions in the body as a major substrate for renal synthesis of ammonia and for renal and hepatic gluconeogenesis.

24

Nucleic Acids

Summary

Nucleic acids are polymers of nucleoside monophosphates, or nucleotides. A purine or pyrimidine ring is attached to ribose or 2'-deoxyribose by a β-N-glycosidic bond. The *anti* conformation of nucleosides predominates in nucleic acids. The furanose rings can pucker, with the C-2' *endo* and C-3' *endo* conformations appearing most often. The bases in DNA are often methylated. Hydrogen bonding by the bases of nucleotides is important for the three-dimensional structure and functions of nucleic acids.

DNA is a double-stranded nucleotide polymer whose monomeric units are linked as 3'–5' phosphodiesters. As originally proposed by Watson and Crick, the two strands are antiparallel. Base pairing between A and G and between C and T allows each strand to serve as a template for the other. Stacking interactions between base pairs are largely responsible for the helical conformation of double-stranded DNA. B-DNA, the most common form, is a right-handed helix stabilized by hydrophobic interactions, stacking interactions, hydrogen bonding, and electrostatic repulsion. The sugar-phosphate backbone winds around the outside of the helix, and the helix surface contains two grooves. DNA can be denatured by heating above its melting temperature, T_m. Separated single strands can renature; nucleation is the rate-limiting step.

Double-stranded DNA can adopt several conformations. A-DNA is right-handed, but compared to B-DNA, the base pairs are tilted and the center of the helix is accessible to solvent. Z-DNA is a left-handed helix with no obvious grooves.

DNA molecules are flexible and assume a variety of shapes. Cruciform structures can form when inverted repeats are present. Naturally occurring DNA is usually underwound, or negatively supercoiled. Topoisomerases catalyze the removal of supercoils and, in some cases, can introduce supercoils.

There are four major classes of RNA: ribosomal RNA, transfer RNA, messenger RNA, and small RNA. RNA is usually single stranded and, like DNA, has considerable secondary structure, including hairpins, or stem-loops. Double-stranded RNA adopts a helical conformation similar to that of A-DNA.

The phosphodiester linkages of nucleic acids are hydrolyzed by nucleases that are classified as DNases or RNases and as exonucleases or endonucleases. Hydrolysis of RNA in alkaline solutions or by the action of RNase A proceeds via a 2',3'-cyclic phosphate intermediate. Restriction endonucleases cleave DNA at specific sequences. Host DNA is protected from hydrolysis by the presence of bases that have been methylated by the restriction enzyme or by another enzyme. When EcoRI binds to DNA, specific residues of the protein contact the nucleotide residues at the recognition site.

In cells, large DNA molecules are condensed and packaged. The histone proteins of eukaryotes bind to DNA to form nucleosomes. Nucleosomes are strung together, and the resulting structure undergoes additional degrees of condensation, forming supercoils that are attached to an RNA-protein scaffold in the nucleus.

Study Information

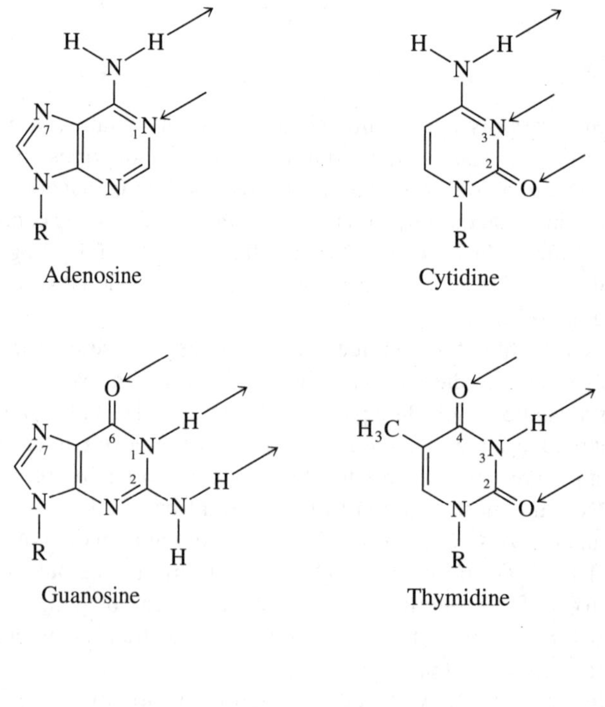

Figure 24·4
Hydrogen bond sites of nucleosides in DNA. Each base contains atoms and functional groups that can serve as hydrogen donor or acceptor sites. The common tautomeric forms of the nucleoside bases are shown. Hydrogen donor and acceptor groups differ in the other tautomers. R represents deoxyribose.

Adenosine

Cytidine

Guanosine

Thymidine

Table 24·3 Major structural features of A-, B-, and Z-DNA

Property	A-DNA	B-DNA	Z-DNA
Helix handedness	right	right	left
Repeating unit	1 base pair	1 base pair	2 base pairs
Rotation per base pair	32.7°	34.6°	30°
Base pairs per turn	~11	~10.4	~12
Inclination of base pair to helix axis	19°	1.2°	9°
Rise per base pair along helix axis	0.23 nm	0.33 nm	0.38 nm
Pitch	2.46 nm	3.40 nm	4.56 nm
Diameter	2.55 nm	2.37 nm	1.84 nm
Conformation of glycosidic bond	*anti*	*anti*	*anti* at C *syn* at G
Sugar pucker	C-3′ *endo*	C-2′ *endo*	C-2′ *endo* at C C-3′ *endo* at G

[Adapted from Dickerson, R. E., Drew, H. R., Conner, B. N., Wing, R. M., Fratini, A. V., and Kopka, M. L. (1982). The anatomy of A-, B-, and Z-DNA. *Science* 216:475–485.]

Table 24·5 Specificities of some common restriction endonucleases

Source	Enzyme[1]	Recognition sequence[2]
Bacillus amyloliquefaciens H	*Bam*HI	G↓GATCC
Bacillus globigii	*Bgl*II	A↓GATCT
Escherichia coli RY13	*Eco*RI	G↓AA*TTC
Escherichia coli R245	*Eco*RII	↓CC*TGG
Haemophilus aegyptius	*Hae*III	GG↓CC
Haemophilus influenzae R$_d$	*Hind*II	GTPy↓PuA*C
Haemophilus influenzae R$_d$	*Hind*III	*A↓AGCTT
Haemophilus parainfluenzae	*Hpa*II	C↓CGG
Nocardia otitidis-caviarum	*Not*I	GC↓GGCCGC
Providencia stuartii 164	*Pst*I	CTGCA↓G
Serratia marcescens S$_b$	*Sma*I	CCC↓GGG

[1] The names of restriction endonucleases are abbreviations of the names of the organisms that produce them. Some abbreviated names are followed by a letter denoting the strain. Roman numerals indicate the order of discovery of the enzyme in that strain.

[2] Recognition sequences are written 5′ to 3′. Only one strand is represented. The arrows indicate cleavage sites. Asterisks represent positions where bases can be methylated. Pu (purine) denotes that either A or G is recognized. Py (pyrimidine) denotes that either C or T is recognized.

Problems

1. Compare hydrogen bonding in the α helix of proteins to hydrogen bonding in the double helix of DNA. Include in the answer the role of hydrogen bonding in stabilizing these two structures.

2. A stretch of double-stranded DNA contains 1000 base pairs, and its base composition is 58% G + C. How many thymine residues are in this region of DNA?

3. The average molecular weight of a base pair in double-stranded DNA is approximately 650. Using the data from Table 24·6, calculate the ratio by weight of protein to DNA in a typical 30-nm chromatin fiber.

4. DNA molecules can be cleaved into fragments by the action of restriction endonucleases that catalyze cleavage at specific palindromic sequences.
   (a) Identify the palindromic sequence in the duplex DNA molecule shown below.

$$5'\cdots\text{T G T A A G C T T C T T G A}\cdots 3'$$
$$3'\cdots\text{A C A T T C G A A G A A C T}\cdots 5'$$

   (b) Consult Table 24·5 to determine which restriction endonuclease cleaves this sequence.
   (c) Indicate the DNA fragments that are generated by the endonuclease.

5. Two molecules of DNA are the same size but differ in base composition. One molecule contains 20% A + T and the other contains 60% A + T. Which molecule has a higher T_m?

6. Solution A contains a 20–base pair DNA molecule at a concentration of 1 M. Solution B contains a 400–base pair DNA molecule at a concentration of 0.05 M. Thus, each solution contains the same total number of nucleotide residues. Assume that the DNA molecules all have the same base composition.
   (a) The temperature of both solutions is increased slowly. Which solution will be the first to contain completely denatured DNA?
   (b) In which solution will the rate of renaturation be greater?

7. (a) What is the linking number (L) of a relaxed, circular B-DNA molecule of 1872 base pairs? Assume 10.4 base pairs per turn.
   (b) *E. coli* topoisomerase II undergoes three catalytic cycles with the DNA substrate in part (a). Recalculate the linking number.
   (c) *E. coli* topoisomerase I then acts once on the molecule in part (b). What is the new linking number?

Solutions

1. In the α helix, hydrogen bonds form between the carbonyl oxygen of one residue and the α-amino nitrogen four residues, or one turn, away. These hydrogen bonds between atoms in the backbone are roughly parallel to the axis of the helix. The amino acid side chains, which point away from the backbone, do not participate in intrahelical hydrogen bonding. In double-stranded DNA, the sugar-phosphate backbone is not involved in hydrogen bonding. Instead, two or three hydrogen bonds, which are roughly perpendicular to the helix axis, form between complementary bases in opposite strands.

In the α helix, the individual hydrogen bonds are weak, but the cumulative forces of these bonds stabilize the helical structure, especially within the hydrophobic interior of a protein where water does not compete for hydrogen bonding. In DNA, the principal role of hydrogen bonding is to allow each strand to act as a template for the other. Although the hydrogen bonds between complementary bases help stabilize the helix, stacking interactions between base pairs in the hydrophobic interior make a greater contribution to helix stability.

2. If 58% of the residues are G + C, 42% of the residues must be A + T. Since every A pairs with a T on the opposite strand, the number of adenine residues equals the number of thymine residues. Therefore, 21%, or 420, of the residues are thymine ($2000 \times 0.21 = 420$).

3. In the 30-nm fiber, DNA is packaged in nucleosomes, each containing about 200 base pairs of DNA; therefore, the DNA in a nucleosome has a molecular weight of 130 000 ($200 \times 650 = 130\,000$). Assuming there is one molecule of histone H1 per nucleosome, the molecular weight of the protein would be 129 800.

Histone H1	21 000
Histone H2A ($\times 2$)	28 000
Histone H2B ($\times 2$)	27 600
Histone H3 ($\times 2$)	30 600
Histone H4 ($\times 2$)	22 600
Total	129 800

Thus, the ratio by weight of protein to DNA is 129 800:130 000, or approximately 1:1.

4. (a) The shading indicates the palindromic sequence of DNA.

    (b) *Hin*dIII recognizes the palindromic sequence and cleaves the phosphodiester backbone at the positions marked by the arrows.

    (c) Two fragments with single-stranded ends are generated.

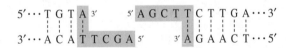

5. The DNA molecule containing 20% A + T has a higher T_m, since it contains 80% G + C. Because stacking interactions are stronger in G/C-rich DNA, more energy is required to denature the DNA.

6. (a) The DNA in solution A will be the first to completely denature, because the stacking interactions in the 20–base pair helix are less extensive than in the 400–base pair helix. Base pairs at the ends of double-stranded DNA molecules tend to be only partially stacked. This "end effect" is more significant in shorter molecules.

    (b) The rate of renaturation will be greater in solution A. Nucleation (the formation of the first base pairs) is rate limiting. The greater the number of single-stranded molecules, the greater their chances of re-forming base pairs. Thus, the DNA in solution A (which contains 2 M single-stranded DNA) will renature more quickly than the DNA in solution B (which contains 0.1 M single-stranded DNA).

7. (a) L = 180 (1872/10.4)

    (b) L = 174. *E. coli* topoisomerase II catalyzes the addition of negative supercoils, thereby decreasing the linking number. Because each catalytic event involves a double-stranded break, three catalytic cycles decrease the linking number by 6.

    (c) L = 175. The linking number increases because *E. coli* topoisomerase I removes a negative supercoil. Because the enzyme catalyzes a single-stranded break in the DNA, the linking number increases by 1.

25

DNA Replication

Summary

DNA replication is, in most cases, semiconservative. Each strand of the parental molecule serves as a template for the synthesis of a complementary daughter strand. After one round of replication, each DNA molecule contains one strand from the parent and one newly synthesized strand.

DNA replication in prokaryotes and eukaryotes occurs by similar mechanisms. In all organisms, DNA replication requires a complex protein machine, although the composition of the machine differs among species. In all organisms, replication is bidirectional and begins at particular sites called origins of replication. In *E. coli,* the origin, *ori*C, is recognized by DnaA, which binds to it and allows the other proteins involved in DNA replication to assemble into an active complex. The complex initiates replication by separating the strands of the helix through the concerted action of DNA topoisomerases, SSB, and DnaB, a helicase. The separated strands are then available to the primosome, which includes primase. DNA polymerase III then joins the complex to complete formation of the replication protein machine. Eukaryotic chromosomes contain multiple origins of replication.

Chain elongation is carried out by DNA polymerases. The polymerase responsible for DNA synthesis in *E. coli* is DNA polymerase III; in eukaryotes it is DNA polymerases α and δ. All DNA polymerases catalyze formation of a phosphodiester linkage between an activated deoxyribonucleoside 5′-triphosphate and a nascent DNA chain. In the replication complex, leading-strand DNA polymerases are processive: they do not dissociate from the DNA until the entire chromosome has been replicated. DNA polymerases require a free 3′-hydroxyl group for activity; all catalyze chain growth in the 5′→3′ direction. Most DNA polymerases are also associated with a 3′→5′ exonuclease activity, which recognizes the distortion caused by an improperly paired base and removes the mismatched nucleotide. This exonuclease activity proofreads newly synthesized DNA and greatly increases the fidelity of DNA replication.

Chain growth differs for the two strands of DNA at the replication fork. One new strand is synthesized continuously in a 5′→3′ direction; it is called the leading strand. The other new strand is synthesized discontinuously in the 5′→3′ direction; it is called the lagging strand. Discontinuous synthesis of the lagging strand produces oligonucleotides called Okazaki fragments. Each Okazaki fragment starts with an RNA primer, which is synthesized by primase. The RNA primer is removed by a 5′→3′ exonuclease, and the resulting gap is filled in by the action of a DNA polymerase. Once the gap is filled, the nick in the phosphodiester backbone is closed by DNA ligase.

Syntheses of the leading and lagging strands are closely coupled in vivo and are carried out simultaneously at the replication fork by the replisome. The replisome contains helicases to open the DNA, primase to synthesize primers on the lagging strand, and two DNA polymerases to lengthen the two new DNA chains. DNA polymerase III synthesizes both leading and lagging strands in *E. coli,* whereas two different polymerases are required in eukaryotes. Some of the polypeptides in the replisome exhibit enzymatic activities that can be detected when the individual polypeptides are isolated; others display no activity in isolation but function by stabilizing the complex or enhancing the activity of the other polypeptides.

DNA replication terminates when two replication forks moving in opposite directions meet. On circular bacterial chromosomes, this occurs opposite the origin in the *ter* region. Tus binds to sites in the *ter* region and prevents replication forks from passing through.

Not all DNA replication is semiconservative. Some bacteriophage DNA is synthesized by a rolling-circle mechanism, which is also used during conjugation in *E. coli* and related species. Mitochondrial DNA and some plasmid DNA molecules are replicated by the D-loop mechanism. The single-stranded RNA genomes of retroviruses are copied into DNA by reverse transcriptase, a special type of DNA polymerase.

Special mechanisms are required to synthesize the ends of linear DNA molecules. In some cases, this involves the formation of linear concatenates, while in other cases, specific proteins bind to the ends of DNA to ensure that entire chromosomes are replicated. Shortening of the ends of eukaryotic chromosomes is prevented by telomerases.

Study Information

Figure 25·11
Schematic diagram of lagging-strand synthesis. A short piece of RNA serves as the primer for synthesis of each Okazaki fragment.

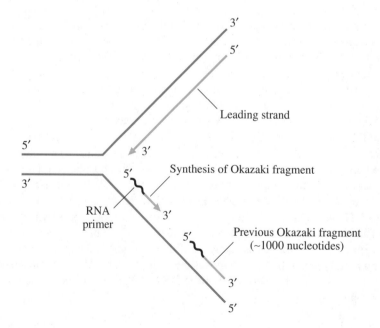

170

Problems

1. The chromosome of a certain bacterium is a circular, double-stranded DNA molecule of 5.2×10^6 base pairs.
   (a) The rate of replication-fork movement is 1000 nucleotides per second. Calculate the time required to replicate the chromosome.
   (b) Under extremely favorable conditions, the bacterial generation time can be as short as 25 minutes. Given that the maximum rate of DNA replication is 1000 nucleotides per second and that the chromosome contains only one origin of replication, explain how the cells are able to divide so rapidly.

2. (a) The addition of SSB to in vitro sequencing reactions often increases the yield of DNA. Explain why.
   (b) Sequencing reactions are usually carried out in vitro at 65°C and usually employ a DNA polymerase isolated from bacteria that grow at high temperatures. What is the advantage of this?

3. How does the use of an RNA primer rather than a DNA primer affect the fidelity of DNA replication in *E. coli?*

4. Both strands of DNA are synthesized in the $5' \rightarrow 3'$ direction.
   (a) Draw a hypothetical reaction mechanism for synthesis of DNA in the $3' \rightarrow 5'$ direction using a 5'-dNTP and a growing chain with a 5'-triphosphate group.
   (b) How would DNA synthesis be affected if the hypothetical enzyme had proofreading activity?

5. Explain the roles of helicases, DNA topoisomerase II, and SSB in DNA replication.

6. In this chapter, we said that processivity helps account for the rapid rate of DNA replication. Describe a situation in which the processivity of an enzyme would have little effect on the overall rate of polymerization.

7. The entire genome of the fruit fly *D. melanogaster* consists of 1.65×10^8 base pairs. If replication at a single replication fork occurs at the rate of 30 base pairs per second, calculate the minimum time it would take to replicate the entire genetic complement if replication is initiated (a) at a single bidirectional origin, or (b) at 2000 bidirectional origins. (c) Replication at its fastest occurs in five to six minutes in the early embryo. What is the minimum number of origins necessary to account for this replication time?

8. The Meselson-Stahl experiment was designed to determine whether replication of DNA is semiconservative, conservative, or dispersive. In this experiment, *E. coli* were grown on media containing $^{15}NH_4Cl$ for one generation and then transferred to media containing $^{14}NH_4Cl$.

   Using solid lines to indicate parental strands and dashed lines to indicate daughter strands, show the DNA molecules that would be produced by two rounds of replication if the process were (a) semiconservative, (b) conservative, or (c) dispersive. For each case, show the DNA banding pattern that would be produced by equilibrium density gradient centrifugation.

Solutions

1. (a) Two replication forks form at the origin of replication and move in opposite directions until they meet at a point opposite the origin. Therefore, the two replisomes each replicate half of the genome (2.6×10^6 base pairs). At each replication fork, two new strands are synthesized (the leading and the lagging strands) at the rate of 1000 nucleotides per second each (2000 nucleotides per second equals 1000 base pairs per second). Replication of the entire chromosome therefore requires

$$\frac{2.6 \times 10^6 \text{ base pairs}}{1000 \text{ base pairs s}^{-1}} = 2600 \text{ s} = 43 \text{ min and } 20 \text{ s}$$

   (b) Although there is only one origin (O), replication can be reinitiated before the previous replication forks have reached the termination site. Thus, more than two replication forks can be present on each double-stranded DNA molecule. Replication of a single chromosome still requires approximately 43 minutes, but completed copies of each chromosome can appear at shorter intervals, depending on the rate of initiation.

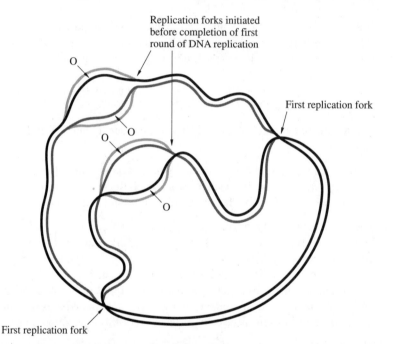

2. (a) The single-stranded DNA template used for DNA synthesis in vitro can form secondary structures such as hairpin loops. SSB prevents the formation of double-stranded structure by binding to the single-stranded template. SSB thus renders the DNA a better substrate for DNA polymerase.
   (b) The yield of DNA in vitro is improved at higher temperatures because formation of secondary structure in the template is less likely. A temperature of 65°C is high enough to prevent formation of secondary structure but not high enough to denature the correctly formed double-stranded regions of newly synthesized DNA. DNA polymerases from bacteria that grow at high temperatures are used in these reactions because they are active at 65°C, a temperature at which DNA polymerases from other bacteria would be inactive.

3. Extremely accurate DNA replication requires a proofreading mechanism to remove errors introduced during the polymerization reaction. Synthesis of an RNA primer by a primase, which does not have proofreading activity, is more

error prone than DNA synthesis. However, because the primer is RNA, it can be removed by the $5' \rightarrow 3'$ exonuclease activity of DNA polymerase I and replaced with accurately synthesized DNA when Okazaki fragments are joined. If the primer were composed of DNA made by a primase without proofreading activity, it would not be removed by DNA polymerase I, and the error rate of DNA replication would be higher at sites of primer synthesis.

4. (a) The hypothetical nucleotidyl-group–transfer reaction would involve nucleophilic attack by the 3'-hydroxyl group of the *incoming* nucleotide on the triphosphate group of the growing chain. Pyrophosphate would be released when a new phosphodiester linkage was formed.

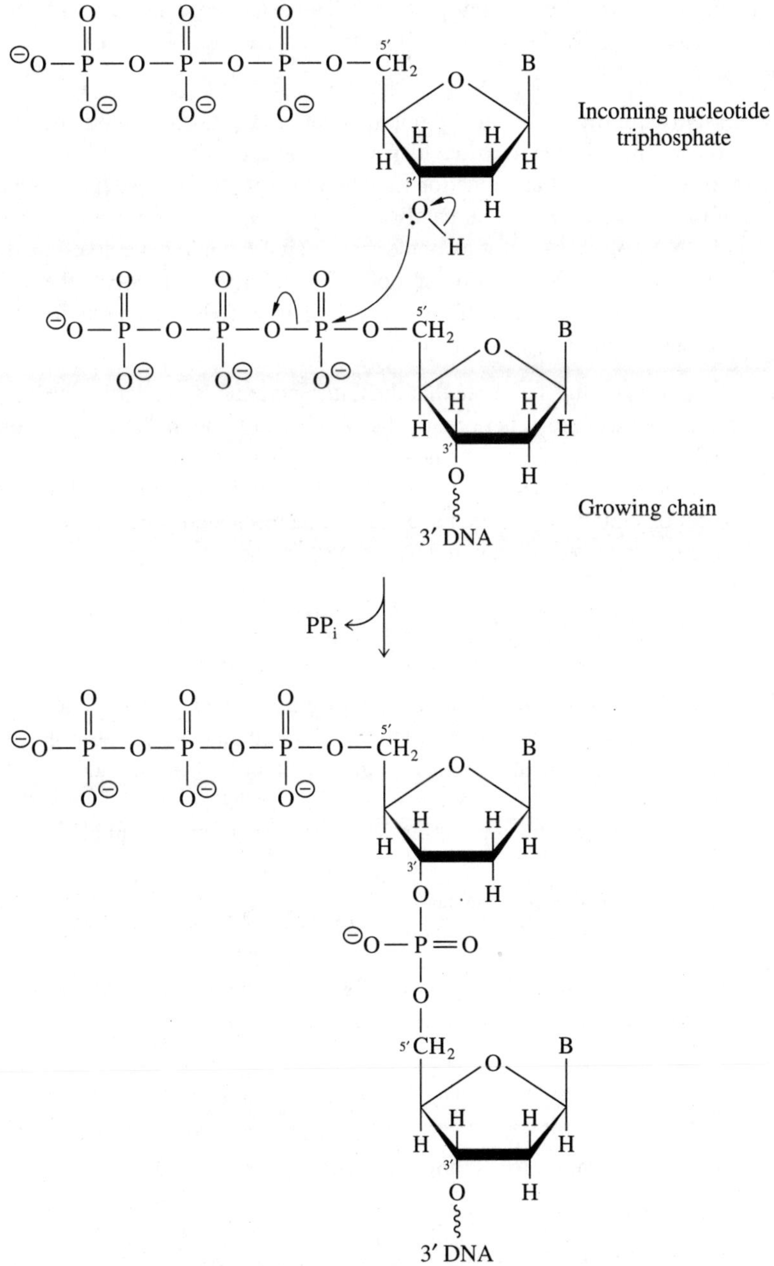

(b) If the hypothetical enzyme had $5' \rightarrow 3'$ proofreading activity, removal of a mismatched nucleotide would leave a 5'-monophosphate group at the end of the growing chain. Further DNA synthesis, which would require a terminal triphosphate group, could not occur.

173

5. Helicases catalyze unwinding of DNA during replication. DnaB, a protein associated with primase in the *E. coli* replisome, is a replication helicase. Unwinding generates supercoils ahead of the replication fork as it advances along the leading strand.

   DNA topoisomerase II, also called DNA gyrase, introduces negative supercoils into DNA ahead of the replication fork in a reaction coupled to ATP hydrolysis. These negative supercoils compensate for the torsional strain created by the unwinding of the helix and promote separation of the parental strands at the replication fork.

   Single-strand binding proteins (SSB) stabilize single-stranded DNA at the replication fork and prevent the DNA molecule from forming secondary structures such as helices or hairpin loops. The extended, relatively inflexible conformation of DNA bound by SSB is the ideal substrate for DNA polymerase.

6. Processivity improves the rates of polymerization reactions by reducing the time it takes the enzyme to locate its substrates. Processivity therefore has little effect on the rate of any reaction in which the rate-limiting step is not location of the substrates but rather catalysis of the polymerization itself. In the case of DNA polymerase, for example, processivity would have no effect on the rate of the reaction if the time it took DNA polymerase to locate the incoming dNTP and the 3′ end of the nascent chain were less than the time it took to catalyze the joining of the two.

7. (a) To begin, you must assume that the entire genome is one large linear molecule of DNA (it isn't really, but this assumption makes the problem easier to answer). Also assume that the origin of replication is at the midpoint of this chromosome. Since the replication forks move in opposite directions, 60 base pairs per second will be replicated. The time required to replicate the entire genome would be

   $$\frac{1.65 \times 10^8 \text{ base pairs}}{60 \text{ base pairs s}^{-1}} = 2.75 \times 10^6 \text{ s} = 764 \text{ h} = 32 \text{ days}$$

   (b) Assuming that the 2000 bidirectional origins are equally spaced along the DNA molecule and that initiation occurs simultaneously at all origins so that each fork replicates the maximum amount of DNA, the rate would be $2000 \times 2 \times 30$ base pairs per second $= 1.2 \times 10^5$ base pairs per second. The time required for replication of the entire genome would be

   $$\frac{1.65 \times 10^8 \text{ base pairs}}{1.2 \times 10^5 \text{ base pairs s}^{-1}} = 1375 \text{ s} = 23 \text{ min}$$

   (c) Again, assume that origins are equally spaced and that initiation at all origins is simultaneous. The required rate of transcription is

   $$\frac{1.65 \times 10^8 \text{ base pairs}}{300 \text{ s}} = 5.5 \times 10^5 \text{ base pairs s}^{-1}$$

   The two replication forks move at a combined rate of 60 base pairs per second.

   $$\frac{5.5 \times 10^5 \text{ base pairs s}^{-1}}{60 \text{ base pairs s}^{-1} \text{origin}^{-1}} = 9170 \text{ origins}$$

   To complete replication in five minutes, therefore, approximately 9200 origins are necessary.

8.

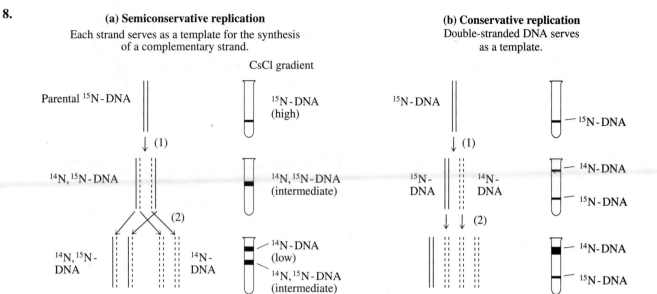

(a) Semiconservative replication
Each strand serves as a template for the synthesis
of a complementary strand.

CsCl gradient

Parental ^{15}N-DNA

^{15}N-DNA
(high)

(1)

^{14}N, ^{15}N-DNA

^{14}N, ^{15}N-DNA
(intermediate)

(2)

^{14}N, ^{15}N-
DNA

^{14}N-
DNA

^{14}N-DNA
(low)
^{14}N, ^{15}N-DNA
(intermediate)

(b) Conservative replication
Double-stranded DNA serves
as a template.

^{15}N-DNA

^{15}N-DNA

(1)

^{15}N-
DNA

^{14}N-
DNA

^{14}N-DNA

^{15}N-DNA

(2)

^{14}N-DNA

^{15}N-DNA

(c) Dispersive replication
Parental and daughter strands are
combined during replication.

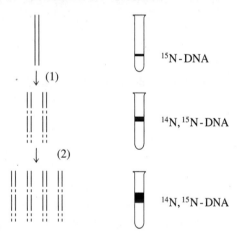

^{15}N-DNA

(1)

^{14}N, ^{15}N-DNA

(2)

^{14}N, ^{15}N-DNA

After one round of replication, centrifugation of DNA produced by semiconservative and dispersive replication would each yield one band (^{14}N,^{15}N-DNA),
whereas conservative replication would yield two distinct bands (^{14}N-DNA
and ^{15}N-DNA). After two rounds, semiconservative replication would yield
two distinct bands (^{14}N-DNA and ^{14}N,^{15}N-DNA), whereas dispersive replication would yield only one band (^{14}N,^{15}N-DNA). The band of DNA produced
by dispersive replication would become less dense with each successive
round as more ^{14}N accumulated in each strand and as ^{15}N was increasingly
diluted.

26

DNA Repair and Recombination

Summary

Mutations in DNA can arise as a result of DNA damage that is not repaired. A variety of different agents can cause damage; these include alkylating agents, intercalating agents, and ionizing radiation. DNA can also be damaged as a result of deamination of bases, incorporation of base analogs, or mistakes during DNA replication. Mutations that are caused by DNA damage can lead to cancer in animals; the Ames test is an assay for the mutagenic effects of various chemicals.

There are four important DNA repair pathways in all organisms: direct repair, excision repair, mismatch repair, and error-prone repair. In direct repair, the normal bases in DNA are restored by enzymes that modify the damaged nucleotides. The excision-repair pathway involves enzymes, such as UvrABC endonuclease, that remove a stretch of damaged DNA. The resulting gap is filled by the repair DNA polymerase. Glycosylases scan DNA for specific types of damage and then excise damaged nucleotides. Mismatch repair is a pathway that allows selective repair of errors introduced during DNA replication. The enzymes involved in this pathway recognize newly synthesized strands of DNA, in some cases because they are undermethylated. Base mismatches are corrected using the parental DNA strand as a template. In error-prone repair, the DNA replication complex is modified to enable it to bypass regions of damage that cannot be readily repaired by the other repair mechanisms.

Homologous recombination involves the exchange of DNA between identical or almost identical DNA strands. Exchange is initiated by the formation of single-stranded tails that invade the double helix on the homologous chromosome. Reciprocal single-strand exchange leads to the formation of a Holliday junction that can be resolved in several different ways by cutting the single strands at the crossover point. In *E. coli,* strand exchange is mediated by RecBC nuclease and RecA, and the Holliday junction is resolved by Ruv proteins.

Recombination probably evolved as a form of repair. The general recombination pathway seen in *E. coli* is similar in all organisms. Recombination is important during conjugation in bacteria and during meiosis in eukaryotes.

Study Information

Figure 26·2
Pathways to nucleotide substitutions. Tautomerization of bases can lead to formation of unusual base pairs and the misincorporation of nucleotides during DNA replication. If the damage is not repaired, a second round of replication leads to a transition or transversion in one of the new strands.

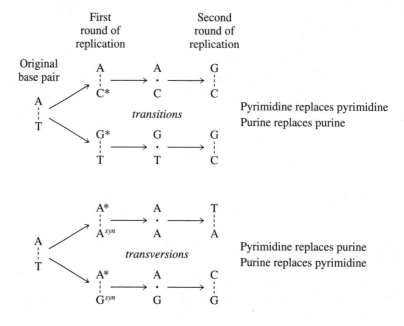

Table 26·1 DNA repair systems in *E. coli*

DNA repair system	Types of damage repaired	Enzymes/proteins involved in repair
Direct repair	Pyrimidine dimers Alkylated bases Other modified bases Mismatched bases	Photoreactivating enzyme Methyltransferases
Excision repair	Alkylated bases Deaminated bases Pyrimidine dimers Other lesions that distort the DNA helix	DNA glycosylases AP endonucleases UvrABC endonuclease Helicase II DNA polymerase I DNA ligase Mfd
Mismatch repair	Mismatched bases on newly synthesized DNA strands	Mut proteins Helicases Exonucleases DNA polymerase III DNA ligase
Error-prone repair	Cross-linked bases Sites where intercalators are bound	UmuC UmuD RecA

Problems

1. Is O^6-methylguanine-DNA methyltransferase an enzyme? Explain.

2. Why do cells exposed to visible light following irradiation with ultraviolet light have a greater survival rate than cells kept in the dark after irradiation with ultraviolet light?

3. Explain why uracil DNA-glycosylase cannot repair the damage when 5-methylcytosine is deaminated to thymine.

4. Why are high rates of mutation observed in regions of DNA that contain methylcytosine?

5. Uridylate is one of the four nucleotides present in RNA, whereas DNA contains a modified uridylate, thymidylate (5-methyluridylate). What is the selective advantage of including thymidylate, which is energetically expensive to synthesize, rather than uridylate in DNA?

6. Why is the hydrolytic deamination of guanine less damaging to the cell than the hydrolytic deamination of adenine?

7. The overall error rate for DNA replication in *E. coli* is approximately 10^{-9}, yet the rate of misincorporation by the replisome is about 10^{-5}. Explain the difference.

Solutions

1. It is common to think of methyltransferases as enzymes since they mediate a reaction that involves the breakage of covalent bonds. The demethylation of O^6-methylguanine would not proceed at a significant rate in the absence of O^6-methylguanine-DNA methyltransferase. However, the methyltransferase is not a true catalyst because it is not regenerated during the reaction. Transfer of the methyl group to the protein inactivates it.

2. Ultraviolet light can damage DNA by causing dimerization of thymine residues. One mechanism for repair of thymine dimers is enzymatic photoreactivation, catalyzed by photoreactivating enzyme. This enzyme uses energy from visible light to cleave the dimer and repair the DNA. Thus, cells that are exposed to visible light following ultraviolet irradiation are better able to repair DNA than cells kept in the dark.

3. The DNA repair enzyme uracil DNA glycosylase removes uracil formed by the hydrolytic deamination of cytosine. Like the other DNA repair enzymes, uracil DNA glycosylase does not recognize thymine or the other three bases normally found in DNA. Thus, it cannot repair the damage when 5-methyl-cytosine is deaminated to thymine.

4. High mutation rates occur at methylcytosine-containing regions because the product of deamination of 5-methylcytosine is thymine, which cannot be recognized as abnormal. The T/G base pair that results from deamination of methylcytosine distorts the double helix, leading to repair. However, the repair enzymes may delete either the incorrect thymine or the correct guanine. When the guanine is replaced by adenine, the resulting A/T base pair is a mutation.

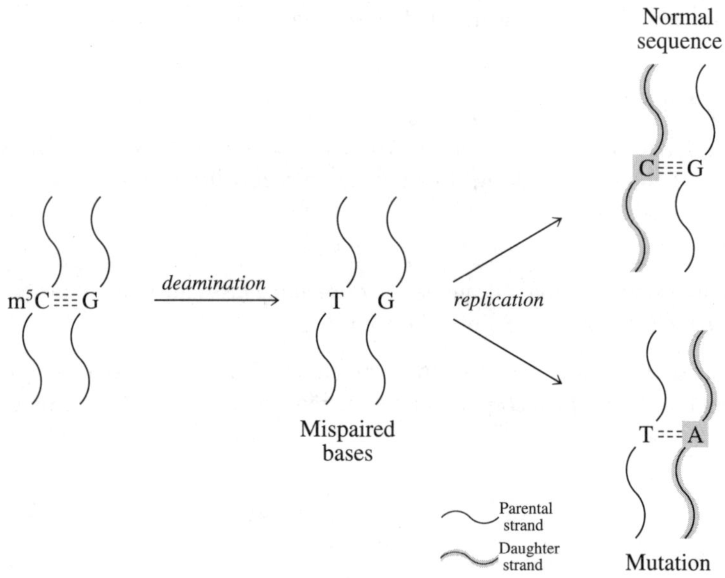

5. If uracil were a naturally occurring base in DNA, repair enzymes would be unable to distinguish normal uracil from uracil that arose by spontaneous deamination of cytosine. Spontaneous deamination of cytosine, which occurs at a relatively high frequency, would lead to high mutation rates in uracil-containing DNA. However, if uracil were tagged with a 5-methyl group before incorporation into DNA (i.e., if thymine were used instead of uracil), then repair enzymes could remove uracil that arose by mutation without disturbing thymine. Thus, the evolution of a mechanism for incorporating thymine instead of uracil into DNA confers a selective advantage because it reduces the mutation rate and increases the stability of genetic information.

6. Deamination of adenine yields hypoxanthine (H), which forms a more stable base pair with cytosine than with thymine. If repair enzymes do not successfully remove hypoxanthine, then the H/T base pair persists. During the next round of DNA replication, the template strand containing T yields a T/A base pair as in the parental DNA, but the strand containing H yields an H/C base pair. When this daughter DNA molecule is repaired or replicated again, the result is a mutant G/C base pair at the site of the original deamination of adenine.

   Deamination of guanine yields xanthine, which still pairs with cytosine, and is therefore less likely to result in a mutation.

7. Proofreading during replication results in excision of 99% of misincorporated nucleotides, thus reducing the overall error rate to 10^{-7}. Of those errors that escape the proofreading step, a further 99% are corrected by repair enzymes. The overall mutation rate is therefore 10^{-9}.

Summary

Bacterial RNA polymerases consist of a multisubunit core and a σ factor that is required for initiation. The *E. coli* core enzyme contains two α subunits, a β subunit, and a β' subunit. The most common σ subunit is σ^{70}. Transcription begins at the 5' end of a gene when the holoenzyme binds to the promoter to form a closed complex. The most common promoter sequence in *E. coli* is the one recognized by the σ^{70}-containing holoenzyme. The consensus sequence of this promoter includes a TATA box 10 base pairs upstream of the transcription start site and a −35 region. The next steps in initiation are unwinding of a small stretch of DNA to form an open complex, synthesis of a short RNA primer, and promoter clearance. The transition to the elongation phase of transcription is accompanied by release of the σ subunit and the association of additional factors such as NusA.

During transcription, RNA polymerase catalyzes a nucleotidyl-group–transfer reaction using nucleoside triphosphates as substrates. The nascent RNA is extended by addition of nucleoside monophosphates to the 3' end, using one of the DNA strands as a template. A short stretch of RNA is base-paired to the template strand within the transcription bubble. Elongation is accompanied by continuous unwinding and rewinding of DNA.

Some nucleotide sequences cause the RNA polymerase to pause. Pausing is exaggerated when the nascent RNA forms a short hairpin structure. This structure can be recognized by NusA, preventing further transcription and leading to termination if the RNA-DNA hybrid is unstable. Termination can also occur when rho binds to free RNA, then comes in contact with the transcription complex. Rho-dependent termination is inhibited by antiterminators during ribosomal RNA synthesis.

There are three different eukaryotic RNA polymerases. RNA polymerase I transcribes large ribosomal RNA genes; RNA polymerase II transcribes most protein-encoding genes; and RNA polymerase III transcribes a variety of small genes. The eukaryotic RNA polymerases are related to each other, and they are more complex than the prokaryotic enzymes. Eukaryotic initiation is also more complex. Initiation on class II genes requires several transcription factors that assist in binding RNA polymerase II to the promoter. The TBP subunit of one of these factors, TFIID, binds to the TATA box. Initiation of class I and class III genes also requires specific transcription factors, but TBP is used by all three RNA polymerases. One of the class III transcription factors, TFIIIA, contains zinc-finger DNA-binding motifs.

Transcription initiation can be regulated negatively or positively. Negative regulation is mediated by repressors that bind DNA near the promoter and prevent initiation. Positive regulation is mediated by activators that bind near a weak promoter and assist initiation. Many repressors and activators are allosteric proteins whose activity is affected by ligand binding. For example, *lac* repressor binds allolactose, and CRP binds cAMP. Both of these proteins are involved in the regulation of the *lac* operon, and both contain a helix-turn-helix DNA-binding motif.

Transcription of some genes is regulated by multiple regulatory proteins that can associate with each other to form a complex at the promoter. In some cases, the regulatory proteins bind to DNA at sites that are far from the promoter, and the resulting complex forms a large loop of DNA. Transcription of many eukaryotic genes is inhibited because they are tightly associated with nucleosomes in heterochromatin. Transcription initiation is accompanied by decondensation of chromatin.

Figure 27·6
Orientation of genes. The sequence of a hypothetical gene and RNA in the process of being transcribed from it are shown. Note that genes are transcribed from the 5′ end to the 3′ end, but the actual template strand of DNA is copied from the 3′ end to the 5′ end. Growth of the ribonucleotide chain proceeds 5′→3′.

Study Information

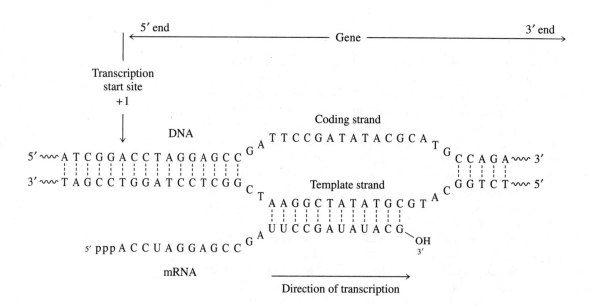

Problems

1. Cordycepin 5′-triphosphate is an analog of ATP.

Assuming cordycepin 5′-triphosphate is similar enough to ATP to be recognized by RNA polymerase, what would happen if a small amount of this drug were present in a cell during RNA transcription?

2. You have used *E. coli* RNA polymerase to transcribe the DNA from a bacteriophage in vitro. In the absence of the termination factor rho, you obtain an RNA transcript with a sedimentation coefficient of 30S after several minutes. In separate experiments, you add rho to the actively transcribing mixture at various times. Several seconds after rho is added, you add an inhibitor to block further transcription initiation. When rho is added immediately after transcription begins, a 10S transcript is obtained. When rho is added after 50 seconds, a 22S transcript is obtained, and when rho is added after 150 seconds, a 30S transcript is obtained. Explain these results.

3. Predict the results when the core polymerase is mixed with double-stranded *E. coli* DNA in the presence of nucleoside triphosphates under optimal conditions for transcription activity.

4. The *E. coli* genome contains hundreds of promoters for mRNA synthesis, but fewer than 10 promoters for rRNA synthesis. Despite this large difference in promoter number, the synthetic capacities for rRNA and mRNA are not very different (Table 27·1). What does this observation suggest about the properties of promoters for protein-encoding genes?

5. (a) RNA polymerase is incubated with a DNA template, all the necessary transcription factors, and the substrates ATP, UTP, GTP, and CTP, which are labelled with ^{18}O in the β-phosphate group and ^{14}C in the ribose. Will ^{18}O and ^{14}C be incorporated into the RNA that is synthesized?

   (b) Predict the results if the nucleoside triphosphates are labelled with ^{14}C in the ribose and ^{18}O in the α-phosphate group.

6. RNA polymerase catalyzes chain elongation by a mechanism similar to that of DNA polymerase. Compare these two enzymes in *E. coli* in terms of the following:

	RNA polymerase	DNA polymerase III
DNA region initially recognized and bound by the polymerase		
Direction of polymerization		
Direction of enzyme movement on template strand		
Processivity (Yes or No)		
Type of nucleotide substrates added to the growing chain		
Mechanism of the reaction		
Products of the reaction		
Rate of polymerization		
Proofreading ability (Yes or No)		

7. Unlike DNA polymerase, RNA polymerase does not have proofreading activity. Explain why the lack of proofreading activity is not detrimental to the cell.

8. A bacterial RNA polymerase elongates RNA at a rate of 70 nucleotides per second, and each transcription complex covers 70 base pairs of DNA.
   (a) What is the maximum number of mRNA molecules that can be produced per minute from a gene of 6000 base pairs?
   (b) What is the maximum number of transcription complexes that can be bound to this gene at one time?

9. *E. coli* cells can grow on a variety of carbon sources. Describe how the rate of transcription of the *lac* operon is affected when bacteria are grown in the presence of the following.
   (a) lactose plus glucose
   (b) glucose alone
   (c) lactose alone

Solutions

1. If cordycepin 5'-triphosphate were mistaken for ATP by RNA polymerase, it would be incorporated into growing RNA chains. However, because cordycepin 5'-triphosphate lacks a 3'-hydroxyl group, it cannot react with the next nucleoside triphosphate during the polymerization reaction. Thus, incorporation of cordycepin 5'-triphosphate during transcription would lead to premature chain termination and, if significant amounts of the drug were present, cell death.

2. These results indicate that the bacteriophage DNA contains one rho-independent termination site and two rho-dependent sites. The third site is the rho-independent site; termination occurs here in the presence or absence of rho. Termination at the other two sites occurs only in the presence of rho. Addition of rho at time 0 causes termination at the first rho-dependent site, yielding 10S RNA. After 50 seconds in the absence of rho, RNA polymerase has read through the first termination site. Addition of rho at this point terminates transcription at the second rho-dependent site, yielding 22S RNA. After 150 seconds in the absence of rho, the RNA polymerase has proceeded past the first and second rho-dependent sites. After this point, termination occurs at the rho-independent site whether or not rho is present.

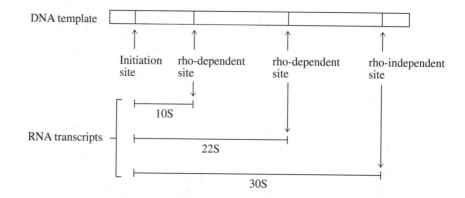

3. Transcription initiation requires the presence of a σ subunit for the polymerase to bind to promoter regions and for formation of an open complex and synthesis of a primer. Because the core enzyme has no σ subunit, no RNA will be synthesized using intact *E. coli* DNA as a template. This reflects the situation in vivo, where only certain genes are expressed, depending on the presence of specific σ subunits and other factors.

However, purified *E. coli* DNA often contains many nicks and gaps that can sometimes serve as initiation sites in vitro. In practice, core polymerase can produce random transcripts of DNA templates even in the absence of additional factors. This aberrant transcription is the basis of the assay for core polymerase activity; without it, the core polymerase would not have been so easily identified as the protein containing the active site for polymerization.

4. The observation suggests that most promoters for protein-encoding genes are weak or are not utilized under normal growth conditions. The frequency of initiation of transcription at the strong rRNA gene promoters must be much higher than the frequency of initiation at weak promoters for mRNA synthesis. Some protein-encoding genes, such as those for the ribosomal proteins, possess strong promoters. Even in these cases, the amount of mRNA that is needed is much less than the amount of rRNA required because each mRNA is translated many times before it is degraded.

5. (a) Since pyrophosphate is released from each nucleoside triphosphate except the first, ^{18}O will be present only at the 5′ end of the newly synthesized RNA, whereas every nucleotide residue will be labelled with ^{14}C-ribose.

   (b) The α-phosphate oxygen is incorporated into RNA. Thus, there will be an ^{18}O and a ^{14}C-ribose in each nucleotide residue of the transcribed RNA.

6.

	RNA polymerase	DNA polymerase III
DNA region initially recognized and bound by the polymerase	*Promoters*	*Replication origin*
Direction of polymerization	$5′ \rightarrow 3′$	$5′ \rightarrow 3′$
Direction of enzyme movement on template strand	$3′ \rightarrow 5′$	$3′ \rightarrow 5′$
Processivity	*Yes*	*Yes*
Type of nucleotide substrates added to the growing chain	*Ribonucleoside triphosphates (NTPs) A, U, G, C*	*Deoxyribonucleoside triphosphates (dNTPs) dA, dT, dG, dC*
Mechanism of the reaction	*Nucleophilic attack by 3′-hydroxyl group of ribose on α-phosphorus of NTP (group-transfer reaction)*	*Nucleophilic attack by 3′-hydroxyl group of deoxyribose on α-phosphorus of dNTP (group-transfer reaction)*
Products of the reaction	$RNA_n + (n-1)\,PP_i$	$DNA_n + (n-1)\,PP_i$
Rate of polymerization	*30–85 nucleotides s^{-1}*	*~1000 nucleotides s^{-1}*
Proofreading ability	*No*	*Yes*

7. The lack of proofreading activity in RNA polymerase makes the error rate of transcription greater than the error rate of DNA replication; however, the defective RNA molecules produced are not likely to affect cell viability because most copies of RNA synthesized from a given gene are normal. In the case of mRNA, the number of defective proteins produced from mRNA containing errors of transcription is only a small percentage of the total number of proteins synthesized. Also, mistakes made during transcription are quickly eliminated since most mRNA molecules have a short half-life.

8. (a) Since the rate of transcription is 70 nucleotides per second and each transcription complex covers 70 base pairs of DNA, an RNA polymerase completes a transcript and leaves the DNA template each second (assuming that the complexes are densely packed). Therefore, when the gene is loaded with transcription complexes, 60 molecules of RNA are produced per minute.

    (b) Since each transcription complex covers 70 base pairs, the maximum number of complexes is

    $$\frac{6000 \text{ base pairs}}{70 \text{ base pairs}/\text{transcription complex}} = 86 \text{ transcription complexes}$$

9. (a) In the presence of both lactose and glucose, the *lac* operon is transcribed at a low level because the repressor forms a complex with *allo*lactose (an isomer of lactose). Because the *allo*lactose-repressor complex cannot bind to the promoter region of the *lac* operon, the repressor does not prevent initiation of transcription.

    (b) In the absence of lactose, no *allo*lactose is formed. Thus, the *lac* repressor binds near the *lac* operon promoter and prevents transcription.

    (c) When lactose is the sole carbon source, the *lac* operon is transcribed at the maximum rate. In the presence of *allo*lactose, transcription is allowed since the *lac* repressor does not bind to the promoter region of the *lac* operon. Also, in the absence of glucose, transcription rates increase because cAMP production increases and thus more CRP-cAMP is available to bind to the promoter region of the *lac* operon. The absence of the repressor and enhancement of transcription initiation by CRP-cAMP allows the cell to synthesize the quantities of enzymes required to support growth when lactose is the only available carbon source.

28

RNA Processing

Summary

RNA processing includes chemical modification of nucleotides and the addition and deletion of nucleotides. Processing appears to be related, in part, to RNA stability. In addition, certain alterations refine RNA structure and allow the RNA to be recognized by other cellular components. Processing is required in all primary transcripts containing tRNA and rRNA and in eukaryotic mRNA molecules.

Precursor tRNA molecules in prokaryotes are part of longer primary transcripts that are cleaved by RNases into immature tRNA monomers. RNase P, which is made up of both protein and a catalytically active RNA molecule, is an endonuclease that cleaves the primary transcript once at each individual tRNA precursor to generate a mature 5′ end. Maturation of the 3′ end of a prokaryotic tRNA precursor involves several additional exonucleases, including RNases D and T. In organisms in which the 3′ terminal sequence CCA is not encoded by the tRNA gene itself, the sequence must be added posttranscriptionally by the enzyme tRNA nucleotidyltransferase.

Many primary transcripts contain sequences called introns that interrupt the functional part of the molecule. During maturation, introns are removed, and the structural portions of the precursor, called exons, are joined together. This process is called RNA splicing. In the case of yeast tRNA splicing, the reaction is mediated by a multifunctional enzyme and an RNA ligase and occurs in two steps. Excision of tRNA introns requires two molecules of ATP.

In both eukaryotes and prokaryotes, certain nucleotides of tRNA molecules are extensively modified. Covalently modified nucleotides tend to lie in regions of the tRNA molecule that are not involved in intrastrand base pairing.

The primary ribosomal RNA transcripts of both prokaryotes and eukaryotes contain more than one rRNA molecule. These individual rRNA species are separated from the long primary transcript by specific endonucleases that recognize double-stranded regions of the precursor. Some rRNA molecules contain introns. Two types of introns have been characterized to date: group I and group II. Some of these introns, called self-splicing introns, excise themselves without the assistance of proteins.

Group I introns are removed as linear molecules. Self-splicing by these introns requires Mg^{2+} and a guanosine cofactor. The guanosine cofactor is added to the 5′ end of the intron during a transesterification reaction, which releases the 5′ exon. Once the 5′ exon has been released, the 3′-hydroxyl group at its 3′ end attacks the 3′ exon boundary in another transesterification reaction. This reaction, like more conventional enzymatic reactions is reversible; these introns are examples of RNA enzymes, or ribozymes.

Self-splicing introns of group II are found in some mitochondrial rRNA and protein-encoding genes. They do not require guanosine cofactors. The mechanism of self-splicing in group II introns involves attack at the 5′ splice site by the 2′-hydroxyl group of an adenylate residue within the intron. Subsequent attack by the 3′-hydroxyl group of the 5′ exon at the 3′ splice site joins the exons and releases the intron as a lariat molecule.

Primary messenger RNA precursors in eukaryotes are processed by nucleotide modification (capping), polyadenylation, and, in some cases, splicing. The cap at the 5′ end protects the RNA from digestion during its long life span and also interacts with ribosomes during translation. mRNA precursors are modified at their 3′ ends by a complex consisting of a cleavage and polyadenylation specificity factor (CPSF), an endonuclease, and poly A polymerase. CPSF binds to a specific site that is 10 to 20 nucleotides upstream of the cleavage site. The endonuclease cleaves the precursor, and poly A polymerase adds a poly A tail consisting of up to 250 adenylate residues. This poly A tail binds to at least one protein that helps stabilize the RNA.

Introns in mRNA precursors range in size from 65 to 10 000 nucleotides. Except for short consensus sequences at the splice sites and the branch point, the sequences of introns share no similarities. Splicing of mRNA precursors is not self-catalyzed but involves a group of ribonucleoproteins (snRNPs) that assemble to form a spliceosome. The spliceosome correctly orients the reactants in the splicing reaction. Assembly of the spliceosome begins when a U1 snRNP binds to the 5′ splice site. This binding is mediated by base pairing between the U1 snRNA and the mRNA precursor. A U2 snRNP then binds at the branch point within the intron, while a U5 snRNP binds at the 3′ splice site. The three bound snRNPs move together, bind to the U4/U6 snRNP, and remove the intron. The actual reaction catalyzed by the spliceosome is similar to group II self-splicing reactions: in one transesterification reaction, nucleophilic attack by the 2′-hydroxyl group at the branch point opens the 5′ splice site; in the second transesterification reaction, the exons are joined and the intron is released as a lariat molecule.

Different mature mRNA molecules can be produced by alternative processing pathways. In some cases, alternative splice sites are used; in other cases, the processing pathways are coupled to the use of alternative promoters or polyadenylation sites.

Study Information

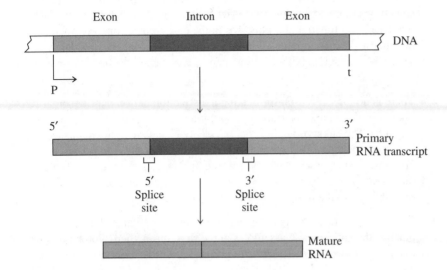

Figure 28·4
Nomenclature of splicing. Introns (dark shade) and exons (light shade) are shown. Splice sites are defined by their positions (5′ or 3′) relative to the intron.

Problems

1. Does the definition of a gene given on Page 27·1 apply to the 16S and tRNA genes shown in Figure 28·10?

2. Clusters of ribosomal RNA genes in *E. coli* are referred to as operons. Are the similar clusters in eukaryotic genomes also operons?

3. Assume that a spliceosome assembles at the first intron of the gene for triose phosphate isomerase in maize (Figure 28·23) almost as soon as the intron is transcribed (i.e., after about 500 nucleotides of RNA have been synthesized). How long must the spliceosome be stable if the splicing reaction cannot occur until transcription terminates? Assume that the rate of transcription by RNA polymerase II in maize is 30 nucleotides per second.

4. The exon sequences of eukaryotic tRNA precursors are joined by an RNA ligase. Why is this ligase not used to splice mRNA precursors?

5. Summarize the covalent modifications to eukaryotic mRNA precursors that may occur prior to the excision of introns.

6. How many high-energy phosphoanhydride bonds are required for the removal of an intron in a yeast tRNA precursor?

7. Explain the difference between a self-splicing reaction (as in the rRNA of *Tetrahymena thermophila*) and RNA acting as a true catalyst—a ribozyme (as in ribonuclease P).

Solutions

1. In Chapter 27, a gene was defined as a DNA sequence that is transcribed. By this definition, the entire ribosomal RNA operon is a gene. However, it is sometimes more convenient to restrict the term *gene* to the segment of RNA that encodes a functional product, as in the case of other operons such as the *lac* operon. The *rrn* operons therefore contain 16S, 23S, 5S, and tRNA genes. The DNA sequences between these genes, although transcribed, are not considered part of any gene.

2. There is no logical reason for making a distinction between a ribosomal RNA operon in *E. coli* and the transcribed region of a ribosomal RNA repeat in eukaryotes. The eukaryotic cluster could easily be called an operon as well. However, the term *operon* usually refers to co-transcribed genes that encode proteins. In such cases, the polycistronic mRNA is not processed. The bacterial *rrn* operons are referred to as operons only because of their superficial resemblance to the protein-encoding bacterial operons.

3. The gene for triose phosphate isomerase in maize contains about 3400 base pairs. If the spliceosome assembles at the first intron, then 2900 base pairs remain to be transcribed. The time required for the transcription of 2900 base pairs is 97 seconds (i.e., 2900 nucleotides ÷ 30 nucleotides per second). If the spliceosome assembles immediately after transcription of the first intron, and if splicing cannot begin until transcription of the entire gene is complete, the spliceosome must be stable for at least 97 seconds.

4. RNA ligase cannot be used to ligate exons in mRNA precursors because the ends of the exons are not adjacent as they are in tRNA splicing. The ends of tRNA exons are next to each other by virtue of the secondary structure of tRNA. In mRNA processing, the spliceosome holds the ends of the exons together.

5. Before they are spliced, mRNA precursors in eukaryotes are often covalently modified at both ends (see facing page, top).
    (1) The γ-phosphate group of the 5′-triphosphate is removed by the activity of phosphohydrolase. The resulting 5′-diphosphate then reacts with GTP in a reaction catalyzed by guanylyltransferase. During this reaction, GMP is attached to the diphosphate, forming a 5′–5′ triphosphate linkage. Subsequent methylation to produce 7-methylguanosine (m^7G) forms a cap 0 structure, depicted in Figure 28·19. (Further methylations of the 2′-hydroxyl groups of the ribose moieties at the 5′ end of the molecule may also occur to form Cap 1 and Cap 2 structures.) Capping occurs soon after transcription is initiated.
    (2) Once RNA polymerase II has transcribed past the endonuclease cleavage site, the endonuclease cleaves the mRNA precursor. The endonuclease recognition sequence occurs approximately 10–20 nucleotides upstream of the endonuclease cleavage site. This cleavage generates a new 3′ end. The newly generated 3′ end of the mRNA is modified in the nucleus by addition of a poly A tail. Poly A polymerase adds up to 250 adenylate residues to the 3′ end of the mRNA chain. This tail may protect against exonucleases. By the time the mature mRNA molecule reaches the cytoplasm, the poly A tail is usually shortened by 50–100 nucleotides..
    (3) Covalent modifications of nucleotides may also occur.

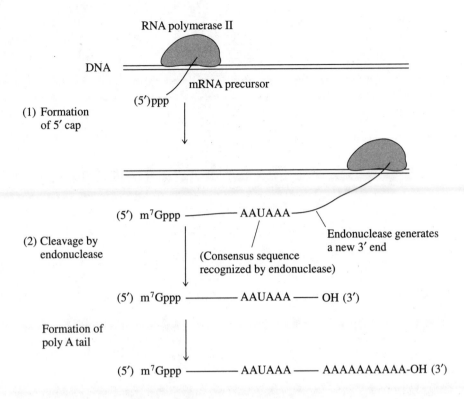

(1) Formation of 5′ cap

(2) Cleavage by endonuclease

Formation of poly A tail

6. Three high-energy bonds are required for the splicing of two exons in a yeast tRNA precursor. One ATP is converted to ADP by the action of the kinase; PP_i is released from another ATP in the step catalyzed by RNA ligase; and the PP_i is subsequently hydrolyzed to $2 P_i$.

7. The primary transcript for the rRNA of *Tetrahymena* undergoes a self-splicing reaction, with loss of its intervening sequence. Because the transcript is permanently modified during the reaction, it is not a true catalyst.

   The RNA component of ribonuclease P, however, can cleave tRNA precursor molecules and yet remain unchanged at the end of the reaction. Therefore, it qualifies as a true catalyst.

29

The Genetic Code and Transfer RNA

Summary

The genetic code is nearly universal and consists of 64 nonoverlapping codons. The starting point for the reading of codons defines the reading frame of a gene; shifting the starting point for the reading of codons by a number of bases not divisible by three shifts the reading frame and changes the entire message.

The genetic code has several salient features. First, the code is degenerate; many synonymous codons specify the same amino acid, and the first two positions of a codon are often enough to specify a given amino acid. Mutations in the third position therefore often do not change the sense of the codons. Second, similar codons specify chemically similar amino acids. Third, special codons signal termination and initiation. Finally, each codon has only one meaning.

Mutations can alter the sense of codons. Missense mutations involve nucleotide substitutions that change the meaning of individual codons so that the wrong amino acid is incorporated into the protein during translation. Nonsense mutations produce termination codons called nonsense codons, which can result in premature termination of protein synthesis. Nonsense mutations can be suppressed by suppressor tRNA molecules, which insert amino acids at nonsense codons. Frameshift mutations involve insertions and deletions of nucleotides that change the reading frame of a gene.

tRNA molecules form the interface between proteins and nucleic acids; they carry activated amino acids to the ribosome for transfer to a growing peptide chain and interact with the mRNA by complementary base pairing. All tRNA molecules share certain structural features. tRNA molecules contain many conserved and many covalently modified nucleotides, some of which help stabilize the secondary and tertiary structures of the molecules. The secondary structure of a tRNA molecule resembles a cloverleaf with four Watson-Crick hydrogen-bonded stems and three loops. The acceptor stem is covalently attached to the amino acid residue, and the anticodon loop interacts with the codon in the mRNA. The tertiary structure of tRNA is L shaped, with the acceptor stem and the anticodon loop at opposite ends of the molecule. Non–Watson-Crick interactions help establish this tertiary structure.

The anticodon consists of three nucleotides that interact with the codon in mRNA by complementary base pairing. This interaction allows flexibility at the 5′ (wobble) position, where certain non–Watson-Crick base pairing is permitted. tRNA molecules that have different anticodons but bind to the same amino acid are called isoacceptor tRNA molecules. The genes that encode isoacceptor tRNA molecules are not equally expressed. In general, the codons used most often are those recognized by the most abundant isoacceptor tRNA molecules.

An amino acid is added to a tRNA molecule in a reaction catalyzed by an aminoacyl-tRNA synthetase. There are two classes of aminoacyl-tRNA synthetases. Class I enzymes catalyze attachment of the amino acid to the 2′-hydroxyl group of the terminal adenylate of the tRNA molecule, whereas class II enzymes catalyze attachment to the 3′-hydroxyl group. The two classes are distinguished on the basis of similarity in amino acid sequence and other aspects of structure.

Aminoacyl-tRNA molecules are synthesized in two steps. First, the amino acid is converted into an energy-rich aminoacyl-adenylate by adenylyl-group transfer from ATP. Second, the aminoacyl group of the energy-rich intermediate is transferred to a tRNA molecule. This aminoacylation of a given tRNA molecule determines which amino acid will be incorporated into the protein at a given codon.

Each aminoacyl-tRNA synthetase is highly specific for one amino acid and its corresponding tRNA molecules. Aminoacyl-tRNA synthetases discriminate among amino acid substrates on the basis of charge, size, and hydrophobicity. Some aminoacyl-tRNA synthetases exhibit a proofreading activity and can catalyze hydrolysis of an inappropriate substrate even after the aminoacyl-adenylate has formed.

Aminoacyl-tRNA synthetases also specifically bind to their corresponding tRNA molecules. The details of this recognition process are still unclear, but recognition is known to involve specific bases in the tRNA, including, in some cases, the anticodon.

Study Information

Figure 29·3

The standard genetic code. The code is composed of 64 triplet codons whose mRNA sequences can be read from this chart. The left-hand column indicates the nucleotide found at the first (5′) position of the codon. The top row indicates the nucleotide found at the second (middle) position of the codon. The right column indicates the nucleotide found at the third (3′) position. Thus, the codon CAG encodes glutamine (Gln), and GGU encodes glycine (Gly). The codon AUG specifies methionine (Met) and is also used to initiate protein synthesis. STOP indicates a termination codon.

First position (5′ end)	Second position				Third position (3′ end)
	U	C	A	G	
U	Phe	Ser	Tyr	Cys	U
	Phe	Ser	Tyr	Cys	C
	Leu	Ser	STOP	STOP	A
	Leu	Ser	STOP	Trp	G
C	Leu	Pro	His	Arg	U
	Leu	Pro	His	Arg	C
	Leu	Pro	Gln	Arg	A
	Leu	Pro	Gln	Arg	G
A	Ile	Thr	Asn	Ser	U
	Ile	Thr	Asn	Ser	C
	Ile	Thr	Lys	Arg	A
	Met	Thr	Lys	Arg	G
G	Val	Ala	Asp	Gly	U
	Val	Ala	Asp	Gly	C
	Val	Ala	Glu	Gly	A
	Val	Ala	Glu	Gly	G

Acceptor stem

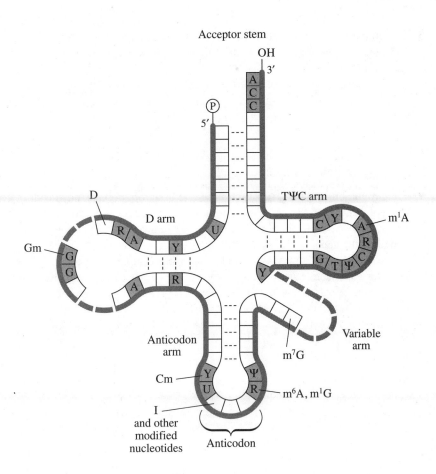

Figure 29·11
Cloverleaf secondary structure of tRNA. Watson-Crick base pairing is indicated by dashed lines between nucleotide residues. The molecule is divided into an acceptor stem and four arms. The acceptor stem is the site of amino acid attachment, and the anti-codon arm is the region of the tRNA molecule that interacts with mRNA. The D and TΨC arms are named for modified nucleotides that are conserved within the arms. The number of nucleotide residues in each arm is more or less constant, except in the variable arm. Conserved bases (colored gray) and positions of commonly modified nucleotides are noted. Abbreviations other than standard nucleotides: R, a purine nucleotide; Y, a pyrimidine nucleotide; m^1A, 1-methyladenylate; m^6A, N^6-methyladenylate; Cm, 2'-O-methylcytidylate; D, dihydrouridylate; Gm, 2'-O-methylguanylate; m^1G, 1-methylguanylate; m^7G, 7-methylguanylate; I, inosinate; Ψ, pseudouridylate; and T, thymidylate.

Problems

1. Point mutations may result in changes in only one amino acid in a protein. Give examples of point mutations that can lead to more severe disruptions in a protein's primary structure.

2. (a) The mRNA segment UUUGAAUGG encodes the tripeptide sequence Phe–Glu–Trp. What other mRNA sequences encode this tripeptide? Explain.

   (b) Many different mRNA sequences can encode the tripeptide sequence Leu–Pro–Gly. Which sequences are most likely to be found in the mRNA of an *E. coli* ribosomal protein that contains this tripeptide sequence?

3. What is the amino acid sequence of the peptide encoded by the following sequence:

   mRNA   5'– C A U G A U A U U C U U C A C U G U A C A A C A U A A A A C A C U A U A A C C – 3'

   (a) in prokaryotes
   (b) in yeast mitochondria?

4. The sequence of a segment of *E. coli* DNA from the *middle* of the coding region of a gene is shown below

   5'〰 C C G G C T A A G C C A T G A C T A G C 〰 3'
   3'〰 G G C C G A T T C G G T A C T G A T C G 〰 5'

   (a) Write the two mRNA sequences that could be transcribed from this sequence.

   (b) Identify the codons in all possible reading frames of this DNA fragment (the standard genetic code is given in Figure 29·3). Which amino acid sequence corresponds to the actual gene product?

5. List all the tRNA species that could be mutated to suppressor tRNA molecules by a single base change in the anticodon. How can a cell with a suppressor tRNA survive?

6. What is the wobble hypothesis? Give examples in your discussion.

7. Why is the accuracy of the aminoacyl-tRNA synthetase reaction so important for protein synthesis?

8. (a) List the common structural features of tRNA molecules.
   (b) How do tRNA molecules differ?

9. Point mutations in DNA (substitution of one base for another) may result in the replacement of one amino acid for another. In some cases, due to the degeneracy of the code, there will be no change in the amino acid sequence encoded by the gene.

   An extracellular protease produced by a bacterium possesses a cysteine residues at the active site (–Gly–Leu–Cys–Arg–). After irradiation, two mutant strains of bacteria were isolated. Strain 1 produced an *inactive* enzyme with serine replacing cysteine at the active site (Gly–Leu–Ser–Arg). In strain 2, a truncated polypeptide was synthesized ending in –Gly–Leu–COO$^{\ominus}$ within the active site sequence shown above. Predict the mutation that might have occurred in each of these strains.

10. What are the advantages of a degenerate rather than a nondegenerate genetic code?

Solutions

1. A point mutation that creates or destroys an initiation codon is likely to severely affect the primary structure of a protein. For example, a mutation such as CTG $\longrightarrow$ ATG that occurred upstream of a normal initiation codon in the DNA sequence that encodes the 5′ leader region could produce a protein with a string of extra amino acid residues in front of the normal N-terminus.

   In eukaryotes, there are other mutations that can severely affect the primary structure of a protein. For example, mutations that eliminate the normal termination codon of a gene, mutations in the splice-junction consensus sequences of genes that contain introns, and mutations that create new splice-junction sequences all affect more than one amino acid in the encoded protein.

2. (a) Although a given mRNA sequence encodes a specific peptide sequence, all but two amino acids can be encoded by more than one codon (Figure 29·3). Thus, a number of mRNA sequences may code for the same tripeptide. There are four mRNA sequences that encode Phe–Glu–Trp.

   Phenylalanine (UUU or UUC)
   Glutamate (GAA or GAG)
   Tryptophan (UGG)

   UUU–GAA–UGG
   UUU–GAG–UGG
   UUC–GAA–UGG
   UUC–GAG–UGG

(b) The genes that encode isoacceptors are expressed unequally, and synonymous codons are used unequally. The frequency of codon usage in *E. coli* ribosomal proteins is shown in Table 29·2. The mRNA sequences encoding Leu–Pro–Gly that are most likely to be found in an *E. coli* ribosomal protein are shown below:

Leu – Pro – Gly
CUG CCG GGU
CUG CCG GGC

3. (a) Figure 29·3 shows the standard genetic code shared by most organisms, including prokaryotes. Starting with the AUG initiation codon and ending with the termination codon UAA, the sequence of the peptide is

Met – Ile – Phe – Phe – Thr – Val – Gln – His – Lys – Thr – Leu

(b) Figure 29·6 shows deviations from the standard genetic code. In yeast mitochondria, AUA codes for methionine (not isoleucine) and CUA codes for threonine (not leucine). The sequence of the peptide is

Met – <u>Met</u> – Phe – Phe – Thr – Val – Gln – His – Lys – Thr – <u>Thr</u>

4. (a) Since either strand could serve as the template for transcription, two possible mRNA sequences could be transcribed.

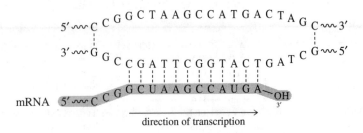

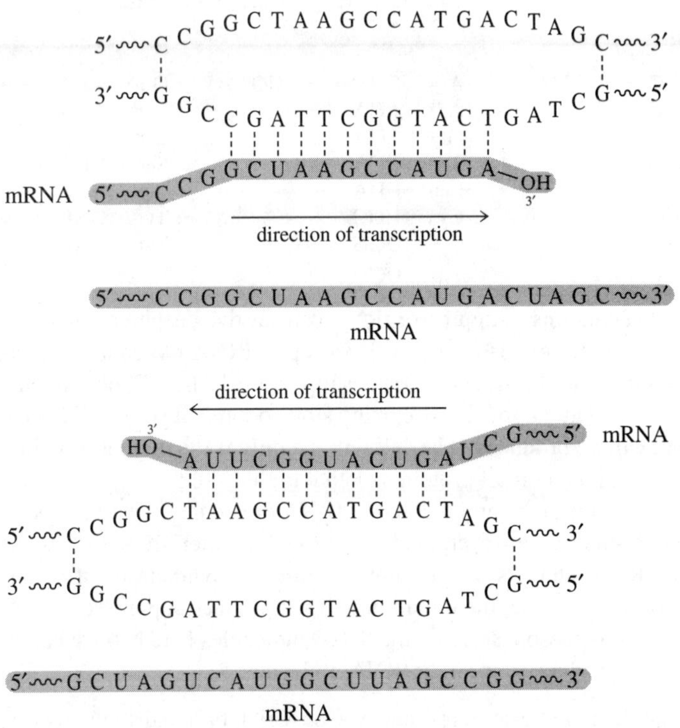

5′ ⌇⌇⌇C C G G C U A A G C C A U G A C U A G C⌇⌇⌇ 3′
mRNA

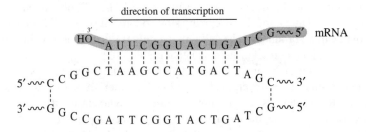

5′⌇⌇⌇G C U A G U C A U G G C U U A G C C G G⌇⌇⌇3′
mRNA

(b) Each mRNA sequence could be translated in three different reading frames. For the first mRNA sequence, the possible codons are

Reading frame 1   5′⌇⌇⌇|C C G|G C U|A A G|C C A|U G A|C U A|G C ⌇⌇⌇3′
           — Pro — Ala — Lys — Pro   STOP

Reading frame 2   5′⌇⌇⌇C|C G G|C U A|A G C|C A U|G A C|U A G|C⌇⌇⌇ 3′
           — Arg — Leu — Ser — His — Asp   STOP

Reading frame 3   5′⌇⌇⌇C C|G G C|U A A|G C C|A U G|A C U|A G C|⌇⌇⌇ 3′
           — Gly   STOP

For the second mRNA sequence, the three possible polypeptide sequences are

Reading frame 1   5′ ∿ |G C U|A G U|C A U|G G C|U U A|G C C|G G ∿ 3′
   — Ala — Ser — His — Gly — Leu — Ala — Gly —

Reading frame 2   5′ ∿ G|C U A|G U C|A U G|G C U|U A G|C C G|G ∿ 3′
   — Leu — Val — Met — Ala   STOP

Reading frame 3   5′ ∿ G C|U A G|U C A|U G G|C U U|A G C|C G G|∿ 3′
   STOP

Since only one of the six possible reading frames does not contain a stop codon, the gene must encode a protein with the internal sequence –Ala–Ser–His–Gly–Leu–Ala–Gly–.

5. Suppressor tRNA molecules contain altered anticodons that allow them to bind the termination codons UAG, UAA, and UGA as if they were codons for amino acids (Section 29·7D). For a single mutation to produce a suppressor tRNA molecule from a normal tRNA molecule, the anticodon of the normal tRNA molecule must be complementary to a codon that differs by a single nucleotide from one of the three termination codons. Such is the case for some of the tRNA molecules specific for glutamine, lysine, glutamate, serine, tryptophan, leucine, tyrosine, arginine, glycine, and cysteine.

Gln	CAG ⟶ UAG	Arg	CGA ⟶ UGA	Gln	CAA ⟶ UAA
Lys	AAG	Arg	AGA	Lys	AAA
Glu	GAG	Gly	GGA	Glu	GAA
Trp	UGG	Ser	UCA	Ser	UCA
Leu	UUG	Leu	UUA	Leu	UUA
Ser	UCG	Cys	UGC	Tyr	UAU
Tyr	UAU	Cys	UGU	Tyr	UAC
Tyr	UAC	Trp	UGG		

A cell that contains a suppressor tRNA can survive despite the loss of a normal tRNA *if* the cell also contains isoacceptor tRNA molecules that carry the same amino acid. In these cases, the suppressor tRNA will occasionally insert an amino acid at a normally occurring stop codon, but some wild-type gene product will be produced. The cell will die only if the mutant protein, which is larger than the normal protein, is lethal to the cell.

A mutation in the anticodon of a tRNA molecule may affect the ability of the cell to survive, however. Such mutations can interfere with how the suppressor tRNA interacts with its aminoacyl-tRNA synthetase during aminoacylation. In this case, the mutation will not produce suppressor tRNA molecules. For this reason, some of the tRNA molecules listed above cannot be mutated to produce suppressor tRNA molecules.

6. The wobble hypothesis states that Watson-Crick base pairing is required for only two of the three base pairs formed when the anticodon interacts with the codon (Section 29·7B). The anticodon must form Watson-Crick base pairs with the 5′ and middle bases of the codon, but other types of base pairs are allowed in the 3′ position of the codon (Table 29·1). This alternative pairing suggests that the 5′ position, or wobble position, of the anticodon is conformationally flexible. Consider the interaction on the next page between the anticodon of the arginyl-tRNA^{Arg} molecule and the two arginine codons.

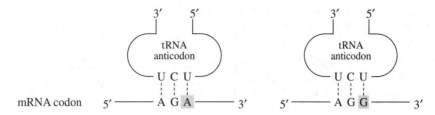

Interactions between U and A and between U and G at the wobble position are both possible.

7. During protein synthesis, the recognition of an mRNA codon by a tRNA anticodon is independent of the amino acid bound to the 3′ terminus of the tRNA; the codon-anticodon interaction determines which charged tRNA binds to the mRNA. Thus, an incorrectly charged tRNA introduces an amino acid substitution in the protein being synthesized, with the possible loss of the biological function of the protein.

8. (a)
   1. All tRNA molecules have –CCA at the 3′ end.
   2. All tRNA molecules fold into L-shaped tertiary structures. The tRNA molecules can also be represented more simply as two-dimensional "cloverleaf" structures. The latter structures show more clearly the four hydrogen-bonded "stems" connecting several "loops."
   3. All tRNA molecules are relatively small in size (about 70–90 nucleotide residues) compared to rRNA and mRNA molecules.
   4. All tRNA molecules contain modified bases.
   5. All tRNA molecules have an acceptor stem, which includes the 5′ and 3′ ends of the tRNA.
   6. There are several loops common to all tRNA molecules (some mitochondrial tRNA molecules are exceptions): the D (dihydrouridine) loop, 7–11 nucleotides; the anticodon loop, 7 nucleotides including three of the anticodon; the pseudouridine loop (TΨC), 7 nucleotides; and the variable loop, 3–21 nucleotides (the size and shape of this loop vary considerably in different tRNA molecules).

   (b) The anticodon for each tRNA differs, thus allowing recognition of one (or a few) mRNA codons. The tertiary structure of each tRNA is slightly different because the nucleotide sequences are not identical. These differences allow aminoacyl-tRNA synthetases to recognize their specific substrates.

9. Cysteine codons (UGU, UGC) differ by only one nucleotide base from several serine codons (AGU, AGC or UCU, UCC). A single point mutation can convert cysteine to serine, for example, UGU (Cys) $\longrightarrow$ AGU (Ser) or UGC $\longrightarrow$ AGC. The mutation in the gene could be either a substitution of an A/T base pair for a T/A base pair or a C/G substitution for G/C.

   The second protein is terminated prematurely. Thus, the mutant gene contains a stop codon following the leucine codon. Assuming a single nucleotide change, there are three possibilities:

$$Ser \longrightarrow Stop$$

$$TCA \longrightarrow TGA$$
$$TCA \longrightarrow TAA$$
$$TCG \longrightarrow TAG$$

10. A nondegenerate genetic code would consist of only 20 codons for amino acids, while the remaining 44 codons would presumably lead to chain termination. Species that utilized such a genetic code would be highly resistant to changes in nucleotide sequence since many substitution mutations would give rise to stop codons, and these would most likely be lethal to the individual organism in which they arose. All other mutations would cause an amino acid substitution in the protein, and such changes are often deleterious. As a consequence, organisms with a nondegenerate genetic code would likely evolve slowly.

With a degenerate genetic code, the interactions between codons and anticodons need not be as precise as would be the case if the code were nondegenerate. This means that fewer tRNA molecules would be needed. Furthermore, if the code were degenerate, some nucleotide substitutions would not result in any amino acid substitution in the protein, and most others would lead to an amino acid substitution rather than premature chain termination. This would increase the probability that new, and better, proteins would be synthesized.

A degenerate genetic code allows many "pathways" leading to the substitution of one amino acid by another as the result of successive mutations. There are many fewer pathways in a nondegenerate code.

30

Protein Synthesis

Summary

The general structure and function of ribosomes are conserved in all organisms. Ribosomes are composed of a large and a small subunit. In *E. coli,* the small subunit is the 30S subunit; it contains 21 proteins and one molecule of 16S rRNA. The large subunit is the 50S subunit; it contains at least 31 different proteins and two rRNA molecules, 5S and 23S. During protein synthesis, the two subunits combine to form an active 70S ribosome. The cytoplasmic ribosomes of eukaryotes are larger than prokaryotic ribosomes (80S, with subunits of 40S and 60S) and contain more proteins. The large subunit of eukaryotic ribosomes contains three molecules of rRNA: 28S, 5S, and 5.8S. The general structures of the rRNA molecules are conserved in all species, even though the exact nucleotide sequences of the molecules vary. Ribosome assembly is coupled to the processing of rRNA precursors in both prokaryotes and eukaryotes.

The formation of peptide bonds on the ribosome occurs in three distinct stages: initiation, elongation, and termination. During initiation in prokaryotes, an initiation complex is formed by the mRNA, the 30S subunit, an aminoacylated initiator tRNA (fMet-tRNA$_f^{Met}$), and the 50S subunit. Formation of the initiation complex is facilitated by three initiation factors that promote binding between the components of the complex and requires the consumption of one molecule of GTP. The proper codon for initiation is selected by an interaction between the initiation codon and the aminoacylated initiator tRNA; this interaction is augmented by base pairing between a region of the 16S rRNA and a complementary Shine-Dalgarno sequence in the mRNA molecule upstream of the initiation codon.

In the elongation stage of protein synthesis, an aminoacyl-tRNA binds initially at the A site, adjacent to a peptidyl-tRNA at the P site. Peptide-bond formation is followed by translocation of the new peptidyl-tRNA from the A site to the P site and ejection of the free tRNA from the E site. This process requires the participation of three elongation factors and the consumption of two molecules of GTP.

Termination of protein synthesis occurs at specific termination codons and requires the participation of release factors and the consumption of one molecule of GTP. Various steps of protein synthesis are inhibited by antibiotics.

Protein synthesis in eukaryotes is similar to protein synthesis in prokaryotes, particularly in chain elongation and termination. Initiation in eukaryotes requires the participation of at least eight initiation factors. One of these initiation factors interacts specifically with the 5′ cap on eukaryotic mRNA molecules. The initiation complex in eukaryotic translation is assembled at the end of the mRNA, and it moves to the first available initiation codon, where translation begins.

Translation can be regulated in several ways. Some proteins, such as ribosomal proteins, bind to the 5′ end of their mRNA and inhibit initiation. Translation can also be inhibited by antisense RNA. Modification of initiation factors can control translation initiation. Additional regulatory mechanisms include translational frameshifting and attenuation.

Many proteins are covalently modified during and after translation. Some of these modifications affect the transport of proteins to different cellular locations. A protein that is destined to be secreted or inserted into a plasma membrane often contains a hydrophobic signal peptide of 16–30 residues at its N-terminus. This peptide is recognized by signal recognition particle (SRP), which blocks further peptide-bond formation until the nascent polypeptide and ribosome are bound to the membrane of the endoplasmic reticulum. The newly synthesized protein is then translocated through a pore in the membrane into the lumen of the endoplasmic reticulum. A signal peptidase cleaves the signal peptide while protein synthesis is still in progress.

Once in the lumen of the endoplasmic reticulum, many secretory proteins are glycosylated. Oligosaccharides may be attached to serine or threonine residues (in *O*-linked glycoproteins) or to asparagine residues (in *N*-linked glycoproteins). In *N*-linked glycoproteins, the oligosaccharide chain is first synthesized as a 14-hexose chain attached to dolichol pyrophosphate. Protein-bound oligosaccharides may be modified in the Golgi apparatus. Some of these modifications direct the transport of glycoproteins to specific cellular locations.

Study Information

Figure 30·5
Comparison of eukaryotic ribosomes and prokaryotic ribosomes. Both consist of two subunits, each of which contains rRNA and proteins. The large subunit of the prokaryotic ribosome contains two molecules of rRNA: 5S and 23S. The large subunit of almost all eukaryotic ribosomes contains three molecules of rRNA: 5S, 5.8S, and 28S. The sequence of the eukaryotic 5.8S rRNA is similar to the sequence of the 5′ end of the prokaryotic 23S rRNA.

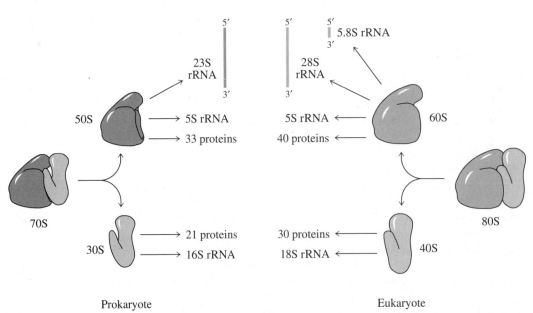

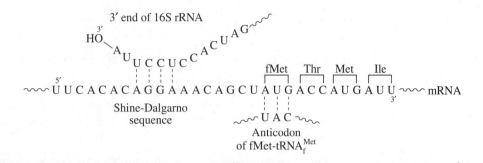

Figure 30·9
Complementary base pairing between the 3′ end of 16S rRNA and the region near the 5′ end of an mRNA. Binding of the 3′ end of the 16S rRNA to the Shine-Dalgarno sequence helps establish the correct reading frame for translation by positioning the initiation codon at the P site.

Problems

1. Bacterial genomes usually contain multiple copies of the genes for rRNA, which are transcribed very efficiently in order to produce large amounts of rRNA for assembly into ribosomes. In contrast, the genes that encode ribosomal proteins are present only as single copies. Explain the difference in the number of genes for rRNA and ribosomal proteins.

2. Calculate the number of high-energy phosphoanhydride bonds that are hydrolyzed during synthesis of a 600–amino acid residue protein in *E. coli*. Do not include the energy required to synthesize the amino acids, mRNA, tRNA, or the ribosome.

3. Many AUG codons may be present in a prokaryotic mRNA segment. How does the ribosome distinguish AUG codons specifying initiation from AUG codons specifying internal methionine?

4. Some prokaryotic mRNA molecules are polycistronic. In contrast, eukaryotic mRNA molecules are monocistronic. Explain this difference in terms of the initiation processes in prokaryotes and eukaryotes.

5. What would be the effect on polypeptide synthesis if EF-Tu recognized and formed a complex with fMet-tRNA$_f^{Met}$?

6. An error rate of about 10^{-4} per amino acid residue is normal during translation. When a slowly hydrolyzed GTP analog, GTPγ-S (Figure 12·45), is used in translation studies in vitro, the accuracy of the protein synthesis process is actually *increased*.
   (a) Explain why the accuracy is increased.
   (b) What effect does GTPγ-S have on the rate of protein synthesis?

7. The mechanism of attenuation requires the presence of a leader region. Predict what effect the following changes would have on regulation of the *trp* operon.
   (a) The entire leader region is deleted.
   (b) The sequence encoding the leader peptide is deleted.
   (c) The leader region does not contain an AUG codon.

8. In the operons that contain genes for isoleucine biosynthesis, the leader regions that precede the genes contain multiple codons that specify not only isoleucine, but valine and leucine as well. Suggest a reason why this is so.

9. Summarize the steps involved in the synthesis and processing of a glycosylated eukaryotic integral membrane protein with a C-terminal cytosolic domain and an N-terminal extracellular domain.

Solutions

1. The transcript of each rRNA gene is an rRNA molecule that is incorporated into a ribosome; therefore, multiple copies of rRNA genes are needed to assemble the large number of ribosomes that the cell requires. In contrast, the transcript of each ribosomal protein gene is an mRNA that can be translated many times. Because of this amplification of RNA to protein, fewer genes are needed for the ribosomal proteins than for rRNA.

2. Two phosphoanhydride bonds are hydrolyzed for each amino acid activated by an aminoacyl-tRNA synthetase.

$$\text{Amino acid} + \text{tRNA} + \text{ATP} \longrightarrow \text{Aminoacyl-tRNA} + \text{AMP} + \text{PP}_i$$
$$\text{PP}_i + \text{H}_2\text{O} \longrightarrow 2\,\text{P}_i$$

The rest of the energy needed for synthesis of the protein is provided by hydrolysis of GTP: one high-energy bond is hydrolyzed in the formation of the 70S initiation complex, another during the insertion of each amino acid into the A site of the ribosome, and another at each translocation step. Since the initial methionyl-tRNA is inserted into the P site, 599 new insertions and 599 translocations occur during the synthesis of a protein of 600 amino acid residues. Finally, one phosphoanhydride bond is hydrolyzed during release of the completed polypeptide chain from the ribosome. The total number of phosphoanhydride bonds hydrolyzed during synthesis of a 600-residue protein is

Activation (600×2)	1200
Initiation	1
Insertion	599
Translocation	599
Termination	1
Total	2400

3. The region of the mRNA molecule upstream of the initiation codon contains the purine-rich Shine-Dalgarno sequence, which is at least partially complementary to a pyrimidine-rich sequence at the 3′ end of the 16S rRNA component of the 30S ribosome subunit (Figure 30·9). By correctly positioning the 30S subunit on the mRNA transcript, the Shine-Dalgarno sequence allows the binding of fMet-tRNA$_f^{Met}$ to the initiation codon. Once protein synthesis begins, subsequent methionine codons are recognized by Met-tRNA^{Met}.

4. In prokaryotes, a 30S ribosomal subunit recognizes and binds the Shine-Dalgarno sequence on an mRNA molecule. Protein synthesis then begins at a nearby initiation codon. There may be multiple Shine-Dalgarno sequences, and they may be found anywhere along the mRNA molecule.

   In eukaryotes, a 40S ribosomal subunit attaches to the 5′ cap of an mRNA molecule. The 40S subunit then slides along the mRNA in the $5' \rightarrow 3'$ direction until the first AUG codon is reached. Usually, assembly of the ribosome is then completed and translation commences.

5. fMet-tRNA$_f^{Met}$ would be inserted into the A site at internal AUG codons. However, formation of a peptide bond could not occur because the amino nitrogen of methionine is blocked by the formyl group and cannot react with the aminoacyl-tRNA in the P site. Protein synthesis would be arrested at internal methionine codons unless the normal Met-tRNA^{Met} were inserted first. The length of time that the translation machinery would stall would depend on the time that fMet-tRNA$_f^{Met}$ remained associated with the complex in the absence of peptide bond formation.

6. (a) The elongation factor EF-Tu–GTP binds an aminoacyl-tRNA molecule and guides it to the A site of the ribosome. Peptide bond synthesis cannot take place until the aminoacyl-tRNA is released, GTP is hydrolyzed, and EF-Tu–GDP dissociates from the translation complex. The conformations of EF-Tu–GTP and EF-Tu–GDP differ; however, both conformations allow the correct aminoacyl-tRNA to form a stable complex with the codon on the mRNA. An incorrectly paired aminoacyl-tRNA will dissociate from the ribosome. The more time available for dissociation of incorrectly paired aminoacyl-tRNA molecules, the less likely it is that an incorrect amino acid will be incorporated into the polypeptide. Since GTPγ-S is hydrolyzed more slowly than GTP, more time is available for an incorrectly paired aminoacyl-tRNA to dissociate, and the accuracy of protein synthesis is increased.

(b) The net rate of protein synthesis is decreased in the presence of GTPγ-S since EF-Tu–GTPγ-S is hydrolyzed more slowly.

7. (a) If the entire leader region were deleted, attenuation would not be possible, and transcription would be controlled exclusively by the *trp* repressor. The overall rate of transcription of the *trp* operon would increase.

(b) If the region encoding the leader peptide were deleted, transcription would be controlled exclusively by the *trp* repressor. Deletion of the sequence encoding the leader peptide would remove sequence 1, thus allowing the stable 2-3 hairpin to form. Since neither the pause site (1-2 hairpin) nor the terminator (3-4 hairpin) could form, initiated transcripts would always continue into the *trp* operon.

(c) If the leader region did not contain an AUG codon, the operon would rarely be transcribed. Due to the absence of the initiation codon, the leader peptide would not be synthesized, and 1-2 hairpins and 3-4 hairpins would almost always form, leading to termination of transcription.

8. The presence of codons specifying valine and leucine in the leader regions of isoleucine operons suggests that a scarcity of these amino acids would prevent attenuation of the genes for isoleucine biosynthesis. Many of the enzymes involved in the biosynthesis of isoleucine are also necessary for the biosynthesis of valine and leucine (Figure 21·20). Thus, even if isoleucine concentrations were high, low concentrations of valine or leucine would ensure that transcription of the isoleucine operon did not terminate prematurely.

9. As the newly synthesized protein is extruded from the ribosome, the N-terminal signal peptide is recognized and bound by a signal-recognition particle (SRP). Further translation is inhibited until the SRP binds to its receptor on the cytosolic face of the endoplasmic reticulum. The ribosome then binds to ribophorins, which anchor the ribosome to the endoplasmic reticulum. When translation resumes, the polypeptide chain is passed through a pore into the lumen. The presence of a stop-transfer sequence in the polypeptide chain signals the transport machinery to stop passing the protein through the membrane. The result is an integral membrane protein with its N-terminus in the lumen of the endoplasmic reticulum and its C-terminus in the cytosol.

Glycosylation of specific residues takes place in the lumen of the endoplasmic reticulum and in the Golgi apparatus. The protein, still embedded in the membrane, is transported between the endoplasmic reticulum and the Golgi apparatus in transfer vesicles that bud off the endoplasmic reticulum.

Secretory vesicles transport the fully glycosylated protein from the *trans* face of the Golgi apparatus to the plasma membrane. When the vesicles fuse with the plasma membrane, the N-terminal portion of the protein, which was in the lumen, is now exposed to the extracellular space, and the C-terminal portion remains in the cytosol.

31

Gene Expression and Development

Summary

The developmental regulation of gene expression involves many of the regulatory mechanisms described in earlier chapters. During infection by bacteriophages and viruses, the temporal expression of many genes is regulated at the level of transcription initiation. The T7 genome encodes a new RNA polymerase that is synthesized early in infection. This polymerase specifically transcribes T7 late genes. Late gene expression in T4 and SP01 is controlled by novel σ factors that are encoded by phage early genes. These σ factors direct host RNA polymerase to phage late-gene promoters. During adenovirus infection, a new transcriptional activator is produced.

Sporulation in *Bacillus subtilis* is an example of differentiation. Regulation of gene expression during sporulation is controlled by the sequential production of σ factors, which is called a σ cascade. In some cases, these σ factors are active either in the mother cell or in the forespore, leading to expression of different genes in the two cells. In these cases, production of active σ factors is coupled to formation of the polar septum.

Infection of *E. coli* by bacteriophage λ is a model of developmental regulation that involves cellular decision making. Bacteriophage λ can enter a lytic or lysogenic pathway after infecting a host cell. The expression of genes during a lytic infection is regulated in part by an antitermination cascade. The products of the *N* and *Q* genes modify the host transcription complex, enabling it to bypass terminators in the λ genome.

The lysogenic pathway of λ development is controlled mainly by cI and cII. The cI repressor activates transcription of its own gene and also represses transcription of phage early genes. cII acts at several promoters to increase transcription of genes required for lysogeny. One of these genes is *c*I. Activation of the *c*I gene by cII depends on the amount of cII present in an infected cell, and this ultimately determines whether lysogeny occurs.

The decision about which developmental pathway to enter is also regulated by competition between Cro and cI as they bind to operators near the early promoters. Both Cro and cI contain helix-turn-helix DNA-binding motifs.

Drosophila development is regulated by a cascade of activators and repressors that are synthesized during embryogenesis. Synthesis of these regulatory proteins is spatially and temporally regulated. The initial pattern of gene expression is defined by gradients of protein encoded by maternally expressed genes. The first round of embryonic gene expression is regulated by the presence or absence of the maternal gene products, and the embryonic gene products are themselves regulatory proteins that influence the expression of other embryonic genes.

Study Information

Figure 31·14
Organization of the phage λ genome. The phage λ control region is enlarged to show some of the immediate-early and early gene promoters and operators. The immediate-early, early, and late genes tend to be clustered into operons. In general, transcription in the clockwise direction leads to the production of proteins involved in lytic growth—proteins for phage λ DNA replication, cell lysis, and phage capsids and tails. Transcription in the counterclockwise direction leads to the production of proteins involved in lysogeny—proteins for integration of the phage λ genome.

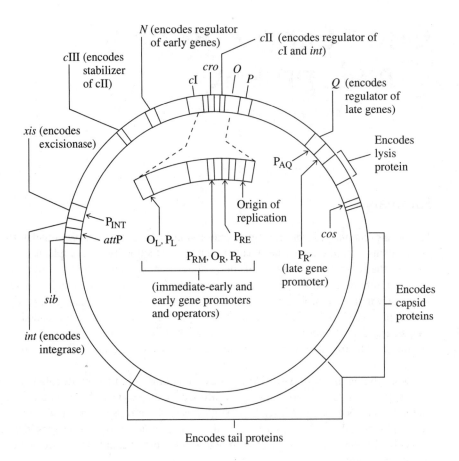

Problems

1. σ cascades in prokaryotic cells ensure that genes are expressed in a certain order. Describe how a eukaryotic cell could achieve the same effect.

2. A student has proposed a study of the SP01 σ^{gp28} gene by cloning the entire gene, including the promoter region, in an *E. coli* strain. Do you anticipate any problems when the σ^{gp28} gene from SP01 is expressed in *E. coli?*

3. The short transcript produced from P_{AQ} is not the only antisense RNA produced during the transcription of phage λ early genes. What other transcript could act as an antisense RNA? How might production of this RNA transcript help establish lysogeny?

4. Describe briefly the effects that deletion of each of the following genes or sequences would have on bacteriophage λ development: (a) *N,* (b) *Q,* (c) *c*II, (d) *c*I, (e) *cro,* (f) *c*I and *cro,* (g) *c*III, (h) *sib.*

5. It has been estimated that about 10% of the genes expressed in eukaryotes are subject to some form of translational control. Unfertilized eggs (oocytes) possess controls of this type for a number of proteins. What are the major advantages of translational control over transcriptional control in oocytes?

Solutions

1. Regulation of the temporal expression of genes in prokaryotic cells sometimes involves the sequential synthesis of σ factors. The temporal expression of eukaryotic genes could be controlled by a similar mechanism. For example, a cellular event might activate a transcription factor or activator, which could then stimulate transcription of a novel set of genes. One of these genes could in turn encode another activator whose synthesis would result in the transcription of yet another set of novel genes.

2. The gene for σ^{gp28} is efficiently transcribed in *B. subtilis.* Since *B. subtilis* and *E. coli* are similar organisms, their RNA polymerases are probably also similar. It is likely that the SP01 σ^{gp28} gene will be transcribed in *E. coli.* Expression of σ^{gp28} in *E. coli* might be detrimental since the bacteriophage σ factor might compete with σ^{70} for binding to RNA polymerase, thereby reducing the rate of transcription initiation of normal *E. coli* genes.

3. Transcription from P_{RE} produces an RNA transcript with a region that is complementary to a region of the transcript produced from P_R (Figure 31·20). If the 3′ end of the transcript produced from P_{RE} were to hybridize to the 5′ end of the transcript produced from P_R, ribosomes would not be able to initiate translation of the transcript produced from P_R. (Note that translation of the *c*I coding region would not be affected.) In this way, transcription from P_{RE}, which is activated in the presence of *c*II, helps decrease both the transcription and translation of the *cro* gene. Since Cro is involved in the establishment of lytic growth, the production of antisense RNA from P_{RE} might help in the establishment of lysogeny.

4. (a) N is an antiterminator for t_L, t_{R1}, and t_{R2}. In the absence of the *N* gene, bacteriophage λ would not be able to proceed from immediate-early to early gene transcription. The production of cII (and therefore the eventual production of cI) would be greatly impaired, and lysogeny would be difficult to establish. In addition, lytic development would not be possible, due to the absence of Q. (The *Q* gene is located past t_{R2}, and its transcription depends on antitermination in the presence of N.) Therefore, deletion of *N* would probably be lethal to the phage.

    (b) Since Q is an antiterminator necessary for transcription of the late genes from the promoter $P_{R'}$, a phage lacking the *Q* gene would not be able to extend transcription past the terminator $t_{R'}$. Production of the head and tail proteins of the phage would be prevented, and assembly of progeny phages or lysis of the host cell would not occur. However, since Xis and Int could still be produced, the genome of a phage growing lysogenically could still be excised from the *E. coli* chromosome.

(c) *c*II is an activator that stimulates transcription of the gene for the λ repressor (*c*I) and the gene for integrase (*int*). Deletion of *c*II would render the phage incapable of establishing lysogeny, since transcription of *c*I from P_{RE} and transcription of *int* from P_{INT} would be insignificant. All phages lacking *c*II would enter the lytic pathway of development.

(d) *c*I encodes the λ repressor, the protein that establishes and maintains lysogenic growth. All phages lacking *c*I would enter the lytic pathway of development.

(e) Cro requires transcription of the *c*I gene. Deletion of *cro* would allow higher levels of *c*I transcription. This might cause the phage to favor lysogeny and delay lytic growth. The pathway of development followed would depend on the stability of *c*II and the levels of *c*I that were produced before significant transcription of the phage late genes had taken place.

(f) Deletion of *c*I and *cro* would render the phage incapable of establishing lysogeny. However, phages with this defect would enter the lytic pathway more slowly than wild-type phages. The N and Q antiterminators would permit transcription of the late genes from $P_{R'}$.

(g) Deletion of *c*III would result in decreased stability of *c*II, decreased ability of the phage to monitor cellular conditions, and increased likelihood of lytic development.

(h) Elimination of the *sib* region would allow the production of integrase even after induction of the lytic genes and excision of the phage genome from the *E. coli* chromosome. The presence of integrase at this stage of development might cause some reintegration of the phage genome into the *att*B site and therefore might interfere with the final stages of lytic development.

5. There are three reasons why translational regulation of gene expression might be advantageous in oocytes. First, the concentration of a protein under translational control can be quickly changed without the need for rapid transcription or degradation of existing mRNA. Translational regulation is therefore more sensitive than transcriptional regulation.

   Second, translational regulation in oocytes allows rapid DNA replication and mitosis during embryogenesis. Most genes cannot be transcribed when DNA is being replicated or when DNA is condensed in metaphase chromosomes.

   Finally, translational regulation in oocytes enables other cells to synthesize specific mRNA molecules and transport them into the oocyte during oogenesis. These mRNA molecules are translated following fertilization. For example, during oogenesis in *Drosophila,* nurse cells deposit Bicoid mRNA at one end of the oocyte. Translation of Bicoid mRNA during embryogenesis yields Bicoid protein, which regulates genes involved in *Drosophila* development.

32

Genes and Genomes

Summary

Many bacteriophages and viruses can integrate into host chromosomes. Bacteriophage λ recombines with a specific site on the *E. coli* chromosome, but bacteriophage Mu can recombine at many sites. Bacteriophage Mu can also transpose to different sites by a mechanism that requires DNA replication and resolution of a cointegrate. Bacteriophage Mu is a class III transposon.

Retroviruses are RNA viruses that are copied into double-stranded DNA for integration into host chromosomes via nonspecific recombination. The ends of retroviral DNA consist of long terminal repeats (LTR) that contain the promoter for the retroviral genes *gag, pol,* and *env.*

Class I bacterial transposons are either simple *IS* elements or composite transposons with two *IS* elements flanking another gene. Class I transposons integrate at specific target sites recognized by the encoded transposase, leaving a short duplication of the host target sequence. Class II bacterial transposons contain a minimum of two genes: a transposase and a resolvase. Integration requires DNA replication and the resolution of a cointegrate by site-specific recombination.

There are three classes of eukaryotic transposon. The retrovirus-like transposons are probably derived from retroviruses because they contain genes homologous to *gag, pol,* and sometimes *env.* Transposition occurs when the transposon is transcribed, copied into double-stranded DNA, and then recombined with the host chromosome. The *Ac*-like transposons are similar to the bacterial class I transposons. They contain a transposase that promotes recombination. The third class of eukaryotic transposons are the retroposons.

Recombination between host DNA and a bacteriophage, virus, or transposon can result in disruption of a host gene. Prokaryotic genomes may include cryptic phages, the remnants of permanently integrated, mutated phage genomes. Both prokaryotic and eukaryotic chromosomes may contain multiple sequences derived from transposons.

Bacterial genomes are compact. There is little space between genes, and many genes are organized into operons. The sizes of bacterial genomes range from about 600 kb to over 12 000 kb. The *E. coli* genome, with about 3000 genes, is of average size for a prokaryote. There are very few examples of gene families or multiple copies of genes in bacterial chromosomes. Some bacteria contain more than one chromosome.

Eukaryotic genomes, which are larger and more complex than bacterial genomes, are organized into several linear chromosomes. The number of chromosomes varies considerably; even closely related species can have different numbers of chromosomes. There are many repetitive sequences in eukaryotic genomes. Some of these are large blocks of simple sequences found at centromeres and telomeres. Others are derived from transposons, such as the retroposons. Eukaryotic chromosomes often contain families of related genes and multiple copies of genes. In addition, the genomes of many multicellular eukaryotes contain pseudogenes and genes with introns. The available evidence suggests that introns in protein-encoding genes arose late in evolution.

The size of eukaryotic genomes is highly variable. Unicellular eukaryotes have compact genomes, much like those of bacteria. Yeast may have only three times as many genes as *E. coli*. The number of genes in more complex organisms such as *Drosophila* may not be much greater, but the distance between genes is greater. There may be about 50 000 genes in mammals.

Gene rearrangements are associated with expression and regulation of some genes. Recombination controls expression of flagellar genes in *Salmonella* and mating-type switching in the yeast *Saccharomyces cerevisiae*. In some terminally differentiated cells, genomic rearrangements are required for expression; mammalian immunoglobulin genes are an example. Genes can also be selectively amplified during development.

Problems

1. The human immunodeficiency virus (HIV) is a retrovirus. The RNA genome of the virus is converted to a double-stranded DNA molecule by the action of reverse transcriptase. Subsequent integration into host DNA, transcription, splicing, and translation lead to the production of new viruses. Antisense RNA molecules have been investigated as potential antiviral agents. At what steps in the retrovirus cycle would antisense RNA molecules be most effective as inhibitors of viral replication?

2. A scientist is studying two strains (A and B) of bacteria with mutations in the *lac* operon. When the strains are grown for many generations, a few cells with normal *lac* operons are obtained. When the strains are grown in the presence of methyl methanesulfonate (an alkylating agent), reversion of the mutation in strain A occurs much sooner, while the reversion rate in strain B is unchanged. In which strain is the mutation in the *lac* operon more likely to be caused by a transposon?

3. Given what you know about reading frames, explain why the development of introns in protein-encoding genes may have been advantageous to rapidly evolving organisms.

4. The *E. coli* genome is approximately 4600 kb in size and contains about 3000 genes. The mammalian genome is approximately 3.2×10^6 kb in size and contains about 50 000 genes. An average gene in *E. coli* is 1500 base pairs long.
   (a) Calculate the percentage of *E. coli* DNA that is not transcribed.
   (b) Although many mammalian genes are larger than bacterial genes, most mammalian gene products are the same size as bacterial gene products. Calculate the percentage of DNA in exons in the mammalian genome.

Solutions

1. Steps involving single-stranded RNA are vulnerable to inhibition by antisense RNA since antisense RNA hybridizes directly with complementary single-stranded RNA sequences. The action of reverse transcriptase, splicing of viral RNA, and translation all involve single-stranded RNA and could therefore be inhibited by formation of a double-stranded RNA complex.

2. Strain B is more likely to harbor a transposon whose insertion into the *lac* operon caused the mutation. Transposons revert spontaneously but at a low rate that is not increased by agents that damage DNA, such as alkylating agents. Other types of mutation, such as base substitutions or small insertions or deletions, are more easily reversed by agents that induce DNA damage.

3. A new gene can be created by combining the coding regions of two different genes. Shuffling the coding regions of the original genes creates a new coding region only if the two genes have the same orientation after recombination (a 1:2 probability), and if the recombination event does not alter the reading frame of the downstream gene (a 1:3 probability). Thus, only one-sixth of all random recombinations of coding regions would be expected to result in a hybrid gene that could be correctly translated. Of course, no new protein would result if recombination introduced a stop codon.

   The presence of introns increases the success rate of recombination events by decreasing the need for precise alignment of reading frames. Exons can be shuffled and joined at random as long as DNA segments are joined at sites that fall within introns and the exons have the same orientation. The spliceosome can then correctly align the reading frames during RNA processing. The increased rate of successful recombination due to the presence of introns would have allowed more rapid evolution.

4. (a) Since the average *E. coli* gene is 1.5 kb (1500 bp) in length, 3000 genes account for about 4500 kb of DNA (3000 genes × 1.5 kb per gene). The remaining 100 kb of the 4600 kb bacterial genome is not transcribed. This 100 kb represents less than 2% of the genome (100 kb ÷ 4600 kb, or 0.02). Some of the nontranscribed DNA consists of promoters and regions that regulate transcription initiation.
   (b) Since the gene products in mammals and bacteria are similar in size, the amount of DNA in the exons of a typical mammalian gene must also be 1500 bp. The total amount of DNA in exons is thus 50 000 genes × 1.5 kb per gene, or 75 000 kb. This amount of DNA represents about 2% of the mammalian genome (7.5×10^4 kb ÷ 3.2×10^6 kb, or 0.02). The remaining 98% of DNA consists of introns, regulatory sequences, and DNA whose function, if any, is not known.

33

Recombinant DNA Technology

Summary

Recombinant DNA technology allows scientific questions that arise within the classic disciplines of biochemistry, genetics, biology, botany, and physiology to be pursued in the areas of molecular biology and chemistry. A central technique of the technology is that of molecular cloning, whereby DNA sequences of interest from a particular organism are isolated and propagated to provide quantities of the DNA suitable for subsequent studies.

Recombinant DNA technology builds upon a few basic techniques: isolation of DNA, cleavage of DNA at particular sequences, ligation of DNA fragments, introduction of DNA into host cells, replication and expression of DNA, and identification of host cells that contain recombinants.

Cleavage of DNA usually involves the use of restriction endonucleases. Ligation of DNA fragments is usually accomplished using DNA ligase. During the generation of recombinant DNA molecules, DNA fragments of interest are inserted into engineered cloning vectors. Vectors can be plasmids, bacteriophages, cosmids, or genomic DNA molecules. Shuttle vectors are used to move DNA among different hosts, and expression vectors are used when a protein-encoding gene is to be transcribed and translated in a host cell.

The introduction of DNA into host cells is called transformation when the DNA is introduced directly and transfection when the DNA is introduced via a phage vector. Once inside a host cell, recombinant DNA molecules are replicated. Host cells containing recombinant DNA molecules are enriched using selections, and populations of hosts that contain particular recombinant DNA molecules are identified using screens. Marker genes are often used in selections and screens.

DNA libraries are collections of populations of recombinant DNA molecules. DNA libraries can be made from genomic DNA or cDNA. Probes are used to identify particular cells that carry a recombinant DNA molecule of interest. Probes may be composed of nucleic acid, or they may be antibodies.

Chromosome walking is used to order fragments of genomic DNA. The polymerase chain reaction is used to amplify particular sequences of interest in DNA samples. Inducible expression vectors are used to overproduce harmful proteins in host cells. Additional techniques, such as DNA sequencing, chemical DNA synthesis, protein sequencing, and site-directed mutagenesis, allow molecular biologists to characterize virtually any gene or gene product in molecular terms. The ability to manipulate DNA in these ways has led to a rapidly increasing understanding of the basic biology of cells and to the production of materials useful for medicine, industry, and agriculture.

Study Information

Table 33·1 Common cloning vectors

Name	Type	Host cells	Selection/screening methods	Common uses
pBR322	Plasmid	*E. coli*	amp^R, tet^R	Library construction
pUC18	Plasmid	*E. coli*	amp^R, blue/white	Expression vector
M13mp18	Bacteriophage	*E. coli*	Blue/white	DNA sequencing, mutagenesis
pEMBL18	Plasmid	*E. coli*	amp^R, blue/white	Expression vector, DNA sequencing, mutagenesis
pcos2EMBL	Cosmid	*E. coli*	amp^R, kan^R	Library construction
λEMBL3	Bacteriophage	*E. coli*		Library construction
λgt11	Bacteriophage	*E. coli*	Blue/white	Expression vector
pMMB33	Cosmid	Gram-negative bacteria	amp^R, kan^R	Transfer of genomic DNA among gram-negative bacteria
pAH9	Plasmid	*E. coli*, yeast	amp^R, *LEU2*	Yeast expression vector, shuttle vector
pMH2	Plasmid	*E. coli*, *Agrobacterium*	amp^R, $neomycin^R$	Plant expression vector, shuttle vector
pSV2-DHFR	Plasmid	*E. coli*, monkey, Chinese hamster ovary cells	amp^R, $methotrexate^R$	Mammalian expression vector, shuttle vector

Problems

1. The C-terminal amino acid sequence and coding DNA sequence of *E. coli* dihydrofolate reductase is

$$\text{Phe — Glu — Ile — Leu — Glu — Arg — Arg — COO}^{\ominus}$$

$$\text{T T T | G A G | A T T | C T G | G A G | C G G | C G G}$$

Describe the mutations that could be made in this sequence to introduce either a *Bst*BI restriction site

$$\downarrow$$
$$\text{T T C G A A}$$

or a *Bgl*II restriction site

$$\downarrow$$

A G A T C T

without changing the amino acid sequence of the protein. How would you introduce these mutations?

2. Explain why the most useful plasmid vectors contain at least one unique restriction endonuclease recognition site.

3. The N-terminal amino acid sequence of a desired protein has been determined to be Met–Asp–Val–Leu–Tyr–Phe–Ala–His–Asn–Pro–Arg–Trp–. Describe the DNA sequences of the longest possible mixed oligonucleotide probe with a degeneracy under 100 that will hybridize to both the mRNA and cDNA corresponding to this protein.

4. A cloning vector is cut with the restriction endonuclease *Sma*I, whose restriction site is

$$\downarrow$$

C C C G G G

and treated with alkaline phosphatase. The synthetic, double-stranded DNA molecule with sequence

5′ Ⓟ–G G A C T T A C T A C C C A A G T A – OH 3′
3′ OH – C C T G A A T G A T G G G T T C A T – Ⓟ 5′

is inserted into the *Sma*I site, whose location within the β-galactosidase gene in the cloning vector is shown below.

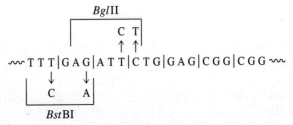

⊕
H₃N – Met — Thr — Met — Ile — Thr — Asn — Ser — Arg — Gly 〰〰〰⟶ functional
β-galactosidase

〰A T G|A C C|A T G|A T T|A C G|A A T|T C C|C G G|G G A〰

*Sma*I

Upon transformation of the ligation mixture into *E. coli* and in the presence of the indicator X-gal, 52% of the plaques on the plate are blue and 48% are clear. Give the most likely explanation for this result.

5. Suppose you want to express the β-globin gene using an expression vector. What factors are important for maximizing the efficiency of transcription and translation if the expression vector is (a) a eukaryotic vector, (b) a prokaryotic vector, (c) a fusion-protein vector?

Solutions

1. The required changes are shown below:

*Bgl*II

C T
↑ ↑
〰T T T|G A G|A T T|C T G|G A G|C G G|C G G〰
↓ ↓
C A

*Bst*BI

To create a *Bst*BI site, replace the TTT phenylalanine codon with the TTC phenylalanine codon and replace the first GAG glutamate codon with the GAA glutamate codon. To create a *Bgl*II site, replace the ATT isoleucine codon with the ATC isoleucine codon and the CTG leucine codon with the

TTG leucine codon. These mutations can be introduced by synthesizing an oligonucleotide containing the changes and using it in oligonucleotide-directed, site-specific mutagenesis as illustrated in Figure 33·26.

2. In order to insert foreign DNA into a circular vector, it is necessary to first linearize the vector. The most efficient way to linearize a plasmid vector is with a restriction endonuclease that catalyzes cleavage at only one site. This ensures that the foreign DNA is inserted at only that site (for example, within a selectable marker gene). Furthermore, since the sequence of the insertion site is known, the resulting recombinant can be manipulated in several ways. For example, if the foreign DNA is cleaved with a different restriction endonuclease before ligation to the vector, loss of the unique restriction site can be used to screen for recombinants.

3. First, write all the mRNA sequences that code for the peptide sequence (refer to Figure 29·3 for the standard genetic code). The coding strand of the cDNA that corresponds to this mRNA sequence has T in place of U.

Next, identify the stretch of mRNA with the least degeneracy. mRNA coding for leucine or arginine is highly degenerate since each of these amino acids is specified by six different codons. In contrast, mRNA coding for tyrosine, phenylalanine, tryptophan, histidine, methionine, aspartate, or asparagine is much less degenerate since these amino acids are specified by only one or two codons.

Finally, determine how many degenerate positions can be included if the degeneracy of the probe is to be kept under 100. Each position for which two nucleotides are possible requires two probes and has a degeneracy of two; each position for which three codons are possible requires three probes and has a degeneracy of three, and so on. The sequence from the tyrosine codon to the asparagine codon includes four positions with two possible nucleotides and one position with four possible nucleotides The degeneracy of a complementary probe would therefore be $2^4 \times 4^1 = 64$. Since extending the probe into the degenerate positions in the proline or leucine codons would increase the degeneracy to more than 100, the longest mixed oligonucleotide probe possible is 17 nucleotides long and extends from the first nucleotide of the tyrosine codon through the second nucleotide of the proline codon.

```
    ⊕
H3N – Met — Asp — Val — Leu — Tyr — Phe — Ala — His — Asn — Pro — Arg — Trp ~

mRNA sequence AUG|GAU|GUU|CUU|UAU|UUU|GCU|CAU|AAU|CCU|AGA|UAU~
                    C   C   C   C   C   C   C   C   C   G   C
                        G   A           A           A CGU
                        A   G           G           G   C
                          UCU                           A
                            C                           G

      Mixed
oligonucleotide      5′ ATA|AAA|CGA|GTA|TTA|GG 3′
     probe                G   G   G   G   G
                                  T
                                  C
```

4. The synthetic DNA is inserted into the *Sma*I site in both orientations. In either orientation, the reading frame of the β-galactosidase gene is maintained. In one orientation, the insert is translated and functional β-galactosidase is produced. Plaques arising from this ligation are blue.

```
                                                                functional
 — Ser — Arg — Thr — Tyr — Tyr — Pro — Ser — Arg — Gly ~~~~>  β-galactosidase
5′~ TCC|CGG|ACT|TAC|TAC|CCA|AGT|AGG|GGA ~ 3′
```

In the opposite orientation, however, two in-frame stop codons (TAG, TAA) halt translation of the β-galactosidase gene; plaques arising from this ligation are clear.

no β-galactosidase

— Ser — Leu — Leu — Gly — Stop — Stop —

5′ ⁓ T C C | C T A | C T T | G G G | T A G | T A A | G T C | C G G | G G A ⁓ 3′

A 50/50 distribution of blue plaques and white plaques is expected. The slight majority of blue plaques most likely arises either from vector DNA that was not cut with *Sma*I or from cleaved vector DNA from which alkaline phosphatase failed to remove the phosphate groups (and which subsequently reclosed without the insert DNA).

5. (a) The vector must contain a strong promoter upstream of the gene. Also, efficient transcription often requires an appropriate transcription termination site and a polyadenylation signal at the 3′ end of the gene.

(b) If the expression vector is to be transformed into a prokaryotic cell, then the DNA fragment containing the β-globin gene must be a cDNA fragment, which contains no introns, since prokaryotes cannot carry out splicing. A promoter that is active in the host cell must also be present. Moreover, efficient translation of the protein requires the presence of a strong ribosome-binding site located at the appropriate distance from the initiation codon. Finally, if β-globin is toxic to the prokaryotic host, the promoter should be inducible.

(c) If the expression vector is a fusion-protein vector, then most of the transcriptional and translational apparatus should already be in place. Efficient expression simply requires that the coding sequence of the β-globin gene be in-frame with and in the same orientation as the coding sequence of the fusion protein.

Appendix

Table A·1 Some physical constants used in biochemistry

R	Universal gas constant	$8.315 \text{ J K}^{-1} \text{ mol}^{-1}$
$\mathcal{F}$	Faraday's constant	$96.48 \text{ kJ V}^{-1} \text{ mol}^{-1}$
N	Avogadro's number	$6.022 \times 10^{23} \text{ mol}^{-1}$

Table A·2 Greek alphabet

A	α	alpha	N	ν	nu
B	β	beta	Ξ	ξ	xi
Γ	γ	gamma	O	o	omicron
Δ	δ	delta	Π	π	pi
E	ε	epsilon	P	ρ	rho
Z	ζ	zeta	Σ	σ	sigma
H	η	eta	T	τ	tau
Θ	θ	theta	Y	υ	upsilon
I	ι	iota	Φ	ϕ	phi
K	κ	kappa	X	χ	chi
Λ	λ	lambda	Ψ	ψ	psi
M	μ	mu	Ω	ω	omega

Table A·3 Atomic numbers and weights of the elements

Element	Symbol	Atomic number	Atomic weight	Element	Symbol	Atomic number	Atomic weight
Actinium	Ac	89	227.03	Mendelevium	Md	101	255.09
Aluminum	Al	13	26.98	Mercury	Hg	80	200.59
Americium	Am	95	243.06	Molybdenum	Mo	42	95.94
Antimony	Sb	51	121.75	Neodymium	Nd	60	144.24
Argon	Ar	18	39.95	Neon	Ne	10	20.18
Arsenic	As	33	74.92	Neptunium	Np	93	237.05
Astatine	At	85	210.99	Nickel	Ni	28	58.71
Barium	Ba	56	137.34	Niobium	Nb	41	92.91
Berkelium	Bk	97	247.07	Nitrogen	N	7	14.01
Beryllium	Be	4	9.01	Nobelium	No	102	255
Bismuth	Bi	83	208.98	Osmium	Os	76	190.20
Boron	B	5	10.81	Oxygen	O	8	16.00
Bromine	Br	35	79.90	Palladium	Pd	46	106.40
Cadmium	Cd	48	112.40	Phosphorus	P	15	30.97
Calcium	Ca	20	40.08	Platinum	Pt	78	195.09
Californium	Cf	98	249.07	Plutonium	Pu	94	242.06
Carbon	C	6	12.01	Polonium	Po	84	208.98
Cerium	Ce	58	140.12	Potassium	K	19	39.10
Cesium	Cs	55	132.91	Praseodymium	Pr	59	140.91
Chlorine	Cl	17	35.45	Promethium	Pm	61	145
Chromium	Cr	24	52.00	Protactinium	Pa	91	231.04
Cobalt	Co	27	58.93	Radium	Ra	88	226.03
Copper	Cu	29	63.55	Radon	Rn	86	222.02
Curium	Cm	96	245.07	Rhenium	Re	75	186.20
Dysprosium	Dy	66	162.50	Rhodium	Rh	45	102.91
Einsteinium	Es	99	254.09	Rubidium	Rb	37	85.47
Erbium	Er	68	167.26	Ruthenium	Ru	44	101.07
Europium	Eu	63	151.96	Samarium	Sm	62	150.40
Fermium	Fm	100	252.08	Scandium	Sc	21	44.96
Fluorine	F	9	18.99	Selenium	Se	34	78.96
Francium	Fr	87	223.02	Silicon	Si	14	28.09
Gadolinium	Gd	64	157.25	Silver	Ag	47	107.87
Gallium	Ga	31	69.72	Sodium	Na	11	22.99
Germanium	Ge	32	72.59	Strontium	Sr	38	87.62
Gold	Au	79	196.97	Sulfur	S	16	32.06
Hafnium	Hf	72	178.49	Tantalum	Ta	73	180.95
Helium	He	2	4.00	Technetium	Tc	43	98.91
Holmium	Ho	67	164.93	Tellurium	Te	52	127.60
Hydrogen	H	1	1.01	Terbium	Tb	65	158.93
Indium	In	49	114.82	Thallium	Tl	81	204.37
Iodine	I	53	126.90	Thorium	Th	90	232.04
Iridium	Ir	77	192.22	Thulium	Tm	69	168.93
Iron	Fe	26	55.85	Tin	Sn	50	118.69
Khurchatovium	Kh	104	260	Titanium	Ti	22	47.90
Krypton	Kr	36	83.80	Tungsten	W	74	183.85
Lanthanum	La	57	138.91	Uranium	U	92	238.03
Lawrencium	Lr	103	256	Vanadium	V	23	50.94
Lead	Pb	82	207.20	Xenon	Xe	54	131.30
Lithium	Li	3	6.94	Ytterbium	Yb	70	173.04
Lutetium	Lu	71	174.97	Yttrium	Y	39	88.91
Magnesium	Mg	12	24.31	Zinc	Zn	30	65.37
Manganese	Mn	25	54.94	Zirconium	Zr	40	91.22

Table A·4 General formulas of organic compounds, important functional groups, and linkages common in biochemistry

(a) *Organic compounds*

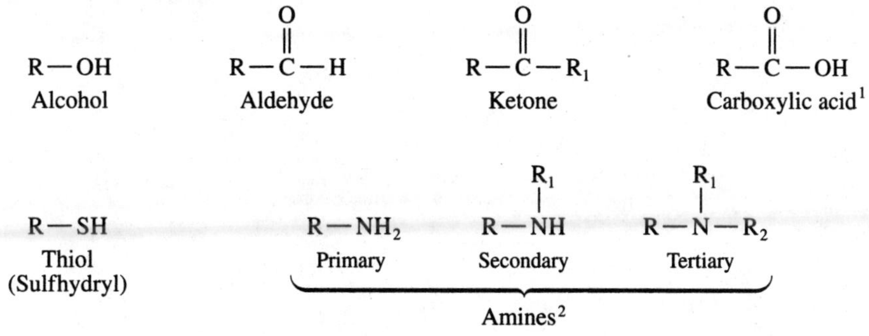

R—OH — Alcohol

R—C(=O)—H — Aldehyde

R—C(=O)—R₁ — Ketone

R—C(=O)—OH — Carboxylic acid[1]

R—SH — Thiol (Sulfhydryl)

R—NH₂ — Primary

R—NH—R₁ — Secondary

R—N(R₁)—R₂ — Tertiary

Amines[2]

(b) *Functional groups*

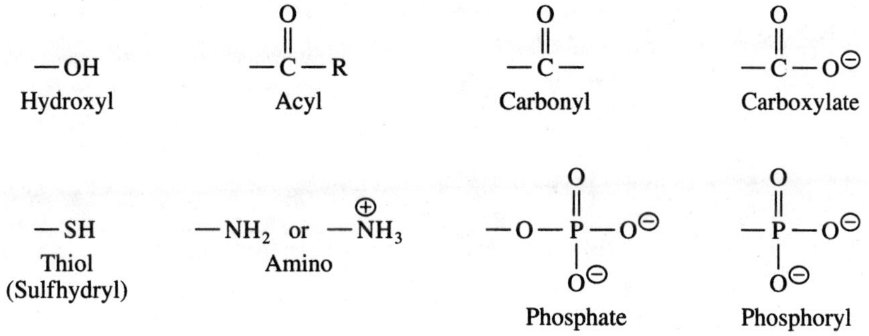

—OH — Hydroxyl

—C(=O)—R — Acyl

—C(=O)— — Carbonyl

—C(=O)—O⁻ — Carboxylate

—SH — Thiol (Sulfhydryl)

—NH₂ or —NH₃⁺ — Amino

—O—P(=O)(O⁻)—O⁻ — Phosphate

—P(=O)(O⁻)—O⁻ — Phosphoryl

(c) *Linkages*

—C—O—C(=O)— — Ester

—C—O—C— — Ether

—N—C(=O)— — Amide

—C—O—P(=O)(O⁻)—O⁻ — Phosphate ester

—O—P(=O)(O⁻)—O—P(=O)(O⁻)—O— — Phosphoanhydride

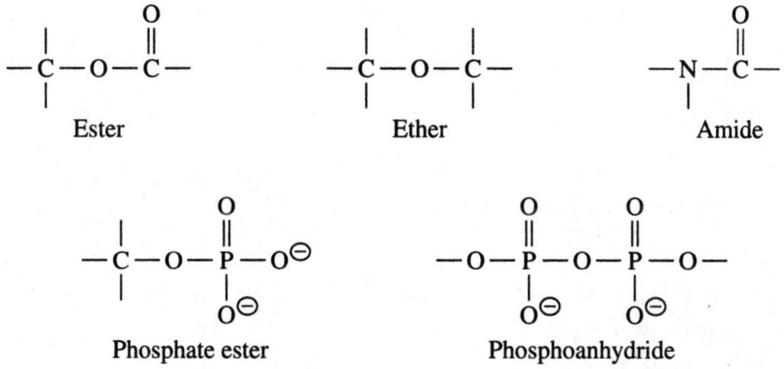

[1] Under most biological conditions, carboxylic acids exist as carboxylate anions: R—C(=O)—O⁻.

[2] Amines can also be protonated: R—NH₃⁺, R—NH₂⁺—R₁, and R—NH⁺—R₂ (with R₁).

Table A·5 Units commonly used in biochemistry

Physical quantity	SI unit	Symbol
Length	Meter	m
Mass	Kilogram	kg
Energy	Joule	J
Electric potential	Volt	V
Time	Second	s
Temperature	Kelvin*	K

*273 K = 0°C.

Table A·6 Prefixes commonly used with SI units

Prefix	Symbol	Multiplication factor
Mega	M	10^6
Kilo	k	10^3
Milli	m	10^{-3}
Micro	μ	10^{-6}
Nano	n	10^{-9}
Pico	p	10^{-12}
Femto	f	10^{-15}

Table A·7 Dissociation constants and pK_a values of weak acids in aqueous solutions at 25°C

Acid	K_a (M)	pK_a
HCOOH (Formic acid)	1.77×10^{-4}	3.8
CH_3COOH (Acetic acid)	1.76×10^{-5}	4.8
$CH_3CHOHCOOH$ (Lactic acid)	1.37×10^{-4}	3.9
H_3PO_4 (Phosphoric acid)	7.52×10^{-3}	2.2
$H_2PO_4^{\ominus}$ (Dihydrogen phosphate ion)	6.23×10^{-8}	7.2
$HPO_4^{2\ominus}$ (Monohydrogen phosphate ion)	2.20×10^{-13}	12.7
H_2CO_3 (Carbonic acid)	4.30×10^{-7}	6.4
$HCO_3^{\ominus}$ (Bicarbonate ion)	5.61×10^{-11}	10.2
$NH_4^{\oplus}$ (Ammonium ion)	5.62×10^{-10}	9.2
$CH_3NH_3^{\oplus}$ (Methylammonium ion)	2.70×10^{-11}	10.7

Table A·8 pK_a values of some commonly used buffers

Buffer	pK_a at 25°C
Phosphate (pK_1)	2.2
Acetate	4.8
MES (2-(N-Morpholino)ethanesulfonic acid)	6.1
Citrate (pK_3)	6.4
PIPES (Piperazine-N,N'-bis(2-ethanesulfonic acid))	6.8
Phosphate (pK_2)	7.2
HEPES (N-2-Hydroxyethylpiperazine-N'-2-ethanesulfonic acid)	7.5
Tris (Tris(hydroxymethyl)aminomethane)	8.1
Glycylglycine	8.2
Glycine (pK_2)	9.8

[Adapted from Stoll, V. S., and Blanchard, J. S. (1990). Buffers: principles and practice. *Methods Enzymol.* 182:24–38.]

Table A·9 One- and three-letter abbreviations for amino acids

A	Ala	Alanine
B	Asx	Asparagine or aspartate
C	Cys	Cysteine
D	Asp	Aspartate
E	Glu	Glutamate
F	Phe	Phenylalanine
G	Gly	Glycine
H	His	Histidine
I	Ile	Isoleucine
K	Lys	Lysine
L	Leu	Leucine
M	Met	Methionine
N	Asn	Asparagine
P	Pro	Proline
Q	Gln	Glutamine
R	Arg	Arginine
S	Ser	Serine
T	Thr	Threonine
V	Val	Valine
W	Trp	Tryptophan
Y	Tyr	Tyrosine
Z	Glx	Glutamate or glutamine

Table A·10 Abbreviations for some monosaccharides and their derivatives

Monosaccharide or derivative	Abbreviation
Pentoses	
Arabinose	Ara
Ribose	Rib
Xylose	Xyl
Hexoses	
Fructose	Fru
Galactose	Gal
Glucose	Glc
Mannose	Man
Deoxy sugars	
Abequose	Abe
Fucose	Fuc
Rhamnose	Rha
Amino sugars	
Glucosamine	GlcN
Galactosamine	GalN
N-Acetylglucosamine	GlcNAc
N-Acetylgalactosamine	GalNAc
N-Acetylneuraminic acid	NeuNAc
N-Acetylmuramic acid	MurNAc
N-Acetylglucosamine 6-sulfate	GlcNAc-6-SO_4
Sugar acids	
Glucuronic acid	GlcA
Iduronic acid	IdoA

Table A·11 Nomenclature of bases, nucleosides, and nucleotides

Base	Ribonucleoside	Ribonucleotide (5′-monophosphate)
Adenine (A)	Adenosine	Adenosine 5′-monophosphate (AMP); adenylate*
Guanine (G)	Guanosine	Guanosine 5′-monophosphate (GMP); guanylate*
Cytosine (C)	Cytidine	Cytidine 5′-monophosphate (CMP); cytidylate*
Uracil (U)	Uridine	Uridine 5′-monophosphate (UMP); uridylate*

Base	Deoxyribonucleoside	Deoxyribonucleotide (5′-monophosphate)
Adenine (A)	Deoxyadenosine	Deoxyadenosine 5′-monophosphate (dAMP); deoxyadenylate*
Guanine (G)	Deoxyguanosine	Deoxyguanosine 5′-monophosphate (dGMP); deoxyguanylate*
Cytosine (C)	Deoxycytidine	Deoxycytidine 5′-monophosphate (dCMP); deoxycytidylate*
Thymine (T)	Deoxythymidine or thymidine	Deoxythymidine 5′-monophosphate (dTMP); deoxythymidylate* or thymidylate*

*Anionic forms of phosphate esters predominant at pH 7.4

Table A·12 Some common fatty acids (anionic forms) incorporated in membrane lipids

Number of carbons	Number of double bonds	Common name	IUPAC name	Melting point, °C	Molecular formula
12	0	Laurate	Dodecanoate	44	$CH_3(CH_2)_{10}COO^{\ominus}$
14	0	Myristate	Tetradecanoate	52	$CH_3(CH_2)_{12}COO^{\ominus}$
16	0	Palmitate	Hexadecanoate	63	$CH_3(CH_2)_{14}COO^{\ominus}$
18	0	Stearate	Octadecanoate	70	$CH_3(CH_2)_{16}COO^{\ominus}$
20	0	Arachidate	Eicosanoate	75	$CH_3(CH_2)_{18}COO^{\ominus}$
22	0	Behenate	Docosanoate	81	$CH_3(CH_2)_{20}COO^{\ominus}$
24	0	Lignocerate	Tetracosanoate	84	$CH_3(CH_2)_{22}COO^{\ominus}$
16	1	Palmitoleate	$cis\text{-}\Delta^9$-Hexadecenoate	− 0.5	$CH_3(CH_2)_5CH=CH(CH_2)_7COO^{\ominus}$
18	1	Oleate	$cis\text{-}\Delta^9$-Octadecenoate	13	$CH_3(CH_2)_7CH=CH(CH_2)_7COO^{\ominus}$
18	2	Linoleate	$cis, cis\text{-}\Delta^{9,12}$-Octadecadienoate	− 9	$CH_3(CH_2)_4(CH=CHCH_2)_2(CH_2)_6COO^{\ominus}$
18	3	Linolenate	all $cis\text{-}\Delta^{9,12,15}$-Octadecatrienoate	−17	$CH_3CH_2(CH=CHCH_2)_3(CH_2)_6COO^{\ominus}$
20	4	Arachidonate	all $cis\text{-}\Delta^{5,8,11,14}$-Eicosatetraenoate	−49	$CH_3(CH_2)_4(CH=CHCH_2)_4(CH_2)_2COO^{\ominus}$

[Values from Dawson, R. M. C., Elliott, D. C., Elliott, W. H., and Jones, K. M. (1986). *Data for Biochemical Research,* 3rd ed. (Oxford: Clarendon Press).]

Common Abbreviations

[Abbreviations for amino acids, monosaccharides, purine and pyrimidine bases, and elements appear in the appendix.]

ACP	acyl carrier protein
ADP	adenosine 5'-diphosphate
AIDS	acquired immune deficiency syndrome
AMP	adenosine 5'-monophosphate
APS	adenosine 5'-phosphosulfate
ARS	autonomously replicating sequence
ATP	adenosine 5'-triphosphate
AZT	3'-azido-2',3'-deoxythymidine
bp	base pair
2,3BPG	2,3-*bis*phosphoglycerate
CAM	Crassulacean acid metabolism
cAMP	3',5'-cyclic adenosine monophosphate
CAT	carnitine acyltransferase
CBP	cap-binding protein
cDNA	complementary DNA
CDP	cytidine 5'-diphosphate
CMP	cytidine 5'-monophosphate
CoA	coenzyme A
CRP	cAMP regulatory protein

CTF	CCAAT box transcription factor
CTP	cytidine 5′-triphosphate
D	dihydrouridylate
ddNTP	dideoxynucleoside triphosphate
DFP	diisopropyl fluorophosphate
DHAP	dihydroxyacetone phosphate
DNA	deoxyribonucleic acid
dNTP	deoxynucleoside triphosphate
DTT	dithiothreitol
E	reduction potential
$E^{\circ\prime}$	standard reduction potential
EF	elongation factor
emf	electromotive force
$\mathcal{F}$	Faraday's constant
F1,6BP	fructose 1,6-*bis*phosphate
F6P	fructose 6-phosphate
FAD	flavin adenine dinucleotide
$FADH_2$	flavin adenine dinucleotide (reduced form)
Fd	ferredoxin
FDNB	1-fluoro-2,4-dinitrobenzene
FMN	flavin mononucleotide
$FMNH_2$	flavin mononucleotide (reduced form)
ΔG	free-energy change
$\Delta G^{\circ\prime}$	standard free-energy change
G3P	glyceraldehyde 3-phosphate
G6P	glucose 6-phosphate
GDP	guanosine 5′-diphosphate
GMP	guanosine 5′-monophosphate
GPI	glycosylphosphatidylinositol
GSH	glutathione
GSSH	glutathione disulfide
GTP	guanosine 5′-triphosphate
H	enthalpy
Hb	hemoglobin
HDL	high density lipoprotein
HIV	human immunodeficiency virus
HMG CoA	3-hydroxy 3-methylglutaryl CoA
HPLC	high-pressure liquid chromatography
I	inosinate
ICR	internal control region
IF	initiation factor
IGS	internal guide sequence

IP_3	inositol 1,4,5-*tris*phosphate
K_a	acid dissociation constant
kb	kilobase pair
k_{cat}	catalytic constant
K_{eq}	equilibrium constant
K_i	inhibition constant
K_m	Michaelis constant
K_{tr}	transport constant
K_w	ion product of water
L	linking number
LDL	low density lipoprotein
LHC	light harvesting complex
LTR	long terminal repeat
Mb	myoglobin
MOI	multiplicity of infection
M_r	relative molecular mass
mRNA	messenger ribonucleic acid
$NAD^{\oplus}$	nicotinamide adenine dinucleotide
NADH	nicotinamide adenine dinucleotide (reduced form)
$NADP^{\oplus}$	nicotinamide adenine dinucleotide phosphate
NADPH	nicotinamide adenine dinucleotide phosphate (reduced form)
NMN	nicotinamide mononucleotide
NMR	nuclear magnetic resonance
NTP	nucleoside triphosphate
Δp	protonmotive force
PAF	platelet activating factor
PAGE	polyacrylamide gel electrophoresis
PAP	phosphoadenosine phosphate
PAPS	3′-phosphoadenosine 5′-phosphosulfate
PCR	polymerase chain reaction
PEP	phosphoenolpyruvate
PFK	phosphofructokinase
P_i	inorganic phosphate
pI	isoelectric point
PIP	phosphatidylinositol 4-phosphate
PIP_2	phosphatidylinositol 4,5-*bis*phosphate
PITC	phenylisothiocyanate
PLP	pyridoxal phosphate
PP_i	inorganic pyrophosphate
PQ	plastoquinone
PQH_2	plastoquinol
PRPP	5-phosphoribosyl 1-pyrophosphate
Ψ	pseudouridylate

$\Delta\psi$	membrane potential
PSI	photosystem I
PSII	photosystem II
PTC	phenylthiocarbamoyl
PTH	phenylthiohydantoin
Q	ubiquinone
QH_2	ubiquinol
R	universal gas constant
RF	release factor
RFLP	restriction fragment length polymorphism
RNA	ribonucleic acid
RPP	reductive pentose phosphate
rRNA	ribosomal ribonucleic acid
RuBisCO	ribulose 1,5-*bis*phosphate carboxylase-oxygenase
S	entropy
SDS	sodium dodecyl sulfate
snRNA	small nuclear ribonucleic acid
snRNP	small nuclear ribonucleoprotein
SRP	signal recognition particle
SSB	single-strand binding protein
T	absolute temperature
TBP	TATA-binding protein
TDP	thymidine 5′-diphosphate
TF	transcription factor
T_m	melting point *or* phase-transition temperature
TMP	thymidine 5′-monophosphate
TPP	thiamine pyrophosphate
tRNA	transfer ribonucleic acid
TTP	thymidine 5′-triphosphate
UAS	upstream activating sequence
UDP	uridine 5′-diphosphate
UMP	uridine 5′-monophosphate
UTP	uridine 5′-triphosphate
v	velocity
v_0	initial velocity
VLDL	very low density lipoprotein
V_{max}	maximum velocity
YAC	yeast artificial chromosome

Dictionary of Biochemical Terms

A site. *See* aminoacyl site

absorptive phase. The period immediately following a meal. During the absorptive phase, which lasts up to four hours in humans, most tissues utilize glucose as fuel.

acceptor stem. The sequence at the 5′ end and the sequence near the 3′ end of a tRNA molecule that are base paired, forming a stem. The acceptor stem is the site of amino acid attachment. Also known as the amino acid stem.

acid. A substance that can donate protons. An acid is converted to its conjugate base by loss of a proton. (The Lewis theory defines an acid as an electron-pair acceptor [Lewis acid].)

acid anhydride. The product formed by condensation of two acids.

acid dissociation constant (K_a). The equilibrium constant for the dissociation of a proton from an acid.

acid protease. *See* aspartic protease

acid-base catalysis. Catalysis in which the transfer of a proton accelerates a reaction.

acidic solution. A solution that has a pH value less than 7.0.

acidosis. A condition in which the pH of the blood is significantly lower than 7.4.

ACP. *See* acyl carrier protein

actin filament. A protein filament composed of actin molecules arranged in a twisted, double-stranded rope. Actin filaments are components of the cytoskeletal network and play a role in contractile systems of many organisms. Also known as a microfilament.

activation energy. The free energy required to promote reactants from the ground state to the transition state in a chemical reaction.

activator. *See* transcriptional activator

active site. The portion of an enzyme that contains the substrate-binding site and the amino acid residues involved in catalyzing the conversion of substrate(s) to product(s). Active sites are usually located in clefts between domains or subunits of proteins or in indentations on the protein surface.

active transport. The process by which a solute specifically binds to a transport protein and is transported across a membrane against the solute concentration gradient. Energy is required to drive active transport. In primary active transport, the energy source may be light, ATP, or electron transport. Secondary active transport is driven by ion concentration gradients.

acyl carrier protein (ACP). A protein (in prokaryotes) or a domain of a protein (in eukaryotes) that binds activated intermediates of fatty acid synthesis via a thio-ester linkage.

adenosine diphosphate (ADP). A ribonucleoside diphosphate in which two phosphoryl groups are successively linked to the 5′ oxygen atom of adenosine. ADP is formed in reactions in which a phosphoryl group is transferred from adenosine triphosphate (ATP) or to adenosine monophosphate (AMP).

adenosine monophosphate (AMP). A ribonucleoside monophosphate in which a phosphoryl group is linked to the 5′ oxygen atom of adenosine. Phosphorylation of AMP produces adenosine diphosphate (ADP), a precursor of adenosine triphosphate (ATP). Also known as adenylate.

adenosine triphosphate (ATP). A ribonucleoside triphosphate in which three phosphoryl groups are successively linked to the 5′ oxygen atom of adenosine. ATP is the central supplier of energy in living cells, linking exergonic reactions with endergonic reactions. The phosphoanhydride linkages of ATP contain considerable chemical potential energy. By donating phosphoryl groups, ATP can transfer that energy to intermediates that then participate in biosynthetic reactions.

adenylate. *See* adenosine monophosphate

adipocyte. A triacylglycerol-storage cell found in animals. An adipocyte consists of a fat droplet surrounded by a thin shell of cytosol in which the nucleus and other organelles are suspended.

adipose tissue. Animal tissue composed of specialized triacylglycerol-storage cells known as adipocytes.

A-DNA. The conformation of DNA commonly observed when purified DNA is dehydrated. A-DNA is a right-handed double helix containing approximately 11 base pairs per turn, in which the deoxyribose residues exist predominately in the C-3′ *endo* conformation and the nucleotides adopt an *anti* conformation.

ADP. *See* adenosine diphosphate

aerobic. Occurring in the presence of oxygen.

affinity chromatography. A chromatographic technique used to separate a mixture of proteins or other macromolecules in solution based on specific binding to a ligand that is covalently attached to the chromatographic matrix.

affinity labelling. A process by which an enzyme (or other macromolecule) is covalently tagged by another detectable molecule that specifically interacts with the active site (or other binding site).

aglycone. An organic molecule, such as an alcohol, an amine, or a thiol, that forms a glycosidic linkage with the anomeric carbon of a sugar molecule to form a glycoside.

aldimine. *See* Schiff base

aldoses. A class of monosaccharides in which the most oxidized carbon atom, designated C-1, is aldehydic.

alkaline solution. *See* basic solution

alkalosis. A condition in which the pH of the blood is significantly higher than 7.4.

allosteric effector. *See* allosteric modulator

allosteric enzyme. An enzyme whose activity is modulated by the binding of another molecule at a site other than the active site.

allosteric modulator. A biomolecule that binds to the regulatory site of an allosteric enzyme and thereby modulates its catalytic activity. An allosteric modulator may be an activator or an inhibitor. Also known as an allosteric effector.

allosteric site. *See* regulatory site

allosteric transitions. The changes in conformation of a protein between the active (R) state and the inactive (T) state, caused by binding or release of an allosteric modulator.

α helix. A common secondary structure of proteins, in which the carbonyl oxygen of each amino acid residue (residue n) forms a hydrogen bond with the amide hydrogen of the fourth residue further toward the C-terminus of the polypeptide chain (residue $n + 4$). In an ideal right-handed α helix, equivalent positions recur every 0.54 nm, each amino acid residue advances the helix by 0.15 nm along the long axis of the helix, and there are 3.6 amino acid residues per turn.

Ames test. A procedure used to identify potential mutagens. The Ames test measures the ability of a substance to cause reversion of the his^- phenotype in a strain of *Salmonella typhimurium.*

amino acid. An organic acid consisting of an α-carbon atom to which an amino group, a carboxylate group, a hydrogen atom, and a specific side chain (R group) are attached. Amino acids are the building blocks of proteins.

amino acid analysis. A chromatographic procedure used for the separation and quantitation of amino acids in solutions such as protein hydrolysates.

amino acid stem. *See* acceptor stem

amino terminus. *See* N-terminus

aminoacyl site. The site on a ribosome that is occupied during protein synthesis by an aminoacyl-tRNA molecule. Also known as the A site.

aminoacyl-tRNA. A tRNA molecule that contains an amino acid covalently attached to the 3'-adenylate residue of the acceptor stem.

aminoacyl-tRNA synthetase. An enzyme that catalyzes the activation and attachment of a specific amino acid to the 3′ end of a corresponding tRNA molecule.

AMP. *See* adenosine monophosphate

amphibolic pathway. A metabolic pathway that can be both catabolic and anabolic.

amphipathic molecule. A molecule that has both hydrophobic and hydrophilic regions.

anabolic reaction. A metabolic reaction that synthesizes a molecule needed for cell maintenance and growth.

anaerobic. Occurring in the absence of oxygen.

anaplerotic reaction. A reaction that replenishes metabolites removed from a central metabolic pathway.

angstrom (Å). A unit of length equal to 1×10^{-10} m, or 0.1 nm.

anhydride. *See* acid anhydride

anion. An ion with an overall negative charge.

annular lipid. The layer(s) of lipid bound to an integral membrane protein.

anode. A positively charged electrode. In electrophoresis, anions move toward the anode.

anomeric carbon. The most oxidized carbon atom of a cyclized monosaccharide. The anomeric carbon has the chemical reactivity of a carbonyl group.

anomers. Isomers of a sugar molecule that have different configurations only at the most oxidized carbon atom.

antenna pigments. Light-absorbing pigments associated with the reaction center of a photosystem. These pigments may form a separate antenna complex or may be bound directly to the reaction-center proteins.

antibiotic. A compound, produced by one organism, that is toxic to other organisms. Clinically useful antibiotics must be specific for pathogens and not affect the human host.

antibody. A glycoprotein synthesized by certain white blood cells as part of the immunological defense system. Antibodies specifically bind to foreign compounds, called antigens, forming antibody-antigen complexes that mark the antigen for destruction. Also known as an immunoglobulin.

anticodon. A sequence of three nucleotides in the anticodon loop of a tRNA molecule. The anticodon binds to the complementary codon in mRNA during translation.

anticodon arm. The stem-and-loop structure in a tRNA molecule that contains the anticodon.

anticodon loop. The single-stranded loop that contains the anticodon of a tRNA molecule.

antigen. A molecule specifically bound by an antibody.

antiport. The cotransport of two different species of ions or molecules in opposite directions across a membrane by a transport protein.

antisense RNA. An RNA molecule that binds to a complementary mRNA molecule, forming a double-stranded region that inhibits translation of the mRNA.

antitermination cascade. The sequential expression of proteins that prevent termination of transcription. Proteins encoded by transcripts that are not terminated may include additional antiterminators.

antiterminator. A protein that prevents termination of transcription.

apoprotein. A protein whose cofactor(s) is absent. Without the cofactor(s), the apoprotein lacks the biological activity characteristic of the corresponding holoprotein.

aspartic protease. A protease that contains two aspartate residues in the catalytic center, one of which acts as a base catalyst and the other, as an acid catalyst. Aspartic proteases have a pH optimum of about 2–4. Also known as an acid protease.

atomic mass unit. The unit of atomic weight equal to 1/12th the mass of the ^{12}C isotope of carbon. The mass of the ^{12}C nuclide is exactly 12 by definition.

ATP. *See* adenosine triphosphate

ATPase. An enzyme that catalyzes hydrolysis of ATP to ADP + P_i. Ion-transporting ATPases use the energy of ATP to transport $Na^{\oplus}$, $K^{\oplus}$, or $Ca^{2\oplus}$ across a cellular membrane.

attenuation. A mechanism of translational regulation in which the rate of ribosome movement along an mRNA molecule determines whether transcription proceeds or terminates. Attenuation allows prokaryotes to regulate expression of an entire operon by the ability to synthesize a short peptide encoded at the beginning of the operon.

autoimmune disease. A human disorder in which the body produces antibodies against normal tissues and cellular components.

autophosphorylation. Phosphorylation of a protein kinase catalyzed by another molecule of the same kinase.

autoregulation. Regulation of expression of a gene by a product of the gene.

autosome. A chromosome other than a sex chromosome.

autotroph. An organism that can survive using CO_2 as its only source of carbon.

backbone. 1. The repeating $N—C_{\alpha}—C$ units connected by peptide bonds in a polypeptide chain. 2. The repeating sugar-phosphate units connected by phosphodiester linkages in a nucleic acid.

bacteriophage. A virus that infects a bacterial cell.

base. 1. A substance that can accept protons. A base is converted to its conjugate acid by addition of a proton. (The Lewis theory defines a base as an electron-pair donor [Lewis base].) 2. The substituted pyrimidine or purine of a nucleoside or nucleotide. The heterocyclic bases of nucleosides and nucleotides can participate in hydrogen bonding.

base pairing. The interaction between the bases of nucleotides in single-stranded nucleic acids to form double-stranded molecules, such as DNA, or regions of double-stranded secondary structure. The most common base pairs are formed by hydrogen bonding of adenine (A) with thymine (T) or uracil (U) and of guanine (G) with cytosine (C).

basic solution. A solution that has a pH value greater than 7.0. Also known as an alkaline solution.

B-DNA. The most common conformation of DNA and the one proposed by Watson and Crick. B-DNA is a right-handed double helix containing approximately 10.4 base pairs per turn, in which the deoxyribose residues exist predominately in the C-2′ *endo* conformation and the nucleotides adopt an *anti* conformation.

β-oxidation pathway. The metabolic pathway that degrades fatty acids to acetyl CoA, producing NADH and QH_2 and thereby generating large amounts of ATP. Each round of β oxidation of fatty acids consists of four steps: oxidation, hydration, further oxidation, and thiolysis.

β sheet. A common secondary structure of proteins that consists of extended polypeptide chains stabilized by hydrogen bonds between the carbonyl oxygen of one peptide bond and the amide hydrogen of another on the same or an adjacent polypeptide chain. The hydrogen bonds are nearly perpendicular to the extended polypeptide chains, which may be either parallel (running in the same N- to C-terminal direction) or antiparallel (running in opposite directions).

β strand. An extended polypeptide chain within a β sheet secondary structure or having the same conformation as a strand within a β sheet.

bile. A suspension of bile salts, bile pigments, and cholesterol that originates in the liver and is stored in the gall bladder. Bile is secreted into the small intestine during digestion.

binary fission. The process by which prokaryotic cells divide. During binary fission, each strand of the double-stranded DNA molecule of the prokaryotic genome directs the synthesis of a complementary strand, the two newly created duplex molecules separate, and the cell divides, with each daughter cell containing one copy of the genetic material.

binding-change mechanism. A proposed mechanism for the formation and release of ATP from $F_O F_1$ ATP synthase. The mechanism proposes three different binding-site conformations for ATP synthase: an open site from which ATP has been released, an ATP-bearing tight-binding site that is catalytically active, and an ADP and P_i loose-binding site that is catalytically inactive. Inward passage of protons through the ATP synthase complex into the mitochondrial matrix causes the open site to become a loose site; the loose site, already filled with ADP and P_i, to become a tight site; and the ATP-bearing site to become an open site.

bioenergetics. The study of energy changes in biological systems.

biological membrane. *See* membrane

biopolymer. A biological macromolecule in which many identical or similar small molecules are covalently linked to one another to form a long chain. Proteins, polysaccharides, and nucleic acids are biopolymers.

blunt end. An end of a double-stranded DNA molecule with no single-stranded overhang.

Bohr effect. The phenomenon observed when exposure to carbon dioxide, which lowers the pH inside the cells, causes the oxygen affinity of hemoglobin in red blood cells to decrease.

branch migration. The movement of a crossover, or branch point, resulting in further exchange of DNA strands during recombination.

brush border. *See* microvilli

buffer. A solution of a weak acid and its conjugate base that resists changes in pH.

buffer capacity. The ability of a solution to resist changes in pH. For a given buffer, maximum buffer capacity is achieved at the pH at which the concentrations of the weak acid and its conjugate base are equal (i.e., when pH = pK_a).

C_3 pathway. *See* reductive pentose phosphate cycle

C_4 pathway. A pathway for carbon fixation in several plant species that minimizes photorespiration by concentrating CO_2. In this pathway, CO_2 is incorporated into C_4 acids in the mesophyll cells, and the C_4 acids are decarboxylated in the bundle sheath cells, releasing CO_2 for use by the reductive pentose phosphate cycle.

calorie (cal). The amount of energy required to raise the temperature of 1 gram of water by 1°C (from 14.5°C to 15.5°C). One calorie is equal to 4.184 J.

Calvin-Benson cycle. *See* reductive pentose phosphate cycle

CAM. *See* Crassulacean acid metabolism

cap. A 7-methylguanosine residue attached by a pyrophosphate linkage to the 5′ end of a eukaryotic mRNA molecule. The cap is added posttranscriptionally and is required for efficient translation. Further covalent modifications yield alternative cap structures. Small nuclear RNA molecules have a different cap structure.

cap-binding protein (CBP). A eukaryotic translation initiation factor that interacts with the 5′ cap of an mRNA molecule during assembly of the translation initiation complex.

capsid. The head portion of a viral particle that contains the genetic material.

carbanion. A carbon anion that results from the cleavage of a covalent bond between carbon and another atom in which both electrons from the bond remain with the carbon atom.

carbocation. *See* carbonium ion

carbohydrate. Loosely defined as a compound that is a hydrate of carbon in which the ratio of C:H:O is 1:2:1. Carbohydrates include monomeric sugars (i.e., monosaccharides) and their polymers. Also known as a saccharide.

carbohydrate loading. The athlete's practice of depleting muscle glycogen by intense exercise followed by consumption of high-carbohydrate meals, resulting in higher-than-normal amounts of muscle glycogen.

carbonium ion. A carbon cation that results from the cleavage of a covalent bond between carbon and another atom in which the carbon atom loses both electrons from the bond. Also known as a carbocation.

carboxyl terminus. *See* C-terminus

carcinogen. An agent that can cause cancer.

carnitine shuttle system. A cyclic pathway that shuttles acetyl CoA from the cytosol to the mitochondria by formation and transport of acyl carnitine.

cascade. Sequential activation of several components, resulting in rapid signal amplification.

catabolic reaction. A metabolic reaction that degrades a complex molecule to provide smaller molecular building blocks and energy to an organism.

catabolite repression. A regulatory mechanism that results in increased rates of transcription of many bacterial genes and operons when glucose is present. A complex between cAMP and cAMP regulatory protein (CRP) activates transcription.

catalytic center. The polar amino acids in the active site of an enzyme that participate in chemical changes during catalysis.

catalytic constant (k_{cat}). A kinetic constant that is a measure of how rapidly an enzyme can catalyze a reaction when saturated with its substrate(s). The catalytic constant is equal to the maximum velocity, V_{max}, divided by the total concentration of enzyme, $[E]_{total}$, or the number of moles of substrate converted to product per mole of enzyme active sites per second, under saturating conditions. Also known as the turnover number.

catalytic triad. The hydrogen-bonded serine, histidine, and aspartate residues in the active site of serine proteases and some other hydrolases. The serine residue serves as a covalent catalyst; the histidine residue serves as an acid-base catalyst; and the aspartate residue aligns the histidine residue and stabilizes its protonated form.

cathode. A negatively charged electrode. In electrophoresis, cations move toward the cathode.

cation. An ion with an overall positive charge.

CBP. *See* cap-binding protein

CCAAT box. A consensus sequence found between 60 and 80 base pairs upstream of the transcription initiation site in mammalian genes. Several transcriptional regulatory proteins bind to the CCAAT box.

cDNA. *See* complementary DNA

cDNA library. A DNA library constructed from cDNA copies of all the mRNA in a given cell type.

cell wall. A mechanically tough, porous outer coat that surrounds the plasma membrane of nearly all bacterial, fungal, and plant cells. The cell wall provides the cell with mechanical strength and protection against osmotic stress.

cellulose. A linear (unbranched) homopolymer of glucose residues linked by $\beta\text{-}(1 \rightarrow 4)$ glycosidic bonds. A structural polysaccharide of plant cell walls, cellulose accounts for over 50% of the organic matter in the biosphere.

central dogma. The pathway for flow of information from a gene to the corresponding protein. Genetic information is stored in DNA, which can be replicated and passed to daughter cells. Information is copied, or transcribed, from DNA to RNA. RNA is translated during synthesis of a polypeptide chain.

ceramide. A molecule that consists of a fatty acid linked to the C-2 amino group of sphingosine by an amide bond. Ceramides are the metabolic precursors of all sphingolipids.

cerebroside. A glycosphingolipid that contains one monosaccharide residue attached via a β-glycosidic linkage to C-1 of a ceramide. Cerebrosides are abundant in nerve tissue and are found in myelin sheaths.

channel. An integral membrane protein with a central aqueous passage, which allows appropriately sized molecules and ions to transverse the membrane in either direction. Also known as a pore.

channelling. Transfer of the product of one reaction of a multifunctional enzyme or a multienzyme complex directly to the next active site or enzyme without entering the bulk solvent. Channelling increases the rate of a reaction pathway by decreasing the transit time for an intermediate to reach the next enzyme and by producing high local concentrations of the intermediate.

chaotropic agent. A strongly hydrogen-bonding substance that enhances the solubility of nonpolar compounds in water by disrupting regularities in hydrogen bonding among water molecules. Concentrated solutions of chaotropic agents, such as urea and guanidinium salts, decrease the hydrophobic effect and are thus effective protein denaturants.

chaperone. A protein that forms complexes with newly synthesized polypeptide chains and assists in their correct folding into biologically functional conformations. Chaperones may also prevent the formation of incorrectly folded intermediates, prevent incorrect aggregation of unassembled protein subunits, assist in translocation of polypeptide chains across membranes, and assist in the assembly and disassembly of large multiprotein structures.

chelate effect. The phenomenon by which the constant for binding of a ligand having two or more binding sites to a molecule or atom is greater than the constant for binding of separate ligands to the same molecule or atom.

chemiosmotic theory. A theory proposing that a proton concentration gradient established during oxidation of substrates provides the energy to drive processes such as the formation of ATP from ADP and P_i.

chemotroph. An organism that requires organic compounds other than CO_2 as sources of matter and chemical energy.

chiral atom. An atom with asymmetric substitution that can exist in two different configurations.

chitin. A linear homopolymer of N-acetylglucosamine residues joined by β-(1→4) linkages. Chitin is found in the exoskeletons of insects and crustaceans and in the cell walls of most fungi and many algae and is the second most abundant organic compound on earth.

chlorophyll. A green pigment in photosynthetic membranes that is the principal light-harvesting component in phototrophic organisms.

chlorophyll *a/b* light harvesting complex (LHC). A large pigment complex in the thylakoid membrane that aids the photosystem in gathering light.

chloroplast. A chlorophyll-containing organelle in algae and plant cells that serves as the site of photosynthesis.

chromatin. A DNA-protein complex in the nuclei of eukaryotic cells that serves as the basic genetic material.

chromatography. A technique used to separate components of a mixture based on their partitioning between a mobile phase, which can be gas or liquid, and a stationary phase, which is a liquid or solid.

chromosome. A single DNA molecule containing many genes. An organism may have a genome consisting of a single chromosome or many.

chromosome walking. A technique for ordering DNA fragments in a genomic library. Chromosome walking involves hybridization, restriction mapping, and isolation of progressively overlapping recombinant DNA molecules.

chylomicron. A type of plasma lipoprotein that transports triacylglycerols, cholesterol, and cholesteryl esters from the small intestine to the tissues.

chyme. The mixture of food and secretions that passes from the stomach to the small intestine.

Circe effect. Accelerated diffusion of a charged substrate into the active site of an enzyme due to strong attractive forces of the enzyme.

circular DNA. A DNA molecule in which the two putative ends are covalently linked by 3′–5′ phosphodiester bonds, forming a closed circle.

cisterna. The lumen of a vesicle of the Golgi apparatus.

citrate transport system. A cyclic pathway that shuttles acetyl CoA from the mitochondria to the cytosol, with oxidation of cytosolic NADH to $NAD^{\oplus}$ and reduction of cytosolic $NADP^{\oplus}$ to NADPH. Two molecules of ATP are consumed in each round of the pathway.

citric acid cycle. A metabolic cycle consisting of eight enzyme-catalyzed reactions that completely oxidizes acetyl units to CO_2. The energy released in the oxidation reactions is conserved as reducing power when the coenzymes $NAD^{\oplus}$ and ubiquinone (Q) are reduced. Oxidation of one molecule of acetyl CoA by the citric acid cycle generates three molecules of NADH, one molecule of QH_2, and one molecule of GTP or ATP. Also known as the Krebs cycle and the tricarboxylic acid cycle.

clone. One of the identical copies derived from the replication or reproduction of a single molecule, cell, or organism.

cloning. The generation of many identical copies of a molecule, cell, or organism.

cloning vector. A DNA molecule that carries a segment of foreign DNA. A cloning vector introduces the foreign DNA into a cell where it can be replicated and sometimes expressed.

coding strand. The strand of DNA within a gene whose nucleotide sequence is identical to that of the transcribed RNA (with the replacement of T by U in RNA).

codon. A sequence of three nucleotide residues in mRNA (or DNA) that specifies a particular amino acid according to the genetic code.

cofactor. An inorganic ion or organic molecule required by an apoenzyme to convert it to a holoenzyme. There are two types of cofactors: essential ions and coenzymes.

cointegrate. A single DNA molecule produced from two bacterial chromosomes as a result of replicative transposition. The cointegrate is resolved into two separate chromosomes by site-specific recombination.

competitive inhibition. Inhibition of an enzyme-catalyzed reaction by an inhibitor that prevents the substrate from binding.

complementary DNA (cDNA). Double-stranded DNA synthesized from an mRNA template by the action of reverse transcriptase followed by DNA polymerase.

complementation. A technique for selecting transformed cells by testing for the ability of the recombinant DNA to supply a gene product missing in the host cell.

concerted theory of cooperativity and allosteric regulation. A model of the cooperative binding of identical ligands to oligomeric proteins. According to the simplest form of the concerted theory, the change in conformation of a protein due to the binding of a substrate or an allosteric modulator shifts the equilibrium of the conformation of the protein between T (a low substrate-affinity conformation) and R (a high substrate-affinity conformation). This theory suggests that all subunits of the protein have the same conformation, either all T or all R. Also known as the symmetry-driven theory.

condensation. A reaction involving the joining of two or more molecules accompanied by the elimination of water, alcohol, or other simple substance.

configuration. A spatial arrangement of atoms, which cannot be altered without breaking and re-forming covalent bonds.

conformation. Any three-dimensional structure, or spatial arrangement, of a molecule that results from rotation of functional groups around single bonds. Because there is free rotation around single bonds, a molecule can potentially assume many conformations.

conjugate acid. The product resulting from the gain of a proton by a base.

conjugate base. The product resulting from the loss of a proton by an acid.

conjugation. The passage of genetic material from one bacterium to another through the sex pilus.

consensus sequence. The sequence of nucleotides most commonly found at each position within a region of DNA or RNA.

converter enzyme. An enzyme that catalyzes the covalent modification of another enzyme, thereby changing its catalytic activity.

cooperativity. 1. The phenomenon whereby the binding of one ligand or substrate molecule to a protein influences the affinity of the protein for additional molecules of the same substance. Cooperativity may be positive or negative. 2. The phenomenon whereby formation of structure in one part of a macromolecule promotes the formation of structure in the rest of the molecule.

corepressor. A ligand that binds to and activates a repressor of a gene.

Cori cycle. An interorgan metabolic loop that recycles carbon and transports energy from the liver to the peripheral tissues. Glucose is released from the liver and metabolized to produce ATP in other tissues. The resulting lactate is then returned to the liver for conversion back to glucose by gluconeogenesis.

cosmid. A cloning vector that accommodates large fragments of insert DNA. Cosmids allow efficient transfection but permit propagation of recombinant DNA molecules as plasmids.

cosubstrate. A coenzyme that serves as a substrate in an enzyme-catalyzed reaction. A cosubstrate is altered during the course of the reaction and dissociates from the active site of the enzyme. The original form of the cosubstrate can be regenerated in a subsequent enzyme-catalyzed reaction.

cotranscriptional processing. RNA processing that occurs before transcription is complete.

cotranslational modification. Covalent modification of a protein that occurs before elongation of the polypeptide is complete.

cotransport. The coupled transport of two different species of solutes across a membrane, in the same direction (symport) or the opposite direction (antiport), carried out by a transport protein.

coupled reactions. Two metabolic reactions that share a common intermediate.

covalent catalysis. Catalysis in which one substrate, or part of it, forms a covalent bond with the catalyst and then is transferred to a second substrate. Many enzymatic group-transfer reactions proceed by covalent catalysis.

Crassulacean acid metabolism (CAM). A modified sequence of carbon-assimilation reactions used primarily by plants in arid environments to reduce water loss during photosynthesis. In these reactions, CO_2 is taken up at night, resulting in the formation of malate. During the day, malate is decarboxylated, releasing CO_2 for use by the reductive pentose phosphate cycle.

cristae. The folds of the inner mitochondrial membrane.

cruciform structure. The crosslike conformation adopted by double-stranded DNA when inverted repeats form a base-paired structure involving the complementary regions within the same strand.

cryptic phage. A bacteriophage whose genome has mutated to become a permanent resident of the host genome.

C-terminus. The amino acid residue bearing a free α-carboxyl group at one end of a peptide chain. Also known as the carboxyl terminus.

cumulative feedback inhibition. Inhibition of an enzyme that catalyzes an early step in several biosynthetic pathways by intermediates or end products of those pathways. The degree of inhibition increases as more of the inhibitors bind.

C-value paradox. The observation that the DNA content of a cell, the C-value, can vary widely between similar species.

cytochrome. A heme-containing protein that serves as an electron carrier in processes such as respiration and photosynthesis.

cytoplasm. The part of a cell enclosed by the plasma membrane, excluding the nucleus.

cytoskeleton. A network of proteins that contributes to the structure and organization of a eukaryotic cell.

cytosol. The aqueous portion of the cytoplasm minus the subcellular structures.

D arm. The stem-and-loop structure in a tRNA molecule that contains dihydro-uridylate (D) residues.

dalton. A unit of mass equal to one atomic mass unit.

de novo pathway. A metabolic pathway in which a biomolecule is formed from simple precursor molecules.

deaminase. An enzyme that catalyzes the removal of an amino group from a substrate, releasing ammonia.

degeneracy. The existence of several different codons that specify the same amino acid.

dehydrogenase. An enzyme that catalyzes the removal of hydrogen from a substrate or the oxidation of a substrate. Dehydrogenases are members of the IUB class of enzymes known as oxidoreductases.

denaturation. A disruption in the native conformation of a biological macromolecule that results in loss of the biological activity of the macromolecule.

deoxyribonuclease (DNase). An enzyme that catalyzes the hydrolysis of deoxyribonucleic acids to form oligodeoxynucleotides and/or monodeoxynucleotides.

deoxyribonucleic acid (DNA). A polymer consisting of deoxyribonucleotide residues joined by 3′–5′ phosphodiester bonds. The sugar moiety in DNA is 2-deoxyribose. DNA serves as the genetic material of all cells and many viruses.

detergent. An amphipathic molecule consisting of a hydrophobic portion and a hydrophilic end that may be ionic or polar. Detergent molecules can aggregate in aqueous media to form micelles. Also known as a surfactant.

developmentally regulated gene. A gene whose expression is spatially or temporally regulated. Such a gene is expressed only in certain cells or at certain times during the life of an organism.

diabetes mellitus. A metabolic disease characterized by hyperglycemia resulting from abnormal regulation of fuel metabolism by insulin. Diabetes mellitus may arise from the lack of insulin or from poor responsiveness of cells to insulin.

dialysis. A procedure in which low-molecular-weight solutes in a sample are removed by diffusion through a semipermeable barrier and replaced by solutes from the surrounding medium.

differential-scanning calorimetry. A technique used to evaluate phase transitions in an aqueous dispersion of molecules, such as membrane lipids, by measuring the rate of heat absorbed by the sample when it is heated simultaneously with a sample of water. When it undergoes phase transition, the sample absorbs heat energy at a faster rate than the water.

diffusion-controlled reaction. A reaction that occurs with every collision between reactant molecules. In enzyme-catalyzed reactions, the k_{cat}/K_m ratio approaches a value of 10^8–$10^9 \, M^{-1} \, s^{-1}$.

digestion. The process by which dietary macromolecules are hydrolyzed to smaller molecules that can be absorbed by an organism.

diploid. Having two sets of chromosomes or two copies of the genome.

---

dipole. Two equal but opposite charges, separated in space, resulting from the uneven distribution of charge within a molecule or a chemical bond.

direct repair. The removal of DNA damage by proteins that continually scan DNA in order to detect lesions. Some damaged nucleotides and mismatched bases are recognized and repaired without cleaving the DNA or excising the base.

distributive enzyme. An enzyme that dissociates from its growing polymeric product after addition of each monomeric unit and must reassociate with the polymer for polymerization to proceed.

disulfide bond. A covalent linkage formed by oxidation of two thiol groups. Disulfide bonds are important in stabilizing the three-dimensional structures of some proteins.

D-loop. An extended region of unwound DNA in which one of the strands has been copied and the other remains single stranded.

DNA. *See* deoxyribonucleic acid

DNA bank. *See* DNA library

DNA fingerprinting. Analysis of the genetic polymorphism of different individuals.

DNA gyrase. *See* topoisomerase

DNA library. The set of recombinant DNA molecules generated by ligating all the fragments of a sample of DNA into vectors. Also known as a DNA bank.

DNA ligase. The enzyme that joins two DNA polynucleotides by catalyzing the formation of a phosphodiester bond. DNA ligase can also repair gaps in double-stranded DNA.

DNA polymerase. An enzyme that catalyzes the DNA template–directed addition of nucleotide residues to the 3′ end of an existing polynucleotide. Some DNA polymerases contain exonuclease activity used in proofreading newly polymerized sequences.

DNase. *See* deoxyribonuclease

domain. A discrete, independent folding unit within the tertiary structure of a protein. Domains are usually combinations of several units of supersecondary structure.

dosage compensation. Selective inactivation of certain genes present in multiple copies within the cell to control the level of gene expression.

double helix. A nucleic acid conformation in which two antiparallel polynucleotide strands wrap around each other to form a two-stranded helical structure stabilized largely by stacking interactions between adjacent hydrogen-bonded base pairs.

double-reciprocal plot. A plot of the reciprocal of initial velocity versus the reciprocal of substrate concentration for an enzyme-catalyzed reaction. The x and y intercepts indicate the values of the reciprocals of the Michaelis constant and the maximum velocity, respectively. A double-reciprocal plot is a linear transformation of the Michaelis-Menten equation. Also known as a Lineweaver-Burk plot.

E. See reduction potential

246

E°′. *See* standard reduction potential

E site. *See* exit site

early gene. A gene that is expressed relatively early in infection by a bacteriophage or virus.

Edman degradation. A procedure used to determine the sequence of amino acid residues from a free N-terminus of a polypeptide chain. The N-terminal residue is chemically modified, cleaved from the chain, and identified by chromatographic procedures, and the rest of the polypeptide is recovered. Multiple reaction cycles allow identification of the new N-terminal residue generated by each cleavage step.

effective molarity. An expression of the rate acceleration of a nonenzymatic reaction within a single compound in terms of the rate relative to the rate of a reaction of two compounds possessing the same reactive groups as the single compound.

effector enzyme. A membrane-associated protein that produces an intracellular second messenger in response to a signal from a transducer.

eicosanoid. A physiologically potent, oxygenated derivative of a 20-carbon polyunsaturated fatty acid. Eicosanoids function intracellularly as short-range messengers in the regulation of various cellular activities.

electrical potential. *See* membrane potential

electrogenic transport. The transfer of ionic solutes across a membrane by a process that results in a net transfer of charge, thereby creating changes in membrane potential.

electromotive force (emf). A measure of the difference between the reduction potentials of the reactions on the two sides of an electrochemical cell (i.e., the voltage difference produced by the reactions).

electroneutral transport. The transfer of ionic solutes across a membrane by a process that results in no net transfer of charge.

electron-transport chain. *See* respiratory electron-transport chain

electrophile. A positively charged or electron-deficient species that is attracted to chemical species that are negatively charged or contain unshared electron pairs (nucleophiles).

electrophoresis. A technique used to separate molecules by their migration in an electric field, primarily on the basis of their net charge.

electrostatic interaction. A noncovalent force between two charged particles, usually between ions of opposite charge.

elongation factors. Proteins that are involved in extending the peptide chain during protein synthesis.

emf. *See* electromotive force

enantiomers. Stereoisomers that are nonsuperimposable mirror images.

endergonic reaction. A chemical reaction that is characterized by a positive free-energy change. Such a reaction cannot occur spontaneously and requires the input of energy from outside the system to proceed.

endocytosis. The process by which matter is engulfed by a plasma membrane and brought into the cell within a lipid vesicle derived from the membrane.

endonuclease. An enzyme that catalyzes the hydrolysis of phosphodiester linkages at various sites within polynucleotide chains.

endoplasmic reticulum. A membranous network of tubules and sheets continuous with the outer nuclear membrane of eukaryotic cells. Regions of the endoplasmic reticulum coated with ribosomes are called the rough endoplasmic reticulum; regions having no attached ribosomes are known as the smooth endoplasmic reticulum. The endoplasmic reticulum is involved in the synthesis, sorting, and transport of certain proteins and in the synthesis of lipids.

endosomes. Smooth vesicles inside the cell that serve as receptacles for endocytosed material.

endotoxin. *See* lipopolysaccharide

energy-rich compound. A compound whose hydrolysis is highly exergonic ($\Delta G_{hydrolysis}$ greater than that for ATP $\longrightarrow$ ADP + P_i). The free energy available upon cleavage of an energy-rich compound can be used to drive other reactions.

enhancer. A region of DNA, located some distance from the promoter, to which a transcriptional activator binds, thereby increasing the rate of transcription.

enteric bacteria. The bacteria that inhabit the large intestine of animals. Most enteric bacteria are facultative anaerobes.

enthalpy (H). A thermodynamic state function that describes the heat content of a system.

entropy (S). A thermodynamic state function that describes the randomness or disorder of a system.

enzymatic reaction. A reaction catalyzed by a biological catalyst, an enzyme. Enzymatic reactions are 10^3 to 10^{17} times faster than the corresponding uncatalyzed reactions.

enzyme. A biological catalyst, almost always a protein. Some enzymes may require additional cofactors for activity. Virtually all biochemical reactions are catalyzed by specific enzymes.

enzyme assay. A method used to analyze the activity of a sample of an enzyme. Typically, enzymatic activity is measured under selected conditions such that the rate of conversion of substrate to product is proportional to enzyme concentration.

enzyme-substrate complex (ES). A complex formed when substrate molecules bind noncovalently within the active site of an enzyme.

epimers. Isomers that differ in configuration at only one of several chiral centers.

equilibrium. The state of a system in which the rate of conversion of substrate to product is equal to the rate of conversion of product to substrate. The free-energy change for a reaction or system at equilibrium is zero.

equilibrium constant (K_{eq}). The ratio of the concentrations of products to the concentrations of reactants at equilibrium. The equilibrium constant is related to the standard free-energy change of a reaction by the equation

$$\Delta G^{\circ\prime}_{reaction} = -RT \ln K_{eq}$$

equilibrium density gradient centrifugation. A technique used to separate macromolecules of different densities in an ultracentrifuge based on their buoyancy in an appropriate density gradient.

error-prone repair. The response to DNA damage that cannot be bypassed by the replication machinery. In error-prone repair, RecA protein binds to single-stranded gaps in DNA near the replication fork. Additional proteins allow the replication complex to proceed through the damaged region, haphazardly inserting nucleotides opposite unrecognizable template-strand nucleotides.

ES. *See* enzyme-substrate complex

escape synthesis. Low-level transcription of a repressed gene that occurs even in the absence of induction. Escape synthesis results from spontaneous dissociation of the repressor for brief interludes.

essential amino acid. An amino acid that cannot be synthesized by an animal and must be obtained in the diet.

essential fatty acid. A fatty acid that cannot be synthesized by an animal and must be obtained in the diet.

essential ion. An ion required as a cofactor for the catalytic activity of certain enzymes. Some essential ions, called activator ions, are reversibly bound to enzymes and often participate in the binding of substrates, whereas tightly bound metal ions frequently participate directly in catalytic reactions.

excision repair. The reversal of DNA damage by excision-repair endonucleases. Gross lesions that alter the structure of the DNA helix are repaired by cleavage on each side of the lesion and removal of the damaged DNA. The resulting single-stranded gap is filled by DNA polymerase and sealed by DNA ligase.

exergonic reaction. A chemical reaction that is characterized by a negative free-energy change. Such a reaction is spontaneous in that it does not require the input of additional energy to proceed.

exit site. The site on a ribosome from which a deaminoacylated tRNA molecule is released during protein synthesis. Also known as the E site.

exocytosis. The process by which material destined for secretion from a cell is enclosed in lipid vesicles that are transported to and fuse with the plasma membrane, releasing the material into the extracellular space.

exon. A nucleotide sequence that is present in the primary RNA transcript and in the mature RNA molecule. The term *exon* also refers to the region of the gene that encodes the corresponding RNA exon.

exonuclease. An enzyme that catalyzes the sequential hydrolysis of phosphodiester linkages from one end of a polynucleotide chain.

expression vector. A cloning vector that allows inserted DNA to be transcribed and translated into protein.

extrinsic membrane protein. *See* peripheral membrane protein

facilitated diffusion. *See* passive transport

facultative anaerobe. An organism that can survive in the presence or absence of oxygen.

fat-soluble vitamin. *See* lipid vitamin

fatty acid. A long-chain aliphatic hydrocarbon with a single carboxyl group at one end. Fatty acids are the simplest type of lipid and are components of many more complex lipids, including triacylglycerols, glycerophospholipids, sphingo-lipids, and waxes.

feedback inhibition. Inhibition of an enzyme that catalyzes an early step in a metabolic pathway by an end product of the same pathway.

feed-forward activation. Activation of an enzyme in a metabolic pathway by a metabolite produced earlier in the pathway.

fermentation. The anaerobic catabolism of metabolites for energy production. In alcoholic fermentation, pyruvate is converted to ethanol and carbon dioxide.

fibrous proteins. A major class of water-insoluble proteins that are often built upon a single repetitive structure. Many fibrous proteins are physically tough and provide mechanical support to individual cells or entire organisms.

fingerprinting. *See* DNA fingerprinting

first-order reaction. A reaction whose rate is directly proportional to the concentration of only one reactant.

Fischer projection. A two-dimensional representation of the three-dimensional structures of sugars and related compounds. In a Fischer projection, the carbon skeleton is drawn vertically, with C-1 at the top. At a chiral center, horizontal bonds extend toward the viewer and vertical bonds extend away from the viewer.

fluid mosaic model. A model proposed for the structure of biological membranes. In this model, the membrane is depicted as a dynamic structure in which lipids and membrane proteins (both integral and peripheral) undergo lateral diffusion.

fluorescence. A form of luminescence in which visible radiation is emitted from a molecule as it passes from a higher to a lower electronic state.

flux. The flow of material through a metabolic pathway. Flux depends on the supply of substrates, the removal of products, and the catalytic capabilities of the enzymes involved in the pathway.

footprinting. A technique used to identify the sequence of DNA or RNA bound by a protein. The bound protein protects specific nucleotide sequences from chemical or enzymatic digestion. The protected regions are visualized as gaps, or footprints, when the ladder of nucleic acid fragments is resolved by gel electrophoresis. Also known as nuclease protection.

frameshift mutation. An alteration in DNA caused by the insertion or deletion of a number of nucleotides not divisible by three. A frameshift mutation changes the reading frame of the corresponding mRNA molecule and affects translation of all codons downstream of the mutation.

free-energy change (ΔG). A thermodynamic quantity that defines the equilibrium condition in terms of the changes in enthalpy (H) and entropy (S) of a system at constant pressure. $\Delta G = \Delta H - T\Delta S$, where T is absolute temperature. Free energy is a measure of the energy available within a system to do work.

freeze-fracture electron microscopy. A technique used to visualize the structure of biological membranes. During freeze-fracture, the membrane sample is rapidly frozen and then split along the interface between the leaflets of the lipid bilayer. The exposed membrane surface is then coated with a thin metal film, producing a replica of the leaflet surface which can be visualized using an electron microscope.

furanose. A monosaccharide structure that forms a five-membered ring as a result of intramolecular hemiacetal formation.

ΔG. *See* free-energy change

ΔG°′. *See* standard free-energy change

ganglioside. A glycosphingolipid in which oligosaccharide chains containing *N*-acetylneuraminic acid are attached to a ceramide. Gangliosides are present on cell surfaces and provide cells with distinguishing surface markers that may serve in cellular recognition and cell-to-cell communication.

gap gene. A gene that is expressed in a localized region of the developing *Drosophila* embryo. Mutation of a gap gene usually causes a deletion, or gap, in the structure of the larva.

gas chromatography. A chromatographic technique used to separate components of a mixture based on their partitioning between the gas phase and a stationary phase, which can be a liquid or solid.

gastric juice. A solution of mucus, the zymogen pepsinogen, ions, and 0.15 M HCl secreted by cells of the stomach wall during digestion.

gastrointestinal tract. The digestive system of an animal. In the gastrointestinal tract, dietary biopolymers, such as starch and proteins, are enzymatically hydrolyzed to their monomeric units, and the monomers are absorbed.

gel-filtration chromatography. A chromatographic technique used to separate a mixture of proteins or other macromolecules in solution based on molecular size, using a matrix of porous beads. Also known as molecular-exclusion chromatography.

gene. Loosely defined as a segment of DNA that is transcribed. In some cases, the term *gene* may also be used to refer to a segment of DNA that encodes a functional protein or RNA molecule.

gene amplification. Regulation of gene expression by increasing the number of copies of a gene within a cell.

gene conversion. A consequence of mismatch repair following homologous recombination between DNA molecules that differ at one or more sites. Gene conversion occurs when the mismatch repair enzymes use DNA strands from the same original molecule as the two correct templates. The result is two double-stranded DNA molecules with the same nucleotide sequence.

gene replacement therapy. A technique in which cells are removed from a patient suffering from an enzyme deficiency, the gene for wild-type enzyme is transferred into the cells in culture, and the cells are returned to the patient.

genetic code. The correspondence between the sequence of residues in a nucleic acid and the amino acid sequence in a protein. A sequence of three nucleotide residues, known as a codon, specifies a single amino acid. The standard genetic code, composed of 64 codons, is used by almost all living organisms.

genetic engineering. *See* recombinant DNA technology

genetic load. The overall risk of harmful or lethal mutation. The genetic load is proportional to the size of the genome of the organism.

genome. The total amount of genetic information in an organism. It is equivalent to a single complete chromosome or to a set of chromosomes (haploid). Mitochondria and chloroplasts have genomes separate from that in the nucleus of eukaryotic cells.

genomic library. A DNA library constructed by fragmenting and cloning all the DNA for the genome of an organism.

globular proteins. A major class of proteins, many of which are water soluble. Globular proteins are compact and roughly spherical, containing tightly folded polypeptide chains. Typically, globular proteins include indentations, or clefts, which specifically recognize and transiently bind other compounds.

glucogenic compound. A compound, such as an amino acid, that can be used for gluconeogenesis in animals.

gluconeogenesis. A pathway for synthesis of glucose from a noncarbohydrate precursor. Gluconeogenesis from pyruvate involves the seven near-equilibrium reactions of glycolysis traversed in the reverse direction. The three metabolically irreversible reactions of glycolysis are bypassed by four enzymatic reactions that do not occur in glycolysis.

glucose. A six-carbon monosaccharide ($C_6H_{12}O_6$) that serves as the initial substrate of glycolysis and as the monomeric component of polysaccharides such as glycogen, starch, and cellulose.

glucose homeostasis. The maintenance of constant levels of glucose in the circulation achieved by balancing glucose synthesis or absorption against utilization.

glucose-alanine cycle. An interorgan metabolic loop that transports nitrogen to the liver and transports energy from the liver to the peripheral tissues. Glucose is released from the liver and metabolized in muscle to pyruvate, with concomitant production of ATP. Pyruvate can be converted to alanine in muscle. Alanine is returned to the liver, where it is metabolized to ammonia and pyruvate. The ammonia is incorporated into urea, and the pyruvate is converted back to glucose by gluconeogenesis.

glucose–fatty acid cycle. An interorgan regulatory system that operates to provide fatty acids as an alternative fuel to glucose. When glucose levels are low, insulin levels are low, and free fatty acids are released from adipocytes. The metabolism of fatty acids generates inhibitors of glucose metabolism, thus sparing the use of glucose.

glycan. A general term for an oligosaccharide or a polysaccharide. A homoglycan is a polymer of identical monosaccharide residues; a heteroglycan is a polymer of different monosaccharide residues.

glycerophospholipid. A lipid consisting of two fatty acyl groups bound to C-1 and C-2 of glycerol 3-phosphate and, in most cases, a polar substituent attached to the phosphate moiety. Glycerophospholipids are major components of biological membranes.

glycocalyx. A coat of lipid- and protein-bound carbohydrate that extends from the extracellular surface of the eukaryotic plasma membrane.

glycoconjugate. A carbohydrate derivative in which one or more carbohydrate chains are covalently linked to a peptide chain, protein, or lipid.

glycoforms. Glycoproteins containing identical amino acid sequences but different oligosaccharide-chain compositions.

glycogen. A branched homopolymer of glucose residues joined by α-$(1 \rightarrow 4)$ linkages with α-$(1 \rightarrow 6)$ linkages at branch points. Glycogen serves as a storage polysaccharide in animals and bacteria.

glycogenolysis. The pathway for intracellular degradation of glycogen.

glycolysis. A catabolic pathway consisting of 10 enzyme-catalyzed reactions by which one molecule of glucose is converted to two molecules of pyruvate. In the process, two molecules of ATP are formed from ADP + P_i, and two molecules of NAD$^\oplus$ are reduced to NADH.

glycoprotein. A protein that contains covalently bound carbohydrate residues.

glycosaminoglycan. An unbranched polysaccharide of repeating disaccharide units. One component of the disaccharide is an amino sugar; the other component is usually a uronic acid.

glycoside. A molecule containing a carbohydrate in which the hydroxyl group of the anomeric carbon has been replaced through condensation with an alcohol, an amine, or a thiol.

glycosidic bond. Acetal or ketal linkage formed by condensation of the anomeric carbon atom of a saccharide with a hydroxyl, amino, or thiol group of another molecule. The most commonly encountered glycosidic bonds are formed between the anomeric carbon of one sugar and a hydroxyl group of another sugar. Nucleosidic bonds are *N*-linked glycosidic bonds.

glycosphingolipid. A lipid containing sphingosine and carbohydrate moieties.

glycosylation. *See* protein glycosylation.

glyoxylate cycle. A variation of the citric acid cycle in certain plants, bacteria, and yeast that allows net production of glucose from acetyl CoA via oxaloacetate. The glyoxylate cycle bypasses the two CO_2-producing steps of the citric acid cycle.

Golgi apparatus. A complex of flattened, fluid-filled membranous sacs in eukaryotic cells, often found in proximity to the endoplasmic reticulum. The Golgi apparatus is involved in the modification, sorting, and targeting of proteins.

GPI membrane anchor. A phosphatidylinositol-glycan structure attached to a protein via a phosphoethanolamine residue. The fatty acyl moieties of the GPI anchor are embedded in the lipid bilayer, thereby tethering the protein to the membrane.

granal lamellae. Regions of the thylakoid membrane that are located within grana and are not in contact with the stroma.

granum. A stack of flattened vesicles formed from the thylakoid membrane in chloroplasts.

gRNA. *See* guide RNA

group translocation. A type of primary active transport during which the translocated species is chemically modified.

group-transfer potential. A measure of the ability of a compound to transfer a reactive group, often a phosphoryl group, to another compound. Under standard conditions, group-transfer potentials have the same values as the standard free energies of hydrolysis but are opposite in sign.

group-transfer protein. A protein that does not itself catalyze reactions but is required for the action of certain enzymes. Also known as a protein coenzyme.

group-transfer reaction. A reaction in which a substituent or functional group is transferred from one substrate to another.

growth factor. A protein that regulates cell proliferation by stimulating resting cells to undergo cell division.

guide RNA (gRNA). An RNA molecule that serves as a template for the insertion of additional nucleotide residues during RNA editing.

H. *See* enthalpy

hairpin. A secondary structure adopted by single-stranded polynucleotides that arises when short regions fold back on themselves and hydrogen bonds form between complementary bases. Also known as a stem-loop.

haploid. Having one set of chromosomes or one copy of the genome.

Haworth projection. A representation in which a cyclic sugar molecule is depicted as a flat ring that is projected perpendicular to the plane of the page. Heavy lines represent the part of the molecule that extends toward the viewer.

HDL. *See* high density lipoprotein

heat of vaporization. The amount of heat required to evaporate 1 gram of a liquid.

heat-shock gene. A gene whose transcription is increased in response to stresses such as high temperature. Many heat-shock genes that encode chaperones are also expressed in the absence of stress.

helicase. An enzyme that is involved in unwinding DNA.

helix-destabilizing protein. *See* single-strand binding protein

hemiacetal. The product formed when an alcohol reacts with an aldehyde or a ketone.

hemiketal. *See* hemiacetal

hemoglobin. A tetrameric, heme-containing globular protein in red blood cells that carries oxygen (O_2) to other cells and tissues.

Henderson-Hasselbalch equation. An equation that describes the pH of a solution in terms of the pK_a of a weak acid and the concentrations of the acid and its conjugate base.

$$pH = pK_a + \log \frac{[\text{conjugate base}]}{[\text{weak acid}]}$$

heterochromatin. Regions of chromatin that are highly condensed. The genes in heterochromatin are inaccessible to transcription factors and are not transcribed.

heterocyclic molecule. A molecule that contains a ring structure made up of more than one type of atom.

heterogeneous nuclear ribonucleoprotein complex (hnRNP). The complex formed by a nascent RNA transcript and specific proteins. The hnRNP protects the primary transcript from endogenous nucleases and inhibits the formation of secondary structures.

heterogeneous nuclear RNA (hnRNA). *See* mRNA precursor

heterotroph. An organism that requires at least one organic nutrient, such as glucose, as a carbon source.

hexose monophosphate shunt. *See* pentose phosphate pathway.

high density lipoprotein (HDL). A type of plasma lipoprotein that is enriched in protein and transports cholesterol and cholesteryl esters from tissues to the liver.

high-pressure liquid chromatography (HPLC). A chromatographic technique used to separate components of a mixture by dissolving the mixture in a liquid solvent and forcing it to flow through a chromatographic column under high pressure.

histones. A class of proteins that bind to DNA to form chromatin. The nuclei of most eukaryotic cells contain five histones, known as H1, H2A, H2B, H3, and H4.

hnRNA. *See* mRNA precursor

hnRNP. *See* heterogeneous nuclear ribonucleoprotein complex

Holliday junction. The region of strand crossover resulting from recombination between two molecules of homologous double-stranded DNA.

holoprotein. An active protein possessing all of its cofactors and subunits.

homeodomain. A DNA-binding protein structural motif of about 60 amino acid residues that form four α helices connected by turns.

homeotic gene. A development-regulating gene, mutation of which results in the transformation of one body structure to another.

homologous recombination. The exchange or transfer of DNA between molecules with very similar or identical nucleotide sequences.

homology. The similarity of genes or proteins as a result of evolution from a common ancestor.

hormone. A regulatory molecule synthesized in one cell or endocrine gland and transported to another cell or tissue. The target cell or tissue contains the appropriate hormone receptor and responds to the extracellular hormone, or first messenger, by producing intracellular second messengers, which alter the activity of certain enzymes.

hormone-response element. A DNA sequence that binds a transcriptional activator consisting of a steroid hormone–receptor complex.

housekeeping genes. Genes that encode proteins or RNA molecules that are essential for the normal activities of all living cells.

HPLC. *See* high-pressure liquid chromatography

hydration. A state in which a molecule or ion is surrounded by water.

hydrogen bond. A weak electrostatic interaction formed when a hydrogen atom bonded covalently to a strongly electronegative atom also bonds to the unshared electron pair of another electronegative atom.

hydrolase. An enzyme that catalyzes the hydrolytic cleavage of its substrate(s) (i.e., hydrolysis).

hydrolysis. Cleavage of a bond within a molecule by the addition of the elements of water.

hydropathy. A measure of the hydrophobicity of amino acid side chains. The more positive the hydropathy value, the greater the hydrophobicity.

hydropathy index. A measure of the overall hydrophobicity of a stretch of amino acid residues in a protein.

hydropathy plot. A graph of the hydropathy index of a protein sequence.

hydrophilicity. The degree to which a compound or functional group interacts with water or is preferentially soluble in water.

hydrophobic effect. The force that drives hydrophobic groups or molecules to associate with each other rather than with water. The hydrophobic effect appears to depend on the increase in entropy of solvent water molecules that are released from an ordered arrangement around the hydrophobic group.

hydrophobic interaction. A weak, noncovalent interaction between nonpolar molecules or substituents that results from the strong association of water molecules with one another. Such association leads to the shielding or exclusion of nonpolar molecules from an aqueous environment.

hydrophobicity. The degree to which a compound or functional group that is soluble in nonpolar solvents is insoluble or only sparingly soluble in water.

ICR. *See* internal control region

immunofluorescence microscopy. A technique for visualizing cellular components by first labelling them with specific fluorescent ligands or antibodies.

immunoglobulin. *See* antibody

in vitro. Occurring under artificial conditions, such as those in a laboratory, rather than under physiological conditions or in an intact organism.

in vivo. Occurring within a living cell or organism.

induced fit. A model for activation of an enzyme by a substrate-initiated conformational change.

inducer. A ligand that binds to and inactivates a repressor, thereby increasing the transcription of the gene controlled by the repressor.

inhibition constant (K_i). The equilibrium constant for the dissociation of an inhibitor from an enzyme-inhibitor complex.

initial velocity (v_0). The rate of conversion of substrate to product in the early stages of an enzymatic reaction, before appreciable product has been formed.

initiation codon. A codon that specifies the initiation site for protein synthesis. The methionine codon, AUG, is the most common initiation codon.

initiator-tRNA. The tRNA molecule that is used exclusively at initiation codons. In eukaryotes and archaebacteria, the initiator-tRNA is usually a methionyl-tRNA; in eubacteria, the methionine moiety is formylated.

insert DNA. The DNA fragment that is carried by a cloning vector.

integral membrane protein. A membrane protein that penetrates the hydrophobic core of the lipid bilayer and usually spans the bilayer completely. Also known as an intrinsic membrane protein.

integron. *See* transposon

intercalating agent. A compound containing a planar ring structure that can fit between the stacked base pairs of DNA. Intercalating agents distort the DNA structure, partially unwinding the double helix.

interconvertible enzyme. An enzyme whose activity is regulated by covalent modification. Interconvertible enzymes undergo transitions between active and inactive states but may be frozen in one state or the other by a covalent substitution.

intermediary metabolism. The metabolic reactions by which the small molecules of cells are interconverted.

intermediate filament. A double- or triple-stranded structure composed of different protein subunits, found in the cytoplasm of most eukaryotic cells. Intermediate filaments are components of the cytoskeletal network.

internal control region (ICR). A promoter sequence located within the transcribed region of a gene.

intrinsic membrane protein. *See* integral membrane protein

intron. An internal nucleotide sequence that is removed from the primary RNA transcript during processing. The term *intron* also refers to the region of the gene that encodes the corresponding RNA intron. Group I and group II introns are classified on the basis of conserved sequences and secondary structures.

introns-early hypothesis. The proposal that introns were present in primitive cells. According to this hypothesis, introns have been lost from most prokaryotic protein-encoding genes but have been retained in many eukaryotes.

introns-late hypothesis. The proposal that introns in eukaryotes arose as the result of insertion of DNA segments into protein-encoding genes.

inverted repeat. A sequence of nucleotides that is repeated in the opposite orientation within the same polynucleotide strand. An inverted repeat in double-stranded DNA can give rise to a cruciform structure.

ion pair. An electrostatic interaction between ionic groups of opposite charge within the interior of a macromolecule such as a globular protein.

ion product of water (K_w). The product of the concentrations of hydronium ions and hydroxide ions in an aqueous solution, equal to $1.0 \times 10^{-14} \, M^2$.

ion-exchange chromatography. A chromatographic technique used to separate a mixture of ionic species in solution, using a charged matrix. In anion-exchange chromatography, a positively charged matrix binds negatively charged solutes, and in cation-exchange chromatography, a negatively charged matrix binds positively charged solutes. The bound species can be serially eluted from the matrix by gradually changing the pH or increasing the salt concentration in the solvent.

ionophore. A compound that facilitates the diffusion of ions across bilayers and membranes by serving as a mobile ion carrier or by forming a channel for ion passage.

isoacceptor tRNA molecules. Different tRNA molecules that bind the same amino acid.

isoelectric focusing. A modified form of electrophoresis that uses buffers to create a pH gradient within a polyacrylamide gel. Each protein migrates to its isoelectric point (pI), that is, the pH in the gradient at which it no longer carries a net positive or negative charge.

isoelectric point (pI). The pH at which a zwitterionic molecule does not migrate in an electric field because its net charge is zero.

isoenzymes. *See* isozymes

isomerase. An enzyme that catalyzes an isomerization reaction, a change in geometry or structure within one molecule.

isopeptide bond. A peptide bond between the α-carboxylate group of an amino acid and the ε-amino group of a lysine residue.

isoprene. A branched, unsaturated five-carbon molecule that forms the basic structural unit of all polyprenyl compounds, including the steroids and lipid vitamins.

isoprenoid. *See* polyprenyl compound

isozymes. Different proteins from a single biological species that catalyze the same reaction. Also known as isoenzymes.

junk DNA. Regions of the genome with no apparent function.

K_a. *See* acid dissociation constant

karyotype. A set of chromosomes visualized by staining.

k_{cat}. *See* catalytic constant

k_{cat}/K_m. The second-order rate constant for conversion of enzyme and substrate to enzyme and product at low substrate concentrations. The ratio of k_{cat} to K_m, when used to compare several substrates, is called the specificity constant.

K_{eq}. *See* equilibrium constant

ketimine. *See* Schiff base

ketogenesis. The pathway that synthesizes ketone bodies from acetyl CoA in the mitochondrial matrix in mammals.

ketogenic compound. A compound, such as an amino acid, that can be degraded to form acetyl CoA and can thereby contribute to the synthesis of fatty acids or ketone bodies.

ketone bodies. Fuel molecules (β-hydroxybutyrate, acetoacetate, and acetone) that are synthesized in the liver from acetyl CoA. During starvation, ketone bodies become a major metabolic fuel.

ketoses. A class of monosaccharides in which the most oxidized carbon atom, usually C-2, is ketonic.

K_i. *See* inhibition constant

kinetic order. The sum of the exponents in a rate equation, which reflects how many molecules are reacting in the slowest step of the reaction. Also known as reaction order.

kinetics. The study of rates of change, such as the rates of chemical reactions.

Klenow fragment. The C-terminal 605-residue fragment of *E. coli* DNA polymerase I produced by partial proteolysis. The Klenow fragment contains both the $5' \rightarrow 3'$ polymerase and $3' \rightarrow 5'$ proofreading exonuclease activities of DNA polymerase I but lacks the $5' \rightarrow 3'$ exonuclease activity of the intact enzyme.

K_m. *See* Michaelis constant

Krebs cycle. *See* citric acid cycle

K_{tr}. *See* transport constant

K_w. *See* ion product of water

L. *See* linking number

lagging strand. The newly synthesized DNA strand formed by discontinuous $5' \rightarrow 3'$ polymerization in the direction opposite replication fork movement.

late gene. A gene that is expressed relatively late in infection by a bacteriophage or virus.

lateral diffusion. The rapid motion of lipid or protein molecules within the plane of one leaflet of a lipid bilayer.

LDL. *See* low density lipoprotein

leader peptide. The peptide encoded by a portion of the leader region of an operon. Synthesis of a leader peptide is the basis for regulating transcription of the entire operon by the mechanism of attenuation.

leader region. The sequence of nucleotides that lies between the promoter and the first coding region of an operon.

leading strand. The newly synthesized DNA strand formed by continuous $5' \rightarrow 3'$ polymerization in the same direction as replication fork movement.

leaflet. One layer of a lipid bilayer.

leaving group. The displaced group resulting from cleavage of a covalent bond.

lectin. A plant protein that binds specific saccharides in glycoproteins.

leucine zipper. A structural motif found in DNA-binding proteins and other proteins. The zipper is formed when the hydrophobic faces (frequently containing leucine residues) of two amphipathic α helices from the same or different polypeptide chains interact to form a coiled-coil structure.

LHC. *See* chlorophyll *a/b* light harvesting complex

ligand. A molecule, group, or ion that binds noncovalently to another molecule or atom.

ligand-gated ion channel. A membrane ion channel that opens or closes in response to binding of a specific ligand.

ligand-induced theory. *See* sequential theory of cooperativity and allosteric regulation

ligase. An enzyme that catalyzes the joining, or ligation, of two substrates. Ligation reactions require the input of the chemical potential energy of a nucleoside triphosphate such as ATP. Ligases are commonly referred to as synthetases.

limit dextrin. A branched oligosaccharide derived from a glucose polysaccharide by the hydrolytic action of amylase or the phosphorolytic action of glycogen phosphorylase or starch phosphorylase. Limit dextrins are resistant to further degradation catalyzed by amylase or phosphorylase. Limit dextrins can be further degraded only after hydrolysis of the α-$(1 \rightarrow 6)$ linkages.

Lineweaver-Burk plot. *See* double-reciprocal plot

linker DNA. The stretch of DNA (approximately 54 base pairs) between two adjacent nucleosome core particles.

linking number (L). A topological property of DNA, defined as the number of times that one strand of a circular DNA molecule crosses over another when the molecule is lying flat on a plane. The linking number is positive in right-handed helices.

lipase. An enzyme that catalyzes the hydrolysis of triacylglycerols.

lipid. A water-insoluble (or sparingly soluble) organic compound found in biological systems, which can be extracted by using relatively nonpolar organic solvents.

lipid bilayer. A double layer of lipids in which the hydrophobic tails associate with one another in the interior of the bilayer and the polar head groups face outward into the aqueous environment. Lipid bilayers are the structural basis of biological membranes.

lipid vitamin. A polyprenyl compound composed primarily of a long hydrocarbon chain or fused ring. Unlike water-soluble vitamins, lipid vitamins can be stored by animals. Lipid vitamins include vitamins A, D, E, and K.

lipid-anchored membrane protein. A membrane protein that is tethered to a membrane through covalent linkage to a lipid molecule.

lipolysis. The metabolic hydrolysis of triacylglycerols.

lipopolysaccharide. A macromolecule composed of lipid A (a disaccharide of phosphorylated glucosamine residues with attached fatty acids) and a polysaccharide. Lipopolysaccharides are found in the outer membrane of Gram-negative bacteria. These compounds are released from bacteria undergoing lysis and are toxic to humans and other animals. Also known as an endotoxin.

lipoprotein. A macromolecular assembly of lipid and protein molecules with a hydrophobic core and a hydrophilic surface. Lipids are transported via lipoproteins.

liposome. A synthetic vesicle composed of a phospholipid bilayer that encloses an aqueous compartment.

local regulators. A class of short-lived eicosanoids that exert their regulatory effects near the cells in which they are produced.

long terminal repeats (LTR). Identical sequences of several hundred base pairs found in the same orientation at the ends of a retroviral genome.

long-patch repair. A type of mismatch repair in which a large piece of DNA containing the mismatch is excised and replaced. Also known as methyl-directed repair.

loop. A nonrepetitive polypeptide region that connects secondary structures within a protein molecule and provides directional changes necessary for a globular protein to attain its compact shape. Loops contain from 2 to 16 residues. Short loops of up to 5 residues are often called turns.

low density lipoprotein (LDL). A type of plasma lipoprotein that is formed during the breakdown of VLDL and is enriched in cholesterol and cholesteryl esters.

LTR. *See* long terminal repeats

lyase. An enzyme that catalyzes a nonhydrolytic or nonoxidative elimination reaction, or lysis, of a substrate, with the generation of a double bond. In the reverse direction, a lyase catalyzes addition of one substrate to a double bond of a second substrate.

lysogenic pathway. The events occurring following bacteriophage infection that result in stable integration of the phage genome into the host genome, where it may remain for many generations.

lysophosphoglyceride. An amphipathic lipid that is produced when one of the two fatty acyl moieties of a glycerophospholipid is hydrolytically removed. Low concentrations of lysophosphoglycerides serve as metabolic intermediates, whereas high concentrations disrupt membranes, causing cells to lyse.

lysosome. A specialized digestive organelle in eukaryotic cells. Lysosomes contain a variety of enzymes that catalyze the breakdown of cellular biopolymers, such as proteins, nucleic acids, and polysaccharides, and the digestion of large particles, such as some bacteria ingested by the cell.

lytic pathway. The events occurring following bacteriophage infection that lead to host cell lysis and release of phage progeny.

major groove. The wide groove on the surface of a DNA double helix created by the stacking of base pairs and the resulting twist in the sugar-phosphate backbones.

marker gene. A gene, carried by a cloning vector, that can be used to distinguish between host cells that carry the vector and those that do not.

mass action ratio (Q). The ratio of the concentrations of products to the concentrations of reactants of a reaction. The mass action ratio is related to the free-energy change of a reaction by the equation

$$\Delta G = \Delta G^{\circ\prime} + RT \ln Q$$

maternal gene. A gene, expressed in the female, whose product acts during development of the progeny.

maximum velocity (V_{max}). The initial velocity of a reaction when the enzyme is saturated with substrate, that is, when all the enzyme is in the form of an enzyme-substrate complex.

melting temperature (T_m). The midpoint of the temperature range in which double-stranded DNA is converted to single-stranded DNA.

membrane. A lipid bilayer containing associated proteins that serves to delineate and compartmentalize cells or organelles. Biological membranes are also the site of many important biochemical processes related to energy transduction and intracellular signalling.

membrane potential ($\Delta\psi$). The charge separation across a membrane that results from differences in ionic concentrations on the two sides of the membrane.

messenger ribonucleic acid (mRNA). A class of RNA molecules that serve as templates for protein synthesis.

metabolic fuel. A small compound that can be catabolized to release energy. In multicellular organisms, metabolic fuels may circulate in the body and be taken up by cells.

metabolically irreversible reaction. A reaction in which the value of the mass action ratio is two or more orders of magnitude smaller than the value of the equilibrium constant. The free-energy change for such a reaction is a large negative number; thus, the reaction is irreversible.

metabolism. The sum total of biochemical reactions carried out by an organism.

metabolite. An intermediate in the synthesis or degradation of biopolymers and their component units.

metabolite coenzyme. A coenzyme synthesized from a common metabolite.

metal-activated enzyme. An enzyme that either has an absolute requirement for metal ions or is stimulated by the addition of metal ions.

metalloenzyme. An enzyme that contains one or more firmly bound metal ions. In some cases, such metal ions constitute part of the active site of the enzyme and are active participants in catalysis.

methyl-directed repair. *See* long-patch repair

micelle. An aggregation of amphipathic molecules in which the hydrophilic portions of the molecules project into the aqueous environment and the hydrophobic portions associate with one another in the interior of the structure to minimize contact with water molecules.

Michaelis constant (K_m). The concentration of substrate that results in an initial velocity (v_0) equal to one-half the maximum velocity (V_{max}) for a given reaction.

Michaelis-Menten equation. A rate equation relating the initial velocity (v_0) of an enzymatic reaction to the substrate concentration ([S]) where V_{max} is the maximum velocity and K_m is the Michaelis constant.

$$v_0 = \frac{V_{max}[S]}{K_m + [S]}$$

microfilament. *See* actin filament

microtubule. A protein filament composed of α and β tubulin heterodimers assembled around a hollow core in a helical arrangement, creating a strong, rigid fiber. Microtubules are components of the cytoskeletal network and can form structures capable of directed movement.

microvilli. Projections of the plasma membrane of epithelial cells that increase the surface area of the cells. Microvilli form a convoluted surface called the brush border.

minor groove. The narrow groove on the surface of a DNA double helix created by the stacking of base pairs and the resulting twist in the sugar-phosphate backbones.

mismatch repair. Restoration of the normal nucleotide sequence in a DNA molecule containing mismatched bases. In mismatch repair, the correct strand is recognized, a portion of the incorrect strand is excised, and correctly base-paired, double-stranded DNA is synthesized by the actions of DNA polymerase and DNA ligase.

missense mutation. An alteration in DNA that involves the substitution of one nucleotide for another, resulting in a change in the amino acid specified by that codon.

mitochondrial matrix. The gel-like phase enclosed by the inner membrane of the mitochondrion. The mitochondrial matrix contains many enzymes involved in aerobic energy metabolism.

mitochondrion. An organelle that serves as the main site of oxidative energy metabolism in most eukaryotic cells. Mitochondria contain an outer and an inner membrane, the latter characteristically folded into cristae.

mixed micelle. A micelle containing more than one type of amphipathic molecule.

mixed oligonucleotide probe. Short stretches of DNA corresponding to each of the possible sequences that could encode a given protein sequence. A mixed oligonucleotide probe is used to screen a DNA library for the presence of the protein-encoding clone.

MOI. *See* multiplicity of infection

molar mass. The weight in grams of one mole of a compound.

molecular weight. *See* relative molecular mass

molecular-exclusion chromatography. *See* gel-filtration chromatography

monocistronic mRNA. An mRNA molecule that encodes only a single polypeptide. Most eukaryotic mRNA molecules are monocistronic.

monomer. A small compound that becomes a residue when polymerized with other monomers.

monosaccharide. A simple sugar of three or more carbon atoms with the empirical formula $(CH_2O)_n$.

monotopic protein. An integral membrane protein anchored by a single membrane-spanning segment.

motif. *See* supersecondary structure

M_r. *See* relative molecular mass

mRNA. *See* messenger ribonucleic acid

mRNA precursor. A class of RNA molecules synthesized by eukaryotic RNA polymerase II. mRNA precursors are processed posttranscriptionally to produce messenger RNA. Also known as heterogeneous nuclear RNA (hnRNA).

mucin. A high-molecular-weight *O*-linked glycoprotein containing as much as 80% carbohydrate by mass. Mucins are extended, negatively charged molecules that contribute to the viscosity of mucus, the fluid found on the surfaces of the gastrointestinal, genitourinary, and respiratory tracts.

multiplicity of infection (MOI). The number of phage or virus particles that on average infect a single host cell.

mutagen. An agent that can cause DNA damage.

mutation. A heritable change in the sequence of nucleotides in DNA that causes a permanent alteration of genetic information.

near-equilibrium reaction. A reaction in which the value of the mass action ratio is close to the value of the equilibrium constant. The free-energy change for such a reaction is small; thus, the reaction is reversible.

Nernst equation. An equation that relates the observed change in reduction potential (ΔE) to the change in standard reduction potential ($\Delta E^{\circ\prime}$) of a reaction.

$$\Delta E = \Delta E^{\circ\prime} - \frac{RT}{n\mathcal{F}} \ln \frac{[A_{ox}][B_{red}]}{[A_{red}][B_{ox}]}$$

neutral solution. An aqueous solution that has a pH value of 7.0.

nick translation. The process in which DNA polymerase binds to a gap between the 3′ end of a nascent DNA chain and the 5′ end of the next RNA primer, catalyzes hydrolytic removal of ribonucleotides using the 5′→3′ exonuclease activity, and replaces them with deoxyribonucleotides using the 5′→3′ polymerase activity.

nitrogen fixation. The reduction of atmospheric nitrogen to ammonia. Biological nitrogen fixation occurs in only a few species of bacteria and algae.

N-linked glycoprotein. A glycoprotein in which one or more oligosaccharide chains are attached to the protein through covalent bonds to the amide nitrogen atom of the side chain of asparagine residues. The oligosaccharide chains of *N*-linked glycoproteins contain a core pentasaccharide of two *N*-acetyl-glucosamine residues and three mannose residues.

NMR spectroscopy. *See* nuclear magnetic resonance spectroscopy

noncompetitive inhibition. Inhibition of an enzyme-catalyzed reaction by an inhibitor that binds to either the enzyme or the enzyme-substrate complex.

nonessential amino acid. An amino acid that an animal can produce in sufficient quantity to meet metabolic needs.

nonsense mutation. An alteration in DNA that involves the substitution of one nucleotide for another, changing a codon that specifies an amino acid to a termination codon. A nonsense mutation results in premature termination of a protein's synthesis.

N-terminus. The amino acid residue bearing a free α-amino group at one end of a peptide chain. In some proteins, the N-terminus is blocked by acylation. The N-terminal residue is usually assigned the residue number 1. Also known as the amino terminus.

nuclear envelope. The double membrane that surrounds the nucleus and contains protein-lined nuclear pore complexes that regulate the import and export of material to and from the nucleus. The outer membrane of the nuclear envelope is continuous with the endoplasmic reticulum; the inner membrane is lined with filamentous proteins, constituting the nuclear lamina.

nuclear magnetic resonance spectroscopy (NMR spectroscopy). A technique used to study the structures of molecules in solution. In nuclear magnetic resonance spectroscopy, the absorption of electromagnetic radiation by molecules in magnetic fields of varying frequencies is used to determine the spin states of certain atomic nuclei.

nuclease. An enzyme that catalyzes hydrolysis of the phosphodiester linkages of a polynucleotide chain. Nucleases can be classified as endonucleases and exonucleases.

nuclease protection. *See* footprinting

nucleic acid. A polymer composed of nucleotide residues linked in a linear sequence by 3′–5′ phosphodiester bonds. DNA and RNA are nucleic acids composed of deoxyribonucleotide residues and ribonucleotide residues, respectively.

nucleoid region. The region within a prokaryotic cell that contains the chromosome.

nucleolus. The region of the eukaryotic nucleus where rRNA transcripts are processed and ribosomes are assembled.

nucleophile. An electron-rich species that is negatively charged or contains unshared electron pairs and is attracted to chemical species that are positively charged or electron-deficient (electrophiles).

nucleosome. A DNA-protein complex that forms the fundamental unit of chromatin. A nucleosome consists of a nucleosome core particle (approximately 146 base pairs of DNA plus a histone octamer), linker DNA (approximately 54 base pairs), and histone H1 (which binds the core particle and linker DNA).

nucleosome core particle. A DNA-protein complex composed of approximately 146 base pairs of DNA wrapped around an octamer of histones (two each of H2A, H2B, H3, and H4).

nucleotide. The phosphate ester of a nucleoside, consisting of a nitrogenous base linked to a pentose phosphate. Nucleotides are the monomeric units of nucleic acids.

nucleotide probe. A labelled oligonucleotide used to screen DNA or RNA molecules for the presence of a specific complementary sequence.

nucleus. An organelle that contains the principal genetic material of eukaryotic cells and functions as the major site of RNA synthesis and processing.

obligate aerobe. An organism that requires the presence of oxygen for survival.

obligate anaerobe. An organism that requires an oxygen-free environment for survival.

obligatory glycolytic tissue. A tissue with little or no capacity for oxidative metabolism and for which glucose is the only usable fuel.

Okazaki fragments. Relatively short strands of DNA, about 1000 residues in length, that are produced during discontinuous synthesis of the lagging strand of DNA.

oligomer. A multisubunit molecule whose arrangement of subunits always has a defined stoichiometry and almost always displays symmetry.

oligonucleotide. A polymer of several (up to about 20) nucleotide residues.

oligonucleotide linker. A short, synthetic, double-stranded DNA molecule whose sequence contains a restriction endonuclease cleavage site. An oligonucleotide linker can be ligated to a blunt-ended DNA molecule and then cleaved by the endonuclease to generate a DNA molecule with sticky ends.

oligopeptide. A peptide containing several (up to about 20) amino acid residues.

oligosaccharide. A polymer of 2 to about 20 monosaccharide residues linked by glycosidic bonds.

oligosaccharide processing. The enzyme-catalyzed addition and removal of saccharide residues during the maturation of a glycoprotein.

***O*-linked glycoprotein.** A glycoprotein in which one or more oligosaccharide chains are attached to the protein through covalent bonds, usually to the hydroxyl oxygen atom of serine or threonine residues.

oncogene. A gene whose product has the ability to transform normal eukaryotic cells into cancer cells. Some oncogenes are carried by viruses.

open reading frame. A stretch of nucleotide triplets that contains no termination codons. Protein-encoding regions are examples of open reading frames.

operator. A DNA sequence to which a specific repressor protein binds, thereby blocking transcription of a gene or operon.

operon. A bacterial transcriptional unit consisting of several different genes co-transcribed from one promoter.

ordered kinetic mechanism. A reaction mechanism in which both the binding of substrates to an enzyme and the release of products from the enzyme follow an obligatory order.

organ. An association of one or more tissues that carry out discrete functions within a multicellular organism.

organelle. Any specialized membrane-bounded structure within a eukaryotic cell. Organelles are uniquely organized to perform specific functions.

origin of replication. A DNA sequence at which replication is initiated.

oxidase. An enzyme that catalyzes an oxidation-reduction reaction in which O_2 is the electron acceptor. Oxidases are members of the IUB class of enzymes known as oxidoreductases.

oxidation. The loss of electrons from a substance through transfer to another substance (the oxidizing agent). Oxidations can take several forms, including the addition of oxygen to a compound, the removal of hydrogen from a compound to create a double bond, or an increase in the valence of a metal ion.

oxidative phosphorylation. A set of reactions in which compounds such as NADH and reduced ubiquinone (QH_2) are aerobically oxidized and ATP is generated from ADP and P_i. Oxidative phosphorylation consists of two tightly coupled phenomena: oxidation of substrates by the respiratory electron-transport chain, accompanied by the translocation of protons across the inner mitochondrial membrane to generate a proton concentration gradient; and formation of ATP, driven by the flux of protons into the matrix through a channel in F_OF_1 ATP synthase.

oxidizing agent. A substance that accepts electrons in an oxidation-reduction reaction and thereby becomes reduced.

oxidoreductase. An enzyme that catalyzes an oxidation-reduction reaction. Some oxidoreductases are known as reductases.

oxygenase. An enzyme that catalyzes incorporation of molecular oxygen into a substrate. Oxygenases are members of the IUB class of enzymes known as oxidoreductases.

P site. *See* peptidyl site

Δp. *See* protonmotive force

PAGE. *See* polyacrylamide gel electrophoresis

pair-rule gene. A development-regulating gene that is expressed in every other segment of the *Drosophila* embryo.

pancreatic juice. A solution of sodium chloride, zymogens, and digestive enzymes secreted by the pancreas during digestion. Pancreatic juice may also contain sodium bicarbonate.

passive transport. The process by which a solute specifically binds to a transport protein and is transported across a membrane, moving with the solute concentration gradient. Passive transport occurs without the expenditure of energy. Also known as facilitated diffusion.

Pasteur effect. The slowing of glycolysis in the presence of oxygen.

pathway. A sequence of metabolic reactions.

pause site. A region of a gene where transcription slows. Pausing is exaggerated at palindromic sequences, where newly synthesized RNA can form a hairpin structure.

PCR. *See* polymerase chain reaction

pentose phosphate pathway. A pathway by which glucose 6-phosphate is metabolized to generate NADPH and ribose 5-phosphate. In the oxidative stage of the pathway, glucose 6-phosphate is converted to ribulose 5-phosphate and CO_2, generating two molecules of NADPH. In the nonoxidative stage, ribulose 5-phosphate can be isomerized to ribose 5-phosphate or converted to two intermediates of glycolysis, fructose 6-phosphate and glyceraldehyde 3-phosphate. Also known as the hexose monophosphate shunt.

peptide. Two or more amino acids covalently joined in a linear sequence by peptide bonds.

peptide bond. The covalent secondary amide linkage that joins the carbonyl group of one amino acid residue to the amino nitrogen of another in peptides and proteins.

peptide group. The nitrogen and carbon atoms involved in a peptide bond and their four substituents: the carbonyl oxygen atom, the amide hydrogen atom, and the two adjacent α-carbon atoms.

peptidoglycan. A macromolecule containing a heteroglycan chain of alternating N-acetylglucosamine and N-acetylmuramic acid cross-linked to peptides of varied composition. Peptidoglycans are the major components of the cell walls of many bacteria.

peptidyl site. The site on a ribosome that is occupied during protein synthesis by the tRNA molecule attached to the growing polypeptide chain. Also known as the P site.

peptidyl transferase. The enzymatic activity responsible for the formation of a peptide bond during protein synthesis.

peptidyl-tRNA. The tRNA molecule to which the growing peptide chain is attached during protein synthesis.

peripheral membrane protein. A membrane protein that is weakly bound to the interior or exterior surface of a membrane through ionic interactions and hydrogen bonding with the polar heads of the membrane lipids or with an integral membrane protein. Also known as an extrinsic membrane protein.

periplasmic space. The region between the plasma membrane and the cell wall in bacteria.

permeability coefficient. A measure of the ability of an ion or small molecule to diffuse across a lipid bilayer.

peroxidase. An enzyme that catalyzes a reaction in which hydrogen peroxide (H_2O_2) is the oxidizing agent. Peroxidases are members of the IUB class of enzymes known as oxidoreductases.

peroxisome. An organelle in all animal and most plant cells that carries out oxidation reactions that produce the toxic compound hydrogen peroxide (H_2O_2). Peroxisomes contain the enzyme catalase, which catalyzes the breakdown of toxic H_2O_2 to water and O_2.

pH. A logarithmic quantity that indicates the acidity of a solution, that is, the concentration of hydronium ions in solution. pH is defined as the negative logarithm of the hydronium ion concentration.

pH optimum. In an enzyme-catalyzed reaction, the pH at the point of maximum catalytic activity.

phage. *See* bacteriophage

phase variation. A switch in gene expression between two genes. Phase variation helps invading pathogens evade the host's immune response.

phase-transition temperature (T_m). The midpoint of the temperature range in which lipids or other macromolecular aggregates are converted from a highly ordered phase or state (such as a gel) to a less-ordered state (such as a liquid crystal).

ϕ (phi). The angle of rotation around the bond between the α-carbon and the nitrogen of a peptide group.

phosphagen. A high-energy phosphate-storage molecule found in animal muscle cells. Phosphagens are phosphoamides and have a higher group-transfer potential than ATP.

phosphatidate. A glycerophospholipid that consists of two fatty acyl groups esterified to C-1 and C-2 of glycerol 3-phosphate. Phosphatidates are metabolic intermediates in the biosynthesis or breakdown of more complex glycerophospholipids.

phosphatidylinositol-glycan–linked glycoprotein. A glycoprotein in which the protein is attached to a phosphoethanolamine moiety that is linked to a branched oligosaccharide to which the lipid phosphatidylinositol is also attached. The phosphatidylinositol-glycan structure is known as a GPI membrane anchor.

phosphoanhydride. A compound formed by condensation of two phosphate groups.

phosphodiester linkage. A linkage in nucleic acids and other molecules in which two alcoholic hydroxyl groups are joined through a phosphate group.

phosphoester linkage. The bond by which a phosphoryl group is attached to an alcoholic or phenolic oxygen.

phospholipid. A lipid containing a phosphate moiety.

phosphorolysis. Cleavage of a bond within a molecule by the addition of the elements of inorganic phosphate (P_i).

phosphorylase. An enzyme that catalyzes the cleavage of its substrate(s) via nucleophilic attack by inorganic phosphate (P_i) (i.e., via phosphorolysis).

phosphorylation. A reaction involving the addition of a phosphoryl group to a molecule.

photon. A quantum of light energy.

photophosphorylation. The light-dependent formation of ATP from ADP and P_i catalyzed by chloroplast ATP synthase.

photorespiration. The light-dependent uptake of O_2 and the subsequent metabolism of phosphoglycolate that occurs primarily in C_3 photosynthetic plants. Photorespiration can occur because O_2 competes with CO_2 for the active site of ribulose 1,5-*bis*phosphate carboxylase-oxygenase, the enzyme that catalyzes the first step of the reductive pentose phosphate cycle.

photosynthesis. The process by which carbohydrates are synthesized from atmospheric CO_2 and water using light as the source of energy.

photosynthetic carbon reduction cycle. *See* reductive pentose phosphate cycle

photosystem. A functional unit of the light-dependent reactions of photosynthesis. Each membrane-embedded photosystem contains a reaction center, which forms the core of the photosystem, and a pool of light-absorbing antenna pigments.

phototroph. An organism that can convert light energy into chemical potential energy (i.e., an organism capable of photosynthesis).

physiological pH. The normal pH of human blood, which is 7.4.

pI. *See* isoelectric point

ping-pong reaction. A reaction in which an enzyme binds one substrate and releases a product, leaving a substituted enzyme that then binds a second substrate and releases a second product, thereby restoring the enzyme to its original form.

pitch. The distance required to complete one turn of a helical structure.

pK_a. A logarithmic value that indicates the strength of an acid. pK_a is defined as the negative logarithm of the acid dissociation constant, K_a.

plaque. An area of dead or slowly growing bacterial cells that indicates the presence of bacteriophage-infected cells among uninfected cells.

plasma membrane. The membrane that surrounds the cytoplasm of a cell and thus defines the perimeter of the cell.

plasmalogen. A glycerophospholipid that has a hydrocarbon chain linked to C-1 of glycerol 3-phosphate through a vinyl ether bond. Plasmalogens are found in the central nervous system and in the membranes of cells of peripheral nerves and muscle tissue.

plasmid. A relatively small, extrachromosomal DNA molecule in bacteria and yeast that is capable of autonomous replication. Plasmids are closed, circular, double-stranded DNA molecules.

P:O ratio. The ratio of molecules of ADP phosphorylated to atoms of oxygen reduced during oxidative phosphorylation.

polar. Having uneven distribution of charge. A molecule or functional group is polar if its center of negative charge does not coincide with its center of positive charge.

pole cells. The cells at the posterior end of an insect egg that give rise to germline tissues.

poly A tail. A stretch of polyadenylate, up to 250 nucleotide residues in length, that is added to the 3' end of a eukaryotic mRNA molecule following transcription.

polyacrylamide gel electrophoresis (PAGE). A technique used to separate molecules of different net charge and/or size based on their migration through a highly cross-linked gel matrix in an electric field.

polycistronic mRNA. An mRNA molecule that contains multiple coding regions. Many prokaryotic mRNA molecules are polycistronic.

polymerase chain reaction (PCR). A method for amplifying the amount of DNA in a sample and for enriching a particular DNA sequence in a population of DNA molecules. In the polymerase chain reaction, oligonucleotides complementary to the ends of the desired DNA sequence are used as primers for multiple rounds of DNA synthesis.

polynucleotide. A polymer of many (usually more than 20) nucleotide residues.

polyol pathway. *See* sorbitol pathway

polypeptide. A peptide containing many (usually more than 20) amino acid residues.

polyprenyl compound. A lipid that is structurally related to isoprene. Also known as an isoprenoid.

polyribosome. *See* polysome

polysaccharide. A polymer of many (usually more than 20) monosaccharide residues linked by glycosidic bonds. Polysaccharide chains can be linear or branched.

polysome. The structure formed by the binding of many translation complexes to a large mRNA molecule. Also known as a polyribosome.

polytene chromosome. A chromosome that is replicated many times without separation of the copies. The resulting structure is composed of regions of condensed chromatin and regions expanded to form puffs.

polytopic protein. An integral membrane protein anchored by multiple membrane-spanning segments.

pore. *See* channel

postabsorptive phase. The period following the absorptive phase. During the postabsorptive phase, which lasts about 12 hours in humans, glucose is mobilized from glycogen stores and is synthesized via gluconeogenesis.

posttranscriptional processing. RNA processing that occurs after transcription is complete.

posttranslational modification. Covalent modification of a protein that occurs after elongation of the polypeptide is complete.

prenylated protein. A lipid-anchored protein that is covalently linked to an isoprenoid moiety via the —SH group of a cysteine residue at the C-terminus of the protein.

primary structure. The sequence in which residues are covalently linked to form a polymeric chain.

primary transcript. A newly synthesized RNA molecule before processing.

primase. An enzyme in the primosome that catalyzes the synthesis of short pieces of RNA about 10 residues in length. These oligonucleotides serve as the primers for synthesis of Okazaki fragments.

primosome. A multiprotein complex, including primase and helicase in *E. coli,* that catalyzes the synthesis of the short RNA primers needed for discontinuous DNA synthesis of the lagging strand.

probe. *See* nucleotide probe

processive enzyme. An enzyme that remains bound to its growing polymeric product through many polymerization steps.

prochiral atom. An atom with multiple substituents, two of which are identical. A prochiral atom can become chiral when one of the identical substituents is replaced.

proenzyme. *See* zymogen

promoter. The region of DNA where RNA polymerase binds during transcription initiation.

prophage. A bacteriophage genome that has been integrated into a host chromosome. The prophage is replicated every time the host genome is replicated.

prostaglandin. An eicosanoid that has a cyclopentane ring. Prostaglandins serve as metabolic regulators that act in the immediate neighborhood of the cells in which they are produced.

prosthetic group. A coenzyme that is tightly bound to an enzyme. A prosthetic group, unlike a cosubstrate, remains bound to a specific site of the enzyme throughout the catalytic cycle of the enzyme.

protease. An enzyme that catalyzes hydrolysis of peptide bonds. The physiological substrates of proteases are proteins.

protein. A biopolymer consisting of one or more polypeptide chains. The biological function of each protein molecule depends not only on the sequence of covalently linked amino acid residues, but also on its three-dimensional structure (conformation).

protein coenzyme. *See* group-transfer protein

protein glycosylation. The covalent addition of carbohydrate to proteins. In *N*-glycosylation, the carbohydrate is attached to the amide group of the side chain of an asparagine residue. In *O*-glycosylation, the carbohydrate is attached to the hydroxyl group of the side chain of a serine or threonine residue.

protein kinase. An enzyme that catalyzes the phosphorylation of protein substrates.

protein phosphatase. An enzyme that catalyzes the hydrolytic removal of a phosphoryl group from a phosphoprotein.

proteoglycan. A complex of proteins containing glycosaminoglycan chains covalently bound through their anomeric carbon atoms. Up to 95% of the mass of a proteoglycan may be glycosaminoglycan.

protonmotive force (Δp). The energy stored in a proton concentration gradient across a membrane.

proto-oncogene. A eukaryotic gene that can be mutated to an oncogene.

proximity effect. The increase in the rate of a nonenzymatic or enzymatic reaction attributable to high effective concentrations of reactants, which result in more frequent formation of transition states.

pseudo first-order reaction. A multi-reactant reaction carried out under conditions where the rate depends on the concentration of only one reactant.

pseudogene. A nonexpressed sequence of DNA that evolved from a protein-encoding gene. Pseudogenes often contain mutations in their coding regions and cannot produce functional proteins.

ψ (psi). The angle of rotation around the bond between the α-carbon and the carbonyl carbon of a peptide group.

$\Delta\psi$. *See* membrane potential

purine. A nitrogenous base having a bicyclic structure in which a pyrimidine is fused to an imidazole ring. Adenine and guanine are substituted purines found in both DNA and RNA.

pyranose. A monosaccharide structure that forms a six-membered ring as a result of intramolecular hemiacetal formation.

pyrimidine. A nitrogenous base having a heterocyclic ring that consists of four carbon atoms and two nitrogen atoms. Cytosine, thymine, and uracil are substituted pyrimidines found in nucleic acids (cytosine in DNA and RNA, uracil in RNA, and thymine principally in DNA).

Q. 1. An abbreviation for mass action ratio. 2. An abbreviation for ubiquinone.

Q cycle. A cyclic pathway proposed to explain the sequence of electron transfers and proton movements within Complex III of mitochondria or the cytochrome *bf* complex in chloroplasts. The net result of the two steps of the Q cycle is oxidation of two molecules of QH_2 or plastoquinol, PQH_2; formation of one molecule of QH_2 or PQH_2; transfer of two electrons; and net translocation of four protons across the inner mitochondrial membrane to the intermembrane space or across the thylakoid membrane to the lumen.

quaternary structure. The organization of two or more polypeptide chains within a multisubunit protein.

R group. A part of a molecule not explicitly shown in a chemical structure, such as an alkyl group or the side chain of an amino acid.

R state. The more active conformation of an allosteric enzyme; opposite of T state.

Ramachandran plot. A plot of ψ versus ϕ values for amino acid residues in a polypeptide chain. Certain ϕ and ψ values are characteristic of different conformations.

random kinetic mechanism. A reaction mechanism in which neither the binding of substrates to an enzyme nor the release of products from the enzyme follows an obligatory order.

rate acceleration. The ratio of the rate constant for a reaction in the presence of enzyme (k_{cat}) divided by the rate constant for that reaction in the absence of enzyme (k_n). The rate acceleration value is a measure of the efficiency of an enzyme.

rate-determining step. The slowest step in a chemical reaction. The rate-determining step has the highest activation energy among the steps leading to formation of a product from the substrate.

reaction center. A complex of proteins, specialized electron-transfer molecules, and a special pair of chlorophyll molecules that forms the core of a photosystem. The reaction center is the site of conversion of photochemical energy to electrochemical energy during photosynthesis.

reaction mechanism. The step-by-step atomic or molecular events that occur during chemical reactions.

reaction order. *See* kinetic order

reading frame. The sequence of nonoverlapping codons of an mRNA molecule that specifies the amino acid sequence. The reading frame of an mRNA molecule is determined by the position where translation begins; usually an AUG codon.

receptor. A cell-surface protein that binds a specific ligand, leading to some cellular response.

recombinant DNA. A DNA molecule that includes DNA from different sources.

recombinant DNA technology. The methodologies for isolating, manipulating, and amplifying identifiable sequences of DNA. Also known as genetic engineering.

recombination. The exchange or transfer of DNA from one chromosome to another.

reducing agent. A substance that loses electrons in an oxidation-reduction reaction and thereby becomes oxidized.

reducing end. The residue containing a free anomeric carbon in a polysaccharide. A polysaccharide usually contains no more than one reducing end.

reductase. *See* oxidoreductase

reduction. The gain of electrons by a substance through transfer from another substance (the reducing agent). Reductions can take several forms, including the loss of oxygen from a compound, the addition of hydrogen to a double bond of a compound, or a decrease in the valence of a metal ion.

reduction potential (E). A measure of the tendency of a substance to reduce other substances.

reductive pentose phosphate cycle (RPP cycle). A cycle of reactions that leads to the reductive conversion of carbon dioxide to carbohydrates during photosynthesis. Also known as the C_3 pathway, the photosynthetic carbon reduction cycle, and the Calvin-Benson cycle.

regulatory enzyme. An enzyme located at a critical point within one or more metabolic pathways, whose activity may be increased or decreased based on metabolic demand.

regulatory site. A ligand-binding site in a regulatory enzyme distinct from the active site. Allosteric modulators alter enzyme activity by binding to the regulatory site. Also known as an allosteric site.

regulon. A group of coordinately regulated operons or genes.

relative molecular mass (M_r). The mass of a molecule relative to 1/12th the mass of ^{12}C.

release factor. A protein that is involved in terminating protein synthesis.

renaturation. The restoration of the native conformation of a biological macromolecule, usually resulting in restoration of biological activity.

replication. The duplication of double-stranded DNA, during which parental strands separate and serve as templates for synthesis of new strands. Replication is carried out by DNA polymerase and associated factors.

replication fork. The Y-shaped junction where double-stranded, template DNA is unwound and new DNA strands are synthesized during replication.

replicative transposition. The mechanism whereby bacterial class II transposons integrate into the host chromosome. In replicative transposition, site-specific recombination between two transposons occurs following replication.

replisome. A multiprotein complex that includes DNA polymerase, primase, helicase, single-strand binding protein, and additional components. The replisomes, located at each of the replication forks, carry out the polymerization reactions of bacterial chromosomal DNA replication.

repressor. A regulatory DNA-binding protein that prevents transcription by RNA polymerase.

residue. A single component within a polymer. The chemical formula of a residue is that of the corresponding monomer minus the elements of water.

respiratory electron-transport chain. A series of enzyme complexes and associated cofactors that serve as electron carriers, passing electrons from reduced coenzymes or substrates to molecular oxygen (O_2), the terminal electron acceptor of aerobic metabolism.

restriction endonuclease. An endonuclease that catalyzes the hydrolysis of double-stranded DNA at a specific nucleotide sequence. Type I restriction endonucleases catalyze both the methylation of host DNA and the cleavage of nonmethylated DNA, whereas type II restriction endonucleases catalyze only the cleavage of nonmethylated DNA.

restriction fragment length polymorphism (RFLP). Variation in the length of DNA fragments generated by the action of restriction endonucleases on genomic DNA from different individuals.

restriction map. A diagram showing the size and arrangement of fragments produced from a DNA molecule by the action of various restriction endonucleases.

retroelement. *See* transposon

retroposons. A class of eukaryotic transposons whose members have no terminal repeats but have an A/sT-rich sequence at one end. Also known as retrotransposons.

retroregulation. The endonucleolytic destruction of specific mRNA molecules that depends on secondary structure in the mRNA molecule itself.

retrotransposons. *See* retroposons

retrovirus. An RNA virus that infects a eukaryotic cell.

reverse transcriptase. A type of DNA polymerase that catalyzes the synthesis of a strand of DNA from an RNA template.

RFLP. *See* restriction fragment length polymorphism

ribonuclease (RNase). An enzyme that catalyzes the hydrolysis of ribonucleic acids to form oligonucleotides and/or mononucleotides.

ribonucleic acid (RNA). A polymer consisting of ribonucleotide residues joined by 3′–5′ phosphodiester bonds. The sugar moiety in RNA is ribose. Genetic information contained in DNA is transcribed in the synthesis of RNA, some of which (mRNA) is translated in the synthesis of protein.

ribonucleoprotein. A complex containing both ribonucleic acid and protein.

ribophorin. A ribosome-binding protein that anchors the ribosome to the endoplasmic reticulum surface and through which proteins are translocated into the lumen as they are synthesized.

ribose. A five-carbon monosaccharide ($C_5H_{10}O_5$) that is the carbohydrate component of RNA, ATP, and numerous coenzymes. 2-Deoxyribose is the carbohydrate component of DNA.

ribosomal ribonucleic acid (rRNA). A class of RNA molecules that are integral components of ribosomes. rRNA is the most abundant cellular RNA.

ribosome. A large ribonucleoprotein complex composed of multiple rRNA molecules and proteins. Ribosomes catalyze the formation of peptide bonds during protein synthesis.

ribozyme. An RNA molecule with enzymatic activity.

rise. The distance between one residue and the next along the axis of a helical macromolecule.

R-looping. A technique used to visualize the organization of a gene. In R-looping, mature RNA is hybridized to DNA corresponding to the gene, and stretches of DNA that are not represented in the RNA sequence, such as introns, appear as single-stranded loops in an electron micrograph.

RNA. *See* ribonucleic acid

RNA editing. Posttranscriptional changes in the coding region of an mRNA molecule. Editing may include nucleotide modifications and insertions or deletions.

RNA processing. The reactions that transform a primary RNA transcript into a mature RNA molecule. The three general types of RNA processing include the removal of RNA nucleotides from primary transcripts, the addition of RNA nucleotides not encoded by the gene, and the covalent modification of bases.

RNA splicing. The process of removing introns and joining exons to form a continuous RNA molecule.

RNase. *See* ribonuclease

rolling-circle replication. The mode of replicating circular DNA in which DNA polymerase binds to the 3′ end of one nicked strand and proceeds around the circle synthesizing DNA complementary to the template strand. With each revolution, the newly synthesized DNA displaces the DNA synthesized during the previous cycle.

RPP cycle. *See* reductive pentose phosphate cycle

rRNA. *See* ribosomal ribonucleic acid

S. *See* entropy *and* Svedberg unit

saccharide. *See* carbohydrate

salvage pathway. A pathway in which a major metabolite, such as a purine or pyrimidine nucleotide, can be synthesized from a preformed molecular entity, such as a purine or pyrimidine.

saturated fatty acid. A fatty acid that does not contain a carbon-carbon double bond.

Schiff base. A complex formed by the reversible condensation of a primary amine with an aldehyde (an aldimine) or a ketone (a ketimine).

screen. A test of transformed cells for the presence of a desired recombinant DNA molecule.

SDS-PAGE. *See* sodium dodecyl sulfate–polyacrylamide gel electrophoresis

secondary structure. The regularities in local conformations within macromolecules. In proteins, secondary structure is maintained by hydrogen bonds between carbonyl and amide groups of the backbone. In nucleic acids, secondary structure is maintained by hydrogen bonds and stacking interactions between the bases.

second-order reaction. A reaction whose rate depends on the concentrations of two reactants.

secretory vesicle. A vesicle carrying proteins destined for secretion. Secretory vesicles bud off the Golgi apparatus and travel to the plasma membrane where the vesicle contents are released by exocytosis.

selectin. A membrane glycoprotein containing a lectinlike domain that binds oligosaccharide groups. Ca^{2+} is required for selectin binding to oligosaccharides.

selection. A technique in which only transformed cells survive under a chosen set of experimental conditions.

self-splicing intron. An intron that is excised in a reaction mediated by the RNA precursor itself.

semiconservative replication. The mode of duplicating DNA in which each strand serves as a template for the synthesis of a complementary strand. The result is two molecules of double-stranded DNA, each of which contains one of the parental strands.

sequential reaction. An enzymatic reaction in which all the substrates must be bound to the enzyme before any product is released.

sequential theory of cooperativity and allosteric regulation. A model of the cooperative binding of identical ligands to oligomeric proteins. According to the simplest form of the sequential theory, the binding of a ligand may induce a change in the tertiary structure of the subunit to which it binds and may alter the conformations of neighboring subunits to varying extents. Only one subunit conformation has a high affinity for the ligand. Also known as the ligand-induced theory.

serine protease. A protease with an active-site serine residue that acts as a nucleophile during catalysis.

Shine-Dalgarno sequence. A purine-rich region just upstream of the initiation codon in prokaryotic mRNA molecules. The Shine-Dalgarno sequence binds to a pyrimidine-rich sequence in the ribosomal RNA, thereby positioning the ribosome at the initiation codon.

short-patch repair. A type of mismatch repair in which a region of about 10 base pairs of DNA containing the mismatch is excised and replaced.

shuttle vector. A cloning vector that can replicate in both prokaryotic and eukaryotic cells. Shuttle vectors are used to transfer recombinant DNA molecules between prokaryotic and eukaryotic cells.

σ cascade. The sequential expression of different σ subunits that bind to the core RNA polymerase. A σ cascade can regulate gene expression since different σ subunits can direct RNA polymerase to different promoters, including those for genes encoding additional σ subunits.

σ factor. *See* σ subunit

σ subunit. A subunit of prokaryotic RNA polymerase, which acts as a transcription initiation factor by binding to the promoter. Different σ subunits are specific for different promoters. Also known as a σ factor.

signal peptidase. An integral membrane protein of the endoplasmic reticulum that catalyzes cleavage of the signal peptide of proteins translocated to the lumen.

signal peptide. The N-terminal sequence of residues in a newly synthesized polypeptide that targets the protein for translocation across a membrane.

signal recognition particle (SRP). A eukaryotic protein-RNA complex that binds a newly synthesized peptide as it is extruded from the ribosome. The signal recognition particle is involved in anchoring the ribosome to the cytosolic face of the endoplasmic reticulum so that protein translocation to the lumen can occur.

signal transduction. The process whereby an extracellular signal is converted to an intracellular signal by the action of a membrane-associated receptor, a transducer, and an effector enzyme.

single-strand binding protein (SSB). A protein that binds tightly to single-stranded DNA, preventing the DNA from folding back on itself to form double-stranded regions. Also known as helix-destabilizing protein.

site-directed mutagenesis. An in vitro procedure by which one particular nucleotide residue in a gene is replaced by another, resulting in production of an altered protein sequence.

site-specific integration. Integration of a phage genome into a host genome as a result of recombination between specific nucleotide sequences in the phage and the host.

small nuclear ribonucleoprotein (snRNP). An RNA-protein complex composed of one or two specific snRNA molecules plus a number of proteins. snRNPs are involved in splicing mRNA precursors and in other cellular events.

small nuclear RNA (snRNA). A class of RNA molecules found in eukaryotic nuclei. Many types of snRNA molecules are extensively base paired and contain modified nucleotides. They are involved in splicing mRNA precursors and in other cellular processes.

snRNA. *See* small nuclear RNA

snRNP. *See* small nuclear ribonucleoprotein

soap. An alkali metal salt of a long-chain fatty acid. Soaps are a type of detergent.

sodium dodecyl sulfate–polyacrylamide gel electrophoresis (SDS-PAGE). Polyacrylamide gel electrophoresis performed in the presence of the detergent sodium dodecyl sulfate. SDS-PAGE allows separation of proteins on the basis of size only rather than charge and size.

solvation. A state in which a molecule or ion is surrounded by solvent molecules.

solvation sphere. The shell of solvent molecules that surrounds an ion or solute.

sorbitol pathway. A metabolic pathway consisting of two enzyme-catalyzed reactions that converts glucose to fructose in some mammalian tissues. Also known as the polyol pathway.

special pair. A specialized pair of chlorophyll molecules in reaction centers that serves as the primary electron donor during the light-dependent reactions of photosynthesis.

specific heat. The amount of heat required to raise the temperature of 1 gram of a substance by 1°C.

specificity constant. *See* k_{cat}/K_m

sphingolipids. An amphipathic lipid with a sphingosine (*trans*-4-sphingenine) backbone. Sphingolipids, which include sphingomyelins, cerebrosides, and gangliosides, are present in plant and animal membranes and are particularly abundant in the tissues of the central nervous system.

sphingomyelin. A sphingolipid that consists of phosphocholine attached to the C-1 hydroxyl group of a ceramide. Sphingomyelins are present in the plasma membranes of most mammalian cells and are a major component of myelin sheaths.

splice site. The conserved nucleotide sequence surrounding an exon-intron junction.

spliceosome. The large protein-RNA complex that catalyzes the removal of introns from mRNA precursors. The spliceosome is composed of small nuclear ribonucleoproteins.

splicing. *See* RNA splicing

sporulation. The response of certain organisms to unfavorable growth conditions. During sporulation, special cells, called spores, are formed. Spores remain dormant until favorable growth conditions are restored.

SRP. *See* signal recognition particle

SSB. *See* single-strand binding protein

stacking interactions. The weak noncovalent forces between adjacent bases or base pairs in single-stranded or double-stranded nucleic acids, respectively. Stacking interactions contribute to the helical shape of nucleic acids.

standard free-energy change ($\Delta G°'$). The free-energy change for a reaction under biochemical standard-state conditions.

standard reduction potential ($E°'$). A measure of the tendency of a substance to reduce other substances under biochemical standard-state conditions.

standard state. A set of reference conditions for a chemical reaction. In biochemistry, the standard state is defined as a temperature of 298 K (25°C), a pressure of 1 atmosphere, a solute concentration of 1.0 M, and a pH of 7.0.

starch. A homopolymer of glucose residues that serves as a storage polysaccharide in plants. There are two forms of starch: amylose, an unbranched polymer of glucose residues joined by α-(1→4) linkages; and amylopectin, a branched polymer of glucose residues joined by α-(1→4) linkages with α-(1→6) linkages at branch points.

starvation. A period in which no food is digested or absorbed. During starvation, which begins about 16 hours after consumption of the last meal in humans, blood glucose originates from hepatic and renal gluconeogenesis, and fatty acids and proteins are catabolized for energy.

steady state. A state in which the rate of synthesis of a compound is equal to its rate of utilization or degradation.

stem-loop. *See* hairpin

stereoisomers. Compounds with the same molecular formula but different spatial arrangements of their atoms.

stereospecificity. The ability of an enzyme to recognize and act upon only a single stereoisomer of a substrate.

steroid. A lipid composed of 18 or more carbon atoms and containing a fused, four-ring polyprenyl structure.

sticky end. An end of a double-stranded DNA molecule with a single-stranded extension of several nucleotides.

stomata. Structures on the surface of a leaf through which carbon dioxide diffuses directly into photosynthetic cells.

stop codon. *See* termination codon

strand invasion. The exchange of single strands of DNA from two nicked molecules having homologous nucleotide sequences.

stretch-gated ion channel. A membrane transport protein that allows ions to pass through the bilayer in response to changes in tension or turgor pressure on one side of the membrane.

stringent response. The production of phosphorylated guanylate compounds triggered by the presence of an uncharged tRNA molecule in the aminoacyl site of a ribosome during protein synthesis in prokaryotes. The net effect is a decrease in RNA synthesis when the cell is starved for amino acids.

stroma. The aqueous matrix of the chloroplast. The stroma is the site of the reductive pentose phosphate cycle, which leads to the reduction of carbon dioxide to carbohydrates.

stromal lamellae. Regions of the thylakoid membrane that are in contact with the stroma.

subcloning. Transferring cloned DNA between cloning vectors.

substrate. A reactant in a chemical reaction. In enzymatic reactions, substrates are specifically acted upon by enzymes, which catalyze the conversion of substrates to products.

substrate cycle. A pair of opposing, metabolically irreversible reactions that catalyzes a cycle between two pathway intermediates. Substrate cycles provide sensitive regulatory sites.

substrate-level phosphorylation. Phosphorylation of a nucleoside diphosphate by transfer of a phosphoryl group from a non-nucleotide substrate.

supercoil. A topological arrangement assumed by over- or underwound double-stranded DNA. Underwinding gives rise to negative supercoils; overwinding produces positive supercoils. Positively supercoiled DNA is not found in nature.

supersecondary structure. A combination of secondary structures that appears in a number of different proteins. Also known as a motif.

suppressor tRNA molecule. A mutant tRNA molecule containing an altered anti-codon that permits it to bind to a termination codon. Such binding results in the incorporation of the amino acid carried by the tRNA into the growing polypeptide chain.

surfactant. *See* detergent

Svedberg unit (S). A unit of 10^{-13} seconds used for expressing the sedimentation coefficient, a measure of the rate at which a large molecule or particle sediments in an ultracentrifuge. Large S values usually indicate large masses.

symmetry-driven theory. *See* concerted theory of cooperativity and allosteric regulation

symport. The cotransport of two different species of ions or molecules in the same direction across a membrane by a transport protein.

synonymous codons. Different codons that specify the same amino acid.

synteny. The similar arrangement of genes in different organisms.

synthase. A common name for an enzyme, often a lyase, that catalyzes a synthetic reaction.

synthetase. An enzyme that catalyzes the joining of two substrates and requires the input of the chemical potential energy of a nucleoside triphosphate. Synthetases are members of the IUB class of enzymes known as ligases.

T state. The less active conformation of an allosteric enzyme; opposite of R state.

targeting. The transport of proteins to specific cellular locations based on markers attached to the newly synthesized proteins.

TATA box. An A/T-rich DNA sequence found within the promoter of both prokaryotic and eukaryotic genes. In prokaryotes, the TATA box is located about 10 base pairs upstream of the transcription initiation site; in eukaryotes, it is located about 19 to 27 base pairs upstream of the transcription initiation site.

teichoic acid. A polymer of up to 30 glycerol phosphate or ribitol phosphate moieties joined by phosphodiester linkages. Teichoic acids are components of the cell walls of certain Gram-positive bacteria.

telomerase. A eukaryotic enzyme that extends the 3′ strand of DNA at the ends of chromosomes in order to prevent progressive shortening of the chromosome during successive rounds of replication.

telomere. The terminal region of eukaryotic chromosomes that consists of large numbers of 4–8 base pair repeats in which the nucleotides on one strand are predominately deoxythymidylate and deoxyguanylate.

template. A strand of DNA or RNA whose sequence of nucleotide residues guides the synthesis of a complementary strand.

template strand. The strand of DNA within a gene whose nucleotide sequence is complementary to that of the transcribed RNA. During transcription, RNA polymerase binds to and moves along the template strand in the $3′ \rightarrow 5′$ direction, catalyzing the synthesis of RNA in the $5′ \rightarrow 3′$ direction.

termination codon. A codon that is not normally recognized by any tRNA molecule but is bound by specific proteins that cause newly synthesized peptides to be released from the translation machinery. The three termination codons are referred to as amber (UAG), ochre (UAA), and opal (UGA) codons. Also known as a stop codon.

terpene. One of a number of polyprenyl compounds found in plants.

tertiary structure. The compacting of polymeric chains into one or more globular units within a macromolecule. In proteins, tertiary structure is stabilized mainly by hydrophobic interactions between side chains.

thermodynamics. The branch of physical science that studies transformations of heat and energy.

thin-layer chromatography. A chromatographic technique used to separate components of a mixture on a thin layer of porous material.

−35 region. A sequence found within the promoter of prokaryotic genes about 30–35 base pairs upstream of the transcription initiation site.

30-nm fiber. A chromatin structure in which nucleosomes are coiled into a solenoid 30 nm in diameter.

thylakoid lamella. *See* thylakoid membrane

thylakoid membrane. A highly folded, continuous membrane network suspended in the aqueous matrix of the chloroplast. The thylakoid membrane is the site of the light-dependent reactions of photosynthesis, which lead to the formation of NADPH and ATP. Also known as the thylakoid lamella.

tissue. An association of cells that carry out the same functions.

T_m. *See* melting temperature *and* phase-transition temperature

topoisomerase. An enzyme that changes the linking number of a DNA molecule by cleaving a phosphodiester linkage in either one or both strands, rewinding the DNA, and resealing the break. Type I topoisomerases catalyze the removal of negative supercoils in DNA by breaking one strand and increase the linking number by 1. Type II topoisomerases catalyze the addition of negative supercoils by breaking both strands of DNA and decrease the linking number by 2. Some type II topoisomerases are also known as DNA gyrases.

topology. 1. The arrangement of membrane-spanning segments and connecting loops in an integral membrane protein. 2. The overall morphology of a nucleic acid molecule.

TΨC arm. The stem-and-loop structure in a tRNA molecule that contains the sequence thymidylate–pseudouridylate–cytidylate (TΨC).

trace element. An element required in very small quantities by living organisms. Examples include copper, iron, and zinc.

transaminase. An enzyme that catalyzes the transfer of an amino group from an α-amino acid to an α-keto acid. Transaminases require the coenzyme pyridoxal phosphate.

transcription. The copying of biological information from a double-stranded DNA molecule to a single-stranded RNA molecule, catalyzed by a transcription complex consisting of RNA polymerase and associated factors.

transcription bubble. A region of DNA that is unwound by RNA polymerase during transcription initiation.

transcription factor. A protein that binds to the promoter region and interacts with RNA polymerase during assembly of the transcription initiation complex. Some transcription factors remain bound during RNA chain elongation.

transcription initiation complex. The complex of RNA polymerase and other factors that assembles at the promoter at the start of transcription.

transcriptional activator. A regulatory DNA-binding protein that enhances the rate of transcription by increasing the activity of RNA polymerase.

transducer. The component of a signal-transduction pathway that couples receptor-ligand binding with generation of a second messenger catalyzed by an effector enzyme.

transfection. The introduction of foreign DNA into a cell via a virus or phage vector.

transfer ribonucleic acid (tRNA). A class of RNA molecules that carry activated amino acids to the site of protein synthesis for incorporation into growing peptide chains. tRNA molecules contain an anticodon that recognizes a complementary codon in mRNA.

transferase. An enzyme that catalyzes a group-transfer reaction. Transferases often require a coenzyme.

transformation. A process by which a cell takes up intact DNA from outside the cell.

transgenic organism. An individual organism that carries recombinant DNA stably integrated in all of its cells.

transition. A mutation resulting from a base alteration that occurs when a purine is substituted by the other purine, or when a pyrimidine is substituted by the other pyrimidine.

transition state. An unstable, high-energy arrangement of atoms in which chemical bonds are being formed or broken. Transition states have structures between those of the substrates and the products of a reaction.

transition-state analog. A compound that resembles a transition state. Transition-state analogs characteristically bind extremely tightly to the active sites of appropriate enzymes and thus act as potent inhibitors.

transition-state stabilization. The increased binding of transition states to enzymes relative to the binding of substrates or products. Transition-state stabilization lowers the activation energy and thus contributes to catalysis.

translation. The synthesis of a polypeptide whose sequence reflects the nucleotide sequence of an mRNA molecule. Amino acids are donated by activated tRNA molecules, and peptide bond synthesis is catalyzed by the translation complex, which includes the ribosome and other factors.

translation complex. The complex of a ribosome and protein factors that carries out the translation of mRNA in vivo.

translation initiation complex. The complex of ribosomal subunits, an mRNA template, an initiator tRNA molecule, and initiation factors that assembles at the start of protein synthesis.

translation initiation factor. A protein involved in the formation of the initiation complex at the start of protein synthesis.

translational frameshifting. The shift in reading frame that may occur during the translation of an mRNA molecule. The synthesis of some proteins requires translational frameshifting.

transport constant (K_{tr}). The substrate concentration at which the rate of transport across a membrane via a transport protein is half-maximal. K_{tr} is analogous to the Michaelis constant (K_m) of an enzyme.

transposable element. *See* transposon

transposon. A mobile genetic element that jumps between chromosomes or parts of a chromosome by taking advantage of recombination mechanisms. Also known as a transposable element or an integron. Retrovirus-like transposons are also known as integrating retroelements.

transverse diffusion. The passage of lipid or protein molecules from one leaflet of a lipid bilayer to the other leaflet. Unlike lateral diffusion within one leaflet of a bilayer, transverse diffusion is extremely slow.

transversion. A mutation resulting from a base alteration that occurs when a purine is substituted by a pyrimidine, or when a pyrimidine is substituted by a purine.

triacylglycerol. A lipid containing three fatty acyl residues esterified to glycerol. Fats and oils are mixtures of triacylglycerols. Formerly known as a triglyceride.

tricarboxylic acid cycle. *See* citric acid cycle

triglyceride. *See* triacylglycerol

tRNA. *See* transfer ribonucleic acid

tumor promoter. A compound that greatly increases the development of tumors following exposure of an organism to a carcinogen.

turn. *See* loop

turnover. The dynamic metabolic steady state in which molecules are degraded and replaced by newly synthesized molecules.

turnover number. *See* catalytic constant

twist. The angle of rotation between adjacent residues within a helical macro-molecule.

ubiquitin. A highly conserved protein in eukaryotic cells that is involved in pro-tein degradation. The covalent attachment of one or more ubiquitin moieties to a protein targets that protein for intracellular hydrolysis.

uncompetitive inhibition. Inhibition of an enzyme-catalyzed reaction by an in-hibitor that binds only to the enzyme-substrate complex, not to the free enzyme.

uncoupling agent. A compound that disrupts the usual tight coupling between electron transport and phosphorylation of ADP.

uniport. The transport of a single type of solute across a membrane by a trans-port protein.

unsaturated fatty acid. A fatty acid with at least one carbon-carbon double bond. An unsaturated fatty acid with only one carbon-carbon double bond is called a monounsaturated fatty acid. A fatty acid with two or more carbon-carbon double bonds is called a polyunsaturated fatty acid. In general, the double bonds of unsaturated fatty acids are of the *cis* configuration and are separated from each other by methylene ($—CH_2—$) groups.

urea cycle. A metabolic cycle consisting of four enzyme-catalyzed reactions that converts nitrogen from ammonia and aspartate to urea. Four ATP equivalents are consumed during formation of one molecule of urea.

v. *See* velocity

v_0. *See* initial velocity

van der Waals force. A weak intermolecular force produced between neutral atoms by transient electrostatic interactions. Van der Waals attraction is strongest when atoms are separated by the sum of their van der Waals radii; strong van der Waals repulsion precludes closer approach.

van der Waals radius. The effective size of an atom. The distance between the nuclei of two nonbonded atoms at the point of maximal attraction is the sum of their van der Waals radii.

variable arm. The arm of a tRNA molecule that is located between the anticodon arm and the TΨC arm. The variable arm can range in length from about 3 to 21 nucleotides.

vector. *See* cloning vector

velocity (*v*). The rate of a chemical reaction, expressed as amount of product formed per unit time.

very low density lipoprotein (VLDL). A type of plasma lipoprotein that transports endogenous triacylglycerols, cholesterol, and cholesteryl esters from the liver to the tissues.

virus. A nucleic acid–protein complex that is capable of invading a host cell. A virus takes over the transcription and replication machinery of the host cell to replicate itself, using both its own and the host's gene products.

vitamin. An organic micronutrient that cannot be synthesized by animals and must be obtained in the diet. Many coenzymes are derived from vitamins.

vitamin-derived coenzyme. A coenzyme synthesized from a vitamin.

VLDL. *See* very low density lipoprotein

V_{max}. *See* maximum velocity

voltage-gated ion channel. A membrane transport protein that allows ions to pass through the bilayer in response to changes in the electrical properties of the membrane.

water-soluble vitamin. An organic micronutrient that is soluble in water. Specifically, water-soluble vitamins include the B vitamins and, in the case of primates and a few other organisms, ascorbic acid (vitamin C).

wax. A nonpolar ester that consists of a long-chain monohydroxylic alcohol and a long-chain fatty acid.

wobble position. The 5′ position of an anticodon, where non-Watson-Crick base pairing is permitted. The wobble position makes it possible for a tRNA molecule to recognize more than one codon.

X-ray crystallography. A technique used to determine secondary, tertiary, and quaternary structures of biological macromolecules. In X-ray crystallography, a crystal of the macromolecule is bombarded with X-rays, which are diffracted and then detected electronically or on a film. The atomic structure is deduced by mathematical analysis of the diffraction pattern.

YAC. *See* yeast artificial chromosome

yeast artificial chromosome (YAC). A cloning vector that contains yeast centromeric DNA and a yeast origin of replication. A YAC can accommodate up to 500 kilobases of insert DNA. The artificial chromosome can be introduced into a yeast cell, where it behaves like a normal chromosome during replication.

ylid. An organic molecule that has opposite ionic charges on adjacent covalently bonded atoms.

Z-DNA. The conformation of synthetic oligonucleotides containing alternating deoxycytidylate and deoxyguanylate residues. Z-DNA is a left-handed double helix containing approximately 12 base pairs per turn, in which each G residue adopts a C-3′ *endo* conformation with the base in the *syn* position, and each C residue adopts a C-2′ *endo* conformation with the base in the *anti* position.

zero-order reaction. A reaction whose rate is independent of reactant concentration.

zinc finger. A structural motif often found in DNA-binding proteins. The finger is formed when a stretch of about 30 amino acids forms a loop whose base is anchored by Zn^{2+}. The Zn^{2+} is coordinated by two conserved histidine residues and two conserved cysteine residues at the base of the loop.

Z-scheme. A zigzag scheme that illustrates the reduction potentials associated with electron flow through photosynthetic electron carriers.

zwitterion. A molecule containing negatively and positively charged groups.

zymogen. A catalytically inactive enzyme precursor that must be modified by limited proteolysis to become enzymatically active. Also known as a proenzyme.